Modern Parasitology

A TEXTBOOK OF PARASITOLOGY

EDITED BY

F. E. G. COX

Professor of Parasite Immunology
King's College London

SECOND EDITION

Blackwell
Science

© 1982, 1993, by
Blackwell Science Ltd
Editorial Offices:
Osney Mead, Oxford OX2 0EL
25 John Street, London WC1N 2BL
23 Ainslie Place, Edinburgh EH3 6AJ
350 Main Street, Malden
  MA 02148 5018, USA
54 University Street, Carlton
  Victoria 3053, Australia
10, rue Casimir Delavigne
  75006 Paris, France

Other Editorial Offices:
Blackwell Wissenschafts-Verlag GmbH
Kurfürstendamm 57
10707 Berlin, Germany

Blackwell Science KK
MG Kodenmacho Building
7–10 Kodenmacho Nihombashi
Chuo-ku, Tokyo 104, Japan

The right of the Author to be
identified as the Author of this Work
has been asserted in accordance
with the Copyright, Designs and
Patents Act 1988.

First published 1982
Reprinted 1987
Second edition 1993
Reprinted 1994, 1996, 1998, 1999

Set by Excel Typesetters Company,
Hong Kong
Printed and bound in the United Kingdom
at the University Press, Cambridge

The Blackwell Science logo is a
trade mark of Blackwell Science Ltd,
registered at the United Kingdom
Trade Marks Registry

DISTRIBUTORS

Marston Book Services Ltd
PO Box 269
Abingdon, Oxon OX14 4YN
(Orders: Tel: 01235 465500
         Fax: 01235 465555)

USA
Blackwell Science, Inc.
Commerce Place
350 Main Street
Malden, MA 02148 5018
(Orders: Tel: 800 759 6102
              781 388 8250
         Fax: 781 388 8255)

Canada
Login Brothers Book Company
324 Saulteaux Crescent
Winnipeg, Manitoba R3J 3T2
(Orders: Tel: 204 837-2987)

Australia
Blackwell Science Pty Ltd
54 University Street
Carlton, Victoria 3053
(Orders: Tel: 3 9347 0300
         Fax: 3 9347 5001)

A catalogue record for this title
is available from the British Library

ISBN 0-632-02585-9

Library of Congress
Cataloging-in-Publication Data
Modern parisitology: a textbook of parasitology/
  edited by F. E. G. Cox. – 2nd ed.
    p.   cm.
  Includes bibliographical references.
  ISBN 0-632-02585-9
  1. Medical parasitology.   I. Cox, Francis E. G.
  [DNLM: 1. Parisitic Diseases – drug
  therapy.   2. Parasitic Diseases – prevention &
  control.   3. Parasites – physiology.
  4. Host – Parasite Relations.
  QX 4 M6895 1993]
QR251.M66   1993
616.9'6 – dc20

For further information on
Blackwell Science, visit our website:
www.blackwell-science.com

Contents

List of Contributors

R. M. ANDERSON FRS *Professor of Zoology and Head of the Department of Zoology, University of Oxford, UK*

C. BRYANT *Professor of Zoology, and Head of the Division of Biochemical and Molecular Biology, Australian National University, Canberra, Australia*

L. H. CHAPPELL *Reader in Zoology, University of Aberdeen, UK*

F. E. G. COX *Professor of Parasite Immunology, King's College, University of London, and Professor of Physic, Gresham College, London, UK*

W. E. GUTTERIDGE *Principal Research Scientist, The Wellcome Foundation Ltd, Beckenham, UK, and Visiting Professor of Biochemical Parasitology, University of Kent, UK*

D. T. HART *Lecturer in Molecular Parasitology, King's College, University of London, UK*

D. H. MOLYNEUX *Director, Liverpool School of Tropical Medicine, and Professor of Tropical Health Sciences, University of Liverpool, UK*

P. J. WHITFIELD *Reader in Parasitology and Head of the Division of Life Sciences, King's College, University of London, UK*

Introduction

Parasites present a continual and unacceptable threat to the well-being of millions of people in the tropics and subtropics and to domesticated animals in all parts of the world and the cost of harbouring parasites in terms of human misery and economic loss is incalculable. Parasitology is the study of parasites and their interactions with their hosts. The science of parasitology has a long history and has its roots in zoology, with its emphasis on the identification and classification of parasites and the elucidation of life cycles, and tropical and veterinary medicine with their concern for the diseases caused by parasites. However, the subject is now so intertwined with microbiology, immunology, cell biology, molecular biology and other aspects of biology and medicine that its limits have become increasingly indistinct.

It was once thought that when a parasite had been identified, implicated as a cause of disease and its life cycle elucidated, its control or eradication would follow with the development of drugs, vaccines and antivector measures. Those who thought this, seriously underestimated the complete hold a parasite has on its host and the intimacy of the relationship between them. Parasites exhibit combinations of biochemical, physiological and nutritional adaptations unique in the animal world and also display a range of mechanisms for evading the immune responses of the host unknown to other pathogens. Their ecology is also far more complex than that of free-living organisms.

Parasites, however, are reluctant to yield up their secrets and, over the last two decades, the various challenges presented have attracted the attention not only of parasitologists but also biochemists, molecular biologists, immunologists, physiologists and epidemiologists as well as those working in the more applied fields of medicine, veterinary medicine and the pharmaceutical industry. Above all, parasitology is a field in which the intellectual rewards of basic and applied research are matched by the potential benefits to be gained.

Parasitology is a very active science with over 20 major journals devoted solely to it and a similar number of books and monographs published each year. The first textbooks of parasitology were essentially zoological and were concerned with detailed descriptions of individual species of parasites with little about the diseases caused. These were soon complemented by books that concentrated on the major parasites of humans and were largely concerned with tropical diseases. Today there are several excellent textbooks that cover the important diseases of humans very well but, inevitably, many important aspects are not considered in such textbooks and areas such as biochemistry, molecular biology, physiology, immunology and epidemiology tend to be the subjects of specialized monographs. Few of these specialized monographs bridge the gap between the traditional texts and recent research developments and this gap has been partially filled by a few multiauthor works that present research findings in a more general context. Inevitably, however, such monographs, excellent as most of them are, tend to be pitched at a level too high for undergraduate courses or those just entering the field of parasitology.

It is against this background that this book has been written and the intention is to introduce the whole of modern parasitology in a succinct and comprehensive way. This book is a direct successor to *Modern Parasitology*, published over 10 years ago, during which time the subject has undergone massive changes as the techniques of molecular biology and biochemistry, on the one hand, and the powerful tools of mathematical analysis, on the other, have taken over from more traditional approaches. This is, therefore, not really a second edition but an almost completely new version. The first two chapters cover familiar ground and describe, in a condensed form, the most important protozoan and helminth parasites that infect humans and domesticated animals and their laboratory counterparts and the third chapter deals similarly with the vectors involved and fills a major gap which was apparent in the earlier version. Taken together, these three chapters introduce the reader to the parasites and their vectors and provide a framework upon which more detailed information can be built. The fourth chapter, which has largely stood the test of time, covers parasite ecology and epidemiology in the context of modern mathematical concepts and now includes comparisons with infectious diseases caused by organisms other than parasites. The next three chapters deal with the ways in which parasites live and describe their biochemistry, molecular biology and physiology. The biochemistry chapter has been considerably revised, the molecular biology chapter is new and the physiology chapter now embraces nutrition. The eighth chapter, which has been completely rewritten, is concerned with the immunology of parasitic infections in the context of modern immunological concepts and also includes comparisons with other infectious diseases. The ninth chapter presents an up to date coverage of the basic principles of chemotherapy and has been enlarged to cover basic principles as well as the drugs used. The final chapter, which is completely new, is concerned with integrated control programmes and brings together various topics such as chemotherapy, immunization and vector control covered in earlier chapters.

This book is intended mainly for undergraduate students who are advised to read the first three chapters and then, selectively, the others in whatever order interests them or is compatible with the course they are following. Graduate students, and other postgraduate research workers who wish to put their own work into a more general context, are also advised to follow the same approach but can be more selective than under-graduates although, in reality, no area of parasitology is so distinct from any other that it cannot benefit from some comprehension of what is being done elsewhere and why it is being done.

It must be emphasized that this book is intended to be an *introduction* to modern parasitology and, as such, should be read in conjunction with other books, reviews and scientific papers. Each chapter is followed by a selected list of important and accessible references which will guide the reader towards more specific topics. The reader will also need to refer to the more general literature of the subject and some of the textbooks and monographs in the English language currently available are listed below under three headings, general texts, specialized monographs and 'bridging the gap'. Finally, reference is made to two outstanding series, *Advances in Parasitology*, published annually, and *Parasitology Today*, published monthly, both of which are essential reading for all parasitologists.

F.E.G. COX

FURTHER READING

General texts

Ash, L.R. & Orihel, T.C. (1987) *Parasites: A Guide to Laboratory Procedures and Identification*. New York, Raven Press.

Ash, L.R. & Orihel, T.C. (1990) *Atlas of Human Parasitology*, 3rd edn. Chicago, American Society of Clinical Pathologists Press.

Bell, D.R. (1990) *Lecture Notes on Tropical Medicine*, 3rd edn. Oxford, Blackwell Scientific Publications.

Commonwealth Agricultural Bureaux (1989) *Manual of Tropical Veterinary Parasitology* (translation of *Precis de Parasitologie Veterinaire Tropicale*). Wallingford, CAB International.

Cowan, G.O. & Hammond, W. (1991) *Atlas of Medical Helminthology and Protozoology*, 3rd edn. Edinburgh, Churchill Livingstone.

Despommier, D.D. & Karapelin, J.W. (1987) *Parasite Life Cycles*. Berlin, Springer-Verlag.

Katz, M., Despommier, D.D. & Gwadz, R.W. (1989) *Parasitic Diseases*, 2nd edn. New York, Springer-Verlag.

Mehlhorn, H. (ed.) (1988) *Parasitology in Focus*. Berlin, Springer-Verlag.

Muller, R. & Baker, J.R. (1990) *Medical Parasitology*. London, Gower Medical.

Peters, W. & Gilles, H.M. (1989) *A Colour Atlas of*

Tropical Medicine and Parasitology, 3rd edn. London, Wolfe Medical Publications.

Piekarski, G. (1989) *Medical Parasitology*, 3rd edn. Berlin, Springer-Verlag.

Schmidt, G.D. & Roberts, L.S. (1989) *Foundations of Parasitology*, 4th edn. New York, Williams and Wilkins.

Urquhart, G., Armour, J., Duncan, J.L., Dunn, A.M. & Jennings, F. (1987) *Veterinary Parasitology*. London, Longman.

Warren, K.S. & Mahmoud, A.A.F. (eds) (1989) *Tropical and Geographical Medicine*, 2nd edn. New York, McGraw-Hill

Zaman, V. & Keong, L.A. (1990) *Handbook of Medical Parasitology*, 2nd edn. Edinburgh, Churchill Livingstone.

Monographs

Behnke, J.M. (ed.) (1990) *Parasites: Immunity and Pathology*. London, Taylor and Francis.

Bryant, C., Behm, C. & Howell, M.J. (1989) *Biochemical Adaptation in Parasites*. London, Chapman and Hall.

Coombs, G.H. & North, M. (eds) (1991) *Biochemical Parasitology*. London, Taylor and Francis.

Esch, G.W., Bush, A.O. & Aho, J.M. (eds) *Parasite Communities: Patterns and Processes*. London, Taylor and Francis.

Hyde, J.E. (1990) *Molecular Parasitology*. Milton Keynes, Open University Press.

Bridging the gap

Englund, P.T. & Sher, A. (eds) *The Biology of Parasitism: A Molecular and Immunological Approach*. New York, Alan R. Liss.

McAdam, K.P.W.J. (ed.) (1990) *New Strategies in Parasitology*. Edinburgh, Churchill Livingstone.

Wyler, D. (ed.) (1990) *Modern Parasite Biology: Cellular, Molecular and Immunological Aspects*. New York, W.H. Freeman.

Parasitological journals

Papers on various aspects of parasitology are published in a variety of journals and for some ideas of the range used by parasitologists, reference should be made to one of the abstracting journals listed below. There are a number of journals devoted to parasitological topics and the journals in English most frequently cited are listed below:

Acta Tropica
American Journal of Tropical Medicine and Hygiene
Annals of Tropical Medicine and Parasitology
Experimental Parasitology
International Journal for Parasitology
Journal of Helminthology
Journal of Parasitology
Molecular and Biochemical Parasitology
Parasite Immunology
Parasitology
Parasitology Today
Systematic Parasitology
Transactions of the Royal Society of Tropical Medicine and Hygiene
Tropical Medicine and Parasitology
Veterinary Parasitology
Parasitology Research

Advances in Parasitology (published annually)

Abstracting journals

Protozoological Abstracts
Helminthological Abstracts
Tropical Diseases Bulletin
Between them, these three publications abstract over 1000 journals each year and provide a virtually complete data base of the available parasitological literature. They are also available on various data retrieval systems and on discs.

Chapter 1 / Parasitic Protozoa

F. E. G. COX

1.1 INTRODUCTION

There are over 45 000 named species of protozoa now living of which nearly 10 000 are parasitic in invertebrates and in almost every species of vertebrate. It is, therefore, hardly surprising that humans and their domesticated animals should act as hosts to protozoa, but the diseases thus caused are out of all proportion to the number of species involved. The protozoa that infect humans range from forms that are never pathogenic to those that cause malaria, sleeping sickness, Chagas disease and leishmaniasis, now regarded as being among the major diseases of tropical countries, and which together threaten over one-quarter of the population of the world. In domesticated animals, nagana and theileriosis take a major toll of cattle in Africa and coccidiosis, in its various forms, presents a continual threat to poultry and cattle throughout the world, particularly under conditions of intensive rearing. Even fish and invertebrates suffer from a variety of protozoan infections which create major problems for those trying to raise these animals for food.

Protozoa lie between the prokaryotic and higher eukaryotic organisms and share some of the characteristics of each. They are small, have short generation times, high rates of reproduction and a tendency to induce immunity to reinfection in those hosts that survive. These are features of infections with microparasites such as bacteria. On the other hand, protozoa are undoubtedly eukaryotic cells with organelles and metabolic pathways akin to those of the host. They have also evolved numerous adaptations that allow them to survive in their hosts and, in particular, to counteract or evade the immune response. For this reason, infections with parasitic protozoa are not short-lived as with most bacteria.

1.2 STRUCTURE AND FUNCTION OF PROTOZOA

Protozoa are unicellular eukaryotic cells measuring $1-150\,\mu$m, the parasitic forms tending towards the lower end of this range. Structurally, each protozoan is the equivalent of a single metazoan cell with its plasma membrane, nucleus, nuclear membrane, chromosomes, endoplasmic reticulum, mitochondria, Golgi apparatus, ribosomes and various specialized structures adapted to meet particular needs. Parasitic protozoa are in no way simple or degenerate forms, and their particular adaptions frequently include complex life cycles and specialized ways of entering their hosts and maintaining themselves therein. Their nutrition, physiology and biochemistry are largely geared to the parasitic habit and are specialized rather than degenerate. Sexual reproduction also occurs in some protozoa and, in the parasitic forms, is particularly important in the sporozoans in which it provides for apparently limitless variation and adaptability. These topics are discussed in more detail in Chapters 5 and 7.

1.3 CLASSIFICATION OF THE PROTOZOA

The small size of protozoa coupled with the fact that they consist of single cells with few obvious morphological features means that their classification has had to be based on a wide range of characteristics including variations in life cycles, details of fine structure and, increasingly at the species level, biochemical and molecular differences.

At one time the protozoa were regarded as a phylum within the kingdom Animalia but protozoologists now believe that the group is not a natural one but contains members that could be classified with animals, plants or fungi. Accordingly the concept of Protozoa as a taxon has disappeared and the term is applied to the animal-like members of the kingdom Protista. There are many arguments concerning the relationships between the various taxonomic groups within the protozoa but these do not need to concern parasitologists whose main interest is in having a simple and consistent framework within which to work. The classification outlined here is essentially a simplified version of that used by Sleigh (1989) and Cox (1992) but is compatible with those used in standard parasitological textbooks such as Mehlhorn (1988) and Schmidt and Roberts (1989) (see Introduction).

Traditionally, the protozoa have been divided into four major groups distinguished by their mode of locomotion: the flagellates, which move by means of flagella; the amoebae, by pseudopodia; the ciliates, by cilia; and the sporozoans lacking any obvious means of locomotion. Following the example of Sleigh (1989) these groups will be retained here simply as convenient and traditional categories. An abbreviated outline classification of the protozoa is given in Table 1.1 in which the most important parasites are listed under their phyla and orders.

1.4 KINETOPLASTID FLAGELLATES

The kinetoplastid flagellates are characterized by the possession of a unique organelle called the kinetoplast which contains DNA and is an integral part of the mitochondrial system. The kinetoplast is situated near the base of the flagellum and is easily seen in stained preparations. Kinetoplastid flagellates are found in invertebrates and vertebrates, the genera in mammals being *Leishmania*, *Trypanosoma* and *Endotrypanum* which are transmitted by insects. The typical form is an elongated organism, called a promastigote, with a kinetoplast and a flagellum at the anterior end. Variations of this form are brought about by the migration of the kinetoplast–flagellum complex within the body of the flagellate associated with changes in the mitochondrial system. The forms in the life cycle of a kinetoplastid flagellate are shown in Fig. 1.1.

Table 1.1 An outline classification of the parasitic protozoa

KINGDOM PROTISTA (Single-celled eukaryotic organisms)

Group 1 The flagellated protozoa (Locomotion by flagella)

PHYLUM KINETOPLASTA (1–2 flagella, kinetoplast present)
Order Trypanosomatida, e.g. *Leishmania*, *Trypanosoma*

PHYLUM METAMONADA
Order Retortamonadida, e.g. *Chilomastix*, *Retortamonas*
Order Diplomonadida, e.g. *Enteromonas*, *Giardia*

PHYLUM PARABASALIA
Order Trichomodadida, e.g. *Dientmoeba*, *Histomonas*, *Trichomonas*

Group 2 The amoeboid protozoa (Locomotion by pseudopodia)

PHYLUM RHIZOPODA
Order Euamoebida, e.g. *Entamoeba*

Group 3 The spore forming protozoa (No obvious means of locomotion)

PHYLUM SPOROZOA (= Apicomplexa)
Order Eimeriida, e.g. *Eimeria*, *Isospora*, *Sarcocystis*, *Toxoplasma*
Order Haemosporidida, e.g. *Plasmodium*
Order Piroplasmida, e.g. *Babesia*, *Theileria*

PHYLUM MICROSPORIDIA
Order Microsporidida, e.g. *Encephalitozoon*, *Nosema*

Group 4 The ciliated protozoa (Locomotion by cilia)

PHYLUM CILIOPHORA
Order Trichostomatida e.g. *Balantidium*

1.4.1 Trypanosomes of humans in South America

Trypanosoma cruzi (Figs 1.2a, 1.3a) infects 11–12 million people in South and Central America and is infective to about 100–150 species of wild and domesticated mammals. It is not at all certain how many of these act as reservoirs of human infection but the armadillo is very important as in this host the infections are long-lived. The vectors are bugs belonging to the family Reduvi-

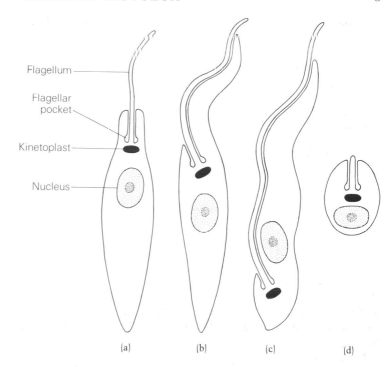

Flagellum

Flagellar pocket

Kinetoplast

Nucleus

Fig. 1.1 Forms in the life cycle of a kinetoplastid flagellate.
(a) Promastigote. (b) Epimastigote.
(c) Trypomastigote.
(d) Amastigote.

(a) (b) (c) (d)

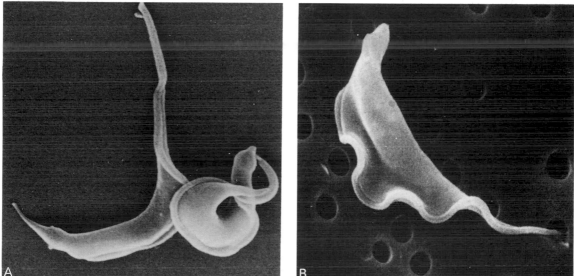

A B

Fig. 1.2 Scanning electronmicrographs of (A): *Trypanosoma cruzi* and (B): *T. brucei*. ×2340. (Photograph (A) kindly given by Dr. D. Snary and (B) by Professor K. Vickerman.)

idae of which three genera are important in the spread of the human disease. When the bug takes up infected blood the trypanosomes multiply in the epimastigote form in the hind gut and infec-

tive or metacyclic forms are passed out with the faeces. These infect the human host if they are rubbed into the bite, another wound or the conjunctiva of the eye. Within the human host, the

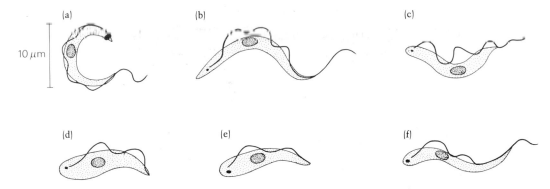

Fig. 1.3 Trypanosomes from the blood of experimentally infected mice. (a) *Trypanosoma cruzi*. (b) *T. brucei rhodesiense* slender from. (c) *T.b. rhodesiense* intermediate form. (d) *T.b. rhodesiense* stumpy form. (e) *T. congolense*. (f) *T. vivax*. All drawn from Giemsa stained slides.

trypanosomes enter various cells, particularly macrophages, muscle and nerve cells, where they round up and multiply in the amastigote form. The amastigotes develop into trypomastigotes that either enter new cells or are taken up when a vector feeds. The disease is called Chagas disease and takes various forms depending on where the amastigotes develop, the most serious consequences being cardiac failure due to parasites in the heart muscles or the loss of the nervous control of the alimentary canal due to parasites in the nervous system.

1.4.2 Trypanosomes of humans in Africa

In contrast to the American trypanosomes the African trypanosomes typically develop in the midgut of the vectors, which are tsetse flies belonging to the family Glossinidae, and are injected from the salivary glands when the fly feeds. Two subspecies infect humans, *T. brucei gambiense* and *T.b. rhodesiense* (Fig. 1.3b–d) the former in riverine conditions in West and Central Africa where it causes chronic sleeping sickness, and the latter in the savanna of East Africa where it causes acute sleeping sickness, although the distinctions between these two forms of the disease are blurred. Both *T.b. gambiense* and *T.b. rhodesiense* can infect a range of mammalian hosts and some of these are important reservoirs of *T.b. rhodesiense*. Natural reservoirs of *T.b. gambiense* seem to be less important because the

infection is essentially a human–human one although pigs may be a source of human infection in some places. The vector of human sleeping sickness is the tsetse fly, *Glossina*; the wet flies of the *G. palpalis* group transmit *T.b. gambiense* and dry flies of the *G. morsitans* group transmit *T.b. rhodesiense* but, again, these distinctions are not absolute. The life cycles of these parasites are the same as that of *T.b. brucei* which is described in Fig. 1.4. The main cause of sleeping sickness as a disease is the invasion of the nervous system by the trypanosomes.

1.4.3 Trypanosomes of domesticated animals

Trypanosomiasis is endemic throughout the tsetse belt of Africa, an area of some 10 million square kilometres. The disease, which is commonly known as nagana, embraces a variety of different manifestations – but usually involves fever, anaemia, lack of appetite and wasting – and causative organisms (see Table 1.2), the three most important of which are *T. congolense*, *T. vivax* (Fig. 1.3e, f) and *T.b. brucei*. *Trypanosoma congolense* is confined to the vascular system and is a particularly important pathogen of cattle although it also affects other animals and can be transmitted by most species of *Glossina*. *Trypanosoma vivax* is also a blood parasite and, like *T. congolense*, is extremely pathogenic in cattle. *Trypanosoma vivax* can be transmitted mechanically by flies that bite and has spread

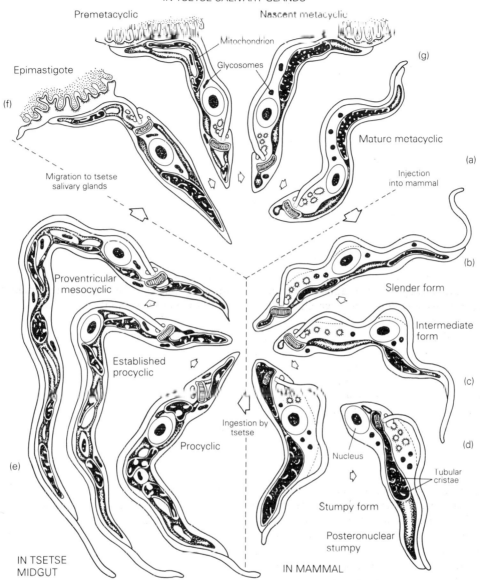

IN TSETSE SALIVARY GLANDS

Premetacyclic

Nascent metacyclic

Mitochondrion

(g)

Glycosomes

Epimastigote

Mature metacyclic

(f)

(a)

Migration to tsetse
salivary glands

Injection
into mammal

(b)

Proventricular
mesocyclic

Slender form

Intermediate
form

Established
procyclic

(c)

Procyclic

Ingestion by
tsetse

(d)

Nucleus

Tubular
cristae

(e)

Stumpy form

Posteronuclear
stumpy

IN TSETSE
MIDGUT

IN MAMMAL

Fig. 1.4 The life cycle of *Trypanosoma brucei*. The infection begins when trypanosomes are injected into the blood of a mammal by a tsetse fly when it feeds (a). The slender forms multiply by binary fission (b) until large numbers build up in the blood and the trypanosomes transform first into intermediate (c) and then stumpy (d) forms that are infective to a tsetse fly. In the slender forms, the mitochondrion is inactive but begins to become active in the stumpy forms. In the midgut of the tsetse fly, the trypanosomes begin to undergo division (e) and then enter the proventriculus and salivary glands where they assume the epimastigote form (f) and undergo further division. The forms in the salivary glands infective to the mammal are known as metacyclic forms (g). In the tsetse fly, the mitochondrion is fully active. The blood stream forms are covered with a glycoprotein coat which is lost in the midgut of the tsetse fly and is reformed in the salivary glands. (After K. Vickerman 1985, *British Medical Bulletin*, **41**, p. 107 and reproduced by permission of the publishers, Churchill Livingstone, Edinburgh.)

Table 1.2 Trypanosomes of humans and domesticated animals

Subgenus	Species	Main hosts	Main reservoirs	Vectors	Disease	Main distribution
STERCORARIA						
Megatrypanum	theileri	Cattle		Tabanid flies	None	Cosmopolitan
Herpetosoma	rangeli	Humans, cat, dogs		Triatomid bugs	None	Central and S. America
Schizotrypanum	cruzi	Humans	150 species of mammals	Triatomid bugs	Chagas disease	Central and S. America
SALIVARIA						
Duttonella	vivax	Cattle, sheep, goats, etc.	Various wild mammals	Tsetse flies	Nagana	Africa
	vivax	Cattle		Tabanid flies	Huequera	S. America
	uniforme	Cattle	Wild ruminants	Tsetse flies	None or mild	Central and E. Africa
Nannomonas	congolense	Cattle, sheep, goats, etc.	Wild ruminants	Tsetse flies	Nagana	Africa
	simiae	Pigs	Warthogs	Tsetse flies	Acute in pigs	Central and E. Africa
Pycnomonas	suis	Pigs	Wild pigs	Tsetse flies	Acute in young	Central and E. Africa
Trypanosoma	brucei brucei	Equines, sheep, goats, etc.	Wild game	Tsetse flies	Nagana	Africa
	brucei rhodesiense	Humans	Wild game	Tsetse flies	Sleeping sickness	E. Africa
	brucei gambiense	Humans	Pigs	Tsetse flies	Sleeping sickness	W. Africa
	evansi	Horses, camels, etc.	Wild mammals	Tabanid flies	Surra	N. Africa, Asia, Middle East
	evansi	Equines	Wild mammals	Tabanid flies	Mal de caderas	Central and S. America
	equiperdum	Equines		Venereal contact	Dourine	Africa, Middle East

outside the tsetse belt not only in Africa but also to South America where it was imported with cattle from the Old World. The third important trypanosome is *T. brucei brucei* which has been extensively studied particularly as a model for human trypanosomiasis (see Fig. 1.4 for the life cycle). *Trypanosoma brucei brucei* is essentially a tissue parasite, living in subcutaneous and connective tissue but also appearing in the blood (Fig. 1.2b). *Trypanosoma brucei brucei* causes serious disease in horses and camels but also infects cattle. Among the other trypanosomes of domesticated animals are *T. uniforme*, which resembles *T. vivax* but generally causes mild infections, *T. simiae* which resembles *T. congolense* in pigs, and *T. suis* also in pigs but a rather rare parasite.

The *T.b. brucei* group of trypanosomes provide an interesting example of evolution in action. Not only has the basic form given rise to the human forms *T.b. gambiense* and *T.b. rhodesiense*, it has also given rise to *T. evansi* which is no longer cyclically transmitted but is passed from animal to animal on the mouthparts of biting flies and even vampire bats in South and Central America. The vampire bat can also be infected with *T. evansi* and thus acts as both host and vector. The parasite formerly called *T. equinum* in South America is now regarded as a synonym for *T. evansi*. *Trypanosoma equiperdum* has dispensed with a vector altogether and is transmitted between horses as a venereal disease. The success of this method of transmission can be seen from the fact that *T. equiperdum* has spread as a disease of horses, donkeys and mules from South and North-West Africa, Syria, Turkey and parts of Asia but has been eradicated from Europe, America, India and most of Asia.

1.4.4 Other trypanosomes of mammals

As well as the trypanosomes described above there are a number of non-pathogenic species that are commonly found in mammals. These include *T. theileri* in cattle, which is transmitted by horse flies, and *T. melophagium* in sheep and sheep keds. *Trypanosoma rangeli*, which occurs in a number of mammals including humans, primates, cats and dogs in Central and South America, is transmitted by bugs and, although the parasite is quite different from *T. cruzi*, the two are occasionally confused. *Trypanosoma rangeli* is harmless in its vetebrate host but may be harmful in the bug. Rodents are frequently infected with trypanosomes transmitted by fleas. Two species, *T. lewisi* in rats and *T. musculi* in mice, are widely used in laboratory studies.

1.4.5 *Leishmania*

The leishmanial parasites exhibit only two forms in their life cycles; amastigotes (Fig. 1.5a) in macrophages of the mammalian host and promastigotes (Fig. 1.5b) in the gut of the vector which is a sandfly (Diptera, Psychodidae). *Leishmania* species cause serious diseases in humans.

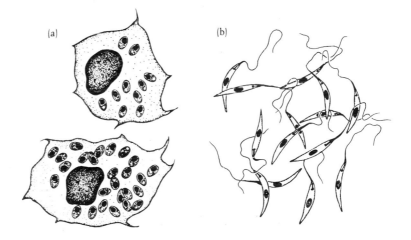

Fig. 1.5 *Leishmania major.*
(a) Amastigote forms in macrophages. (b) Promastigote forms from culture, these forms are equivalent to those that occur in the sandfly. Drawn from Giemsa stained slides. (After, World Health Organization (1984) *The Leishmaniases. Technical Report Series No. 701* and reproduced by permission of the World Health Organization.)

Table 1.3 The main species of *Leishmania* that cause human disease

Species	Disease	Distribution	Reservoir	Vector
OLD WORLD				
L. tropica	Dry cutaneous; urban	Europe, Asia, N. Africa	Dogs	*Phlebotomus*
L. major	Wet cutaneous; rural	Asia, Africa	Rodents	*Phlebotomus*
L. aethiopica	Dry cutaneous; diffuse	Ethiopia, Kenya	*Hyrax*	*Phlebotomus*
L. donovani	Visceral (kala-azar)	Africa, Asia	–	*Phlebotomus*
L. infantum	Infantile visceral	Mediterranean	Dogs, foxes	*Phlebotomus*
NEW WORLD				
L. mexicana	Cutaneous	Central America	Rodents	*Lutzomyia*
L. amazonensis	Cutaneous	Brazil	Rodents, etc.	*Lutzomyia*
L. pifanoi	Cutaneous	Venezuela	Rodents	*Lutzomyia*
L. venezuelensis	Cutaneous	Venezuela	?	*Lutzomyia*
L. braziliensis	Mucocutaneous	Brazil	Rodents	*Lutzomyia*
L. guyanensis	Cutaneous	S. America	?	*Lutzomyia*
L. panamensis	Cutaneous	Panama	Sloths, etc.	*Lutzomyia*
L. peruviana	Cutaneous	S. America	Dogs	*Lutzomyia*
L. chagasi	Visceral	S. America	Foxes	*Lutzomyia*

The typical infection is cutaneous but in many species, and in particular individuals, the parasites may invade subcutaneous or deeper tissues causing hideous and permanent disfiguration. The most serious disease, kala-azar, involves the macrophages of organs such as the liver. Leishmaniasis is now known to be caused by a complex of species and subspecies. As the morphology of all these parasites is similar, the identification of species and subspecies tends to be based on isoenzyme and DNA techniques. The classification of these parasites is still in a state of flux following a change of emphasis away from morphological and disease-associated characteristics towards biochemical and molecular criteria. The most widely used scheme is summarized in Table 1.3. The approach favoured here is to accept a number of species rather than the subspecies used in earlier classifications but the various systems are compatible. There are also several other species recorded from South America, including *L. lainsoni*, *L. naiffi* and *L. shawi* that infect humans and *L. enriettii*, *L. hertigi*, *L. deanei* and *L. aristidesi* that are found only in wild animals. In the Old World, the main species causing cutaneous leishmaniasis are *L. tropica* and *L. major* and the species causing visceral leishmaniasis is *L. donovani*. In the New World

L. chagasi causes visceral leishmaniasis but cutaneous and mucocutaneous leishmaniasis are caused by several species including *L. braziliensis*, *L. mexicana* and *L. peruviana*. A number of species of *Leishmania* have been isolated from hosts other than humans.

1.5 INTESTINAL AND RELATED FLAGELLATES

A number of flagellates occur in the alimentary canals of humans and domesticated animals and similar species are found in laboratory animals. In most cases, the life cycles are very simple and involve the ingestion of food or water contaminated with encysted forms which excyst in the intestine where multiplication by binary fission takes place. Large infestations can build up but the infections are seldom harmful although some may cause gastro-intestinal disorders. Flagellates similar to those in the intestine can occur in other parts of the body, such as the urinogenital system, and these may cause more serious infections.

1.5.1 Intestinal and related forms in humans

Eight species of flagellate are ubiquitous and common parasites of the human gastro-intestinal

Table 1.4 Intestinal amoebae and flagellates and related forms in humans

Species	Cysts	Site	Pathology
Entamoeba histolytica	[C]	Colon/liver	Ulceration, diarrhoea
E. hartmanni	[C]	Colon	None
E. coli	[C]	Colon	None
E. gingivalis	–	Mouth	Gingivitis
Endolimax nana	[C]	Colon/caecum	None
Iodamoeba buetschlii	[C]	Colon/caecum	None
Trichomonas vaginalis	–	Vagina/urethra	Vaginitis, urethritis
T. tenax	–	Mouth	None
T. hominis	–	Colon	None
Dientamoeba fragilis	–	Colon/caecum	None
Giardia duodenalis	[C]	Duodenum	Diarrhoea
Retortamonas intestinalis	[C]	Colon/caecum	None
Enteromonas hominis	[C]	Colon/caecum	None
Chilomastix mesnili	[C]	Colon/caecum	None

tract or urinogenital system (Table 1.4). Few do any real harm but some occasionally give rise to unpleasant symptoms which can usually be easily treated. *Giardia duodenalis* (Fig. 1.6a) which is also known as *G. lamblia*, *G. intestinalis* or *Lamblia intestinalis*, is found in the upper part of the small intestine where large in-festations may cause malabsorption particularly in children. The overall prevalence is usually 1–30% but can reach 70% in unsanitary institutions. It is not known for certain whether there is a reservoir host. Three species of *Trichomonas* are common in all parts of the world: *T. hominis* occurs in the caecum and large

Fig. 1.6 Scanning electronmicrographs of (a) *Giardia duodenalis* and (b) *Trichomonas vaginalis*. The sucking disc of *Giardia* and the undulating membrane of *Trichomonas* are characteristic features. (a) ×2520, (b) ×1260. (Both photographs kindly given by Professor V. Zaman.)

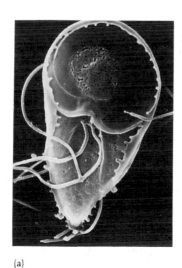

(a)

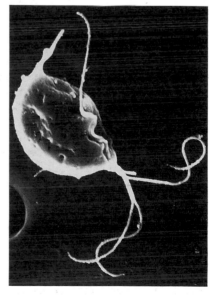

(b)

intestine, *T. tenax* in the mouth and *T. vaginalis* (Fig. 1.6b) in the vagina and urethra of women and in the urethra, seminal vesicles and prostate of men. *Trichomonas vaginalis* may cause inflammation and discharge and is an increasingly important venereal disease afflicting some 180 million women. These trichomonads do not form cysts: *T. hominis* forms rounded resistant stages while *T. vaginalis* and *T. tenax* are transmitted by direct contact. *Retortamonas intestinalis* is a cosmopolitan species, shared with pigs and primates, in the caecum and colon of 1–10% of humans and it is occasionally associated with mild diarrhoea. *Enteromonas hominis* is cosmopolitan and harmless. *Chilomastix mesnili* is rare and harmless. *Dientamoeba fragilis* is transmitted through the eggs of the pinworm *Enterobius vermicularis* and may cause diarrhoea.

1.5.2 Intestinal and related forms in domesticated animals

The most important intestinal flagellates are found in birds and include *Histomonas meleagridis* which occurs as an amoeboid form in the cells of the small intestine and liver and as a flagellated form in the caecum of chickens and turkeys throughout the world. In turkeys it causes a serious disease, blackhead, which may kill 50–100% of young birds. Transmission is via the eggs of the caecal nematode *Heterakis gallinarum*. *Tetratrichomonas gallinarum*, also known as *Trichomonas gallinarum* or *T. pullorum*, occurs in the caecum and liver of chickens, turkeys and other gallinaceous birds throughout the world but is not a serious pathogen. *Spironucleus meleagridis*, also known as *Hexamita meleagridis*, is found in the duodenum and small intestine of turkeys and causes catarrhal enteritis in young birds. *Trichomonas gallinae* occurs in the upper digestive tract of pigeons and occasionally chickens and turkeys where it is usually harmless but may cause serious disease if the parasites invade other parts of the body, such as the brain.

In cattle, the most important, flagellate is *Trichomonas foetus*, sometimes called *Tritrichomonas foetus*. It is a parasite of the genital organs of cattle throughout the world. In the bull it is harmless but in cows may cause early abortion resulting in the loss of 50–100% of calves. Transmission is by direct contact and, as bulls never lose the infection and treatment is unreliable, control can be effected by the use of artificial insemination using uninfected donors.

1.5.3 Other flagellates

A variety of flagellates commonly occur in wild, domesticated and laboratory animals throughout the world. Mice, for example harbour species of *Giardia*, *Spironucleus (Hexamita)* and *Trichomonas* and the first two may be pathogenic in laboratory colonies.

1.6 PARASITIC AMOEBAE

Six species of amoebae are common in humans in most parts of the world but only one, *Entamoeba histolytica*, is an important pathogen.

1.6.1 *Entamoeba histolytica*

This parasite occurs throughout the world in humans, apes, monkeys, dogs, cats and rats. The trophozoite, or feeding stage (Fig. 1.7a) inhabits the lower small intestine and colon where it multiplies by binary fission and forms characteristic four-nucleated cysts (Fig. 1.7b) which are passed

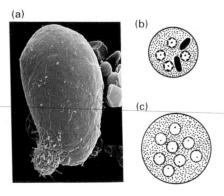

Fig. 1.7 Parasitic amoebae. (a) Trophozoite of *Entamoeba histolytica* ×1325 (scanning electronmicrograph kindly given by Professor V. Zaman). (b) *Entamoeba histolytica* cyst and (c) *E. coli* cyst drawn from stained slides.

out and subsequently ingested in contaminated food or water. Sometimes the amoebae invade the mucosa and submucosa and may be carried via the portal vein to the liver and other parts of the body. Considerable damage may be caused in the wall of the bowel or in the liver. In most people, there is no tissue invasion and the parasite causes no harm. The symptoms following the invasion of the tissues are variable but usually include diarrhoea or dysentery with the loss of blood (amoebic dysentery). In Europe and the USA the prevalence of *E. histolytica* is less than 5% and overt symptoms are rare but in many parts of the tropics and subtropics the prevalence is more than 50% and dysentery and liver involvement may be common.

1.6.2 Other intestinal amoebae of humans

There are four other amoebae commonly found all over the world. *Entamoeba hartmanni*, once regarded as a small form of *E. histolytica*, resembles the pathogenic form but has smaller cysts. *Entamoeba coli* (Fig. 1.7c) is the most common amoeba in humans and has cysts with eight nuclei. *Endolimax nana* inhabits the upper part of the colon and has four-nucleate cysts. *Iodamoeba buetschlii* has cysts with a single nucleus. None of these four parasites is pathogenic.

1.6.3 Facultative amoebae of humans

There have been occasional reports of free-living amoebae infecting humans, sometimes with fatal results. Several species of *Acanthamoeba* can cause upper respiratory tract infections in immunocompromised individuals. *Acanthamoeba* species also act as reservoirs for *Legionella pneumophilia*, the causative agent of legionellosis. *Naegleria fowleri* and other *Naegleria* species, which are strictly speaking flagellates, have been implicated in primary meningoencephalitis in otherwise healthy individuals.

1.7 COCCIDIA

The coccidia are very common parasites mainly of the intestinal tracts of vertebrates. Some are major pathogens of domesticated animals and

losses attributed to them run into millions of pounds or dollars each year. The life cycle usually involves one host and is shown in Fig 1.8. There are a number of minor variations to this basic pattern mainly relating to the site of the infection and the number of schizogonic generations. The genera are identified on the morphology of the infective stage or oocyst. Each oocyst contains a number of sporocysts each containing sporozoites. In the two most important genera, the oocysts of *Eimeria* contain four sporocysts while those of *Isospora* contain two. Certain parasites with isosporan features have two hosts in their life cycles and these will be discussed in Section 1.7.2.

1.7.1 Coccidiosis in domesticated animals

Coccidia are extremely common in domesticated animals and some cause very serious diseases. Coccidia normally have self-limiting infections followed by the acquisition of immunity to reinfection. Under natural conditions, animals become infected with small numbers of oocysts and are only mildly affected by the infection. Under crowded conditions, such as those existing in batteries, large numbers of oocysts may be ingested causing severe or fatal infections, particularly in young animals. The actual pathological effects produced depend to a large extent on the number of oocysts ingested. It is therefore difficult to attribute pathogenicity to particular species and, as the species are extremely difficult to identify, there is considerable controversy about the number of species that exist in any given host and exactly how important one particular species may be. A list of the more important and widely accepted species of *Eimeria* is given in Table 1.5. The species in rodents, *E. nieschulzi* and *E. falciformis*, are widely used in experimental investigations. The life cycle of a typical member of the genus *Eimeria* is shown in Fig. 1.8.

1.7.2 *Toxoplasma* and related coccidia

Until 1970, all coccidians with two sporocysts in the oocyst were classified as *Isospora* species and it was assumed that all had simple life cycles in a single host like that of *Eimeria*. Since 1970, it has

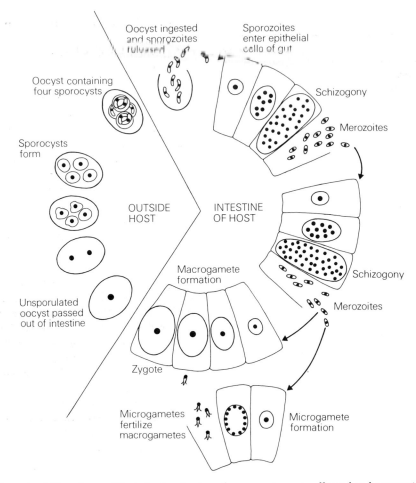

Fig. 1.8 The life cycle of *Eimeria* spp. The infection begins when oocysts are swallowed and sporozoites released in the gut. The sporozoites enter gut cells and undergo a phase of multiplication (schizogony) resulting in the formation of uninucleate merozoites. These merozoites invade other cells and the cycle is repeated two or three times. Eventually merozoites enter other cells where they develop into male and female gametocytes. The nucleus of the male or microgametocyte divides and produces flagellated microgametes which fertilize the female or macrogamete to produce a zygote. A cyst wall or oocyst forms round the zygote within which the nucleus divides twice to produce four uninucleate bodies. Another cyst wall or sporocyst forms round each of these and one further cell division occurs to produce two sporozoites in each. The oocyst containing four sporocysts each with two sporozoites is the infective stage.

There are a number of variations on this life cycle pattern, the main ones being in the number and sites of the schizogonic cycles.

become clear that many of these isosporans develop in an intermediate host which may or may not be obligatory. In such life cycles, the oocysts or sporocysts are passed out from the definitive host and are ingested by an intermediate host within which multiplication in various organs occurs and eventually cysts are formed which, when ingested by the definitive host, initiate the typical coccidian life cycle once again. The problem has been that the parasites in the intermediate hosts were all well-known and had been given valid names. The various parts of each life

Table 1.5 Important *Eimeria* species

Host	Species	Pathogenicity
Chickens	E. acervulina	+
	E. brunetti	++
	E. maxima	+
	E. mitis	+
	E. necatrix	+++
	E. tenella	+++
Turkeys	E. meleagrimitis	++
	E. adenoeides	
Geese	E. anseris	+
	E. nocens	++
	E. truncata	+++
Ducks	E. danailova	++
Cattle	E. bovis	++
	E. zuernii	++
Sheep	E. crandallis	+++
	E. ovina	++
	E. ovinoidalis	+++
	E. arloingi	++
	E. ninakohlyakimovae	++
Pigs	E. debliecki	++
Horses	E. leukarti	++
Rabbits	E. flavescens	++
	E. intestinalis	++
	E. stiedai	+++
Rats	E. nieschulzi	++
Mice	E. falciformis	++

+++, Very pathogenic; ++, moderately pathogenic;
+, slightly pathogenic.

cycle have now been put together and it is possible to identify each parasite from the stages in either the definitive or the intermediate host. The most important species are given in Table 1.6.

This group now includes seven genera, whose characteristics are summarized below. *Isospora* is classified on the basis of stages in the definitive host while all the others are classified by stages in the intermediate host.

● *Isospora*. Direct life cycle.
● *Toxoplasma*. Intermediate host not essential. Development in the lymphoid macrophage system. Cysts thin-walled containing many organisms.
● *Hammondia*. Similar to *Toxoplasma* but intermediate host essential.
● *Cystoisospora*. Intermediate host not essential. Similar to *Toxoplasma* but thin-walled cyst contains only one infective organism.
● *Sarcocystis*. Septate cysts in muscle of intermediate host.
● *Besnoitia*. Thick-walled cysts in connective tissue of intermediate host.
● *Frenkelia*. Cysts in brain of intermediate host.

As a result of these discoveries, the species of *Isospora* in humans, cats and dogs have had to be redefined. The two species originally described from humans were *I. belli* and *I. hominis*. *Isospora belli* remains unchanged while *I. hominis* becomes *S. hominis* and *S. suihominis*. In the dog, *I. canis* becomes *C. canis*, *I. rivolta* becomes *C. rivolta* and *I. bigemina* becomes *S. tenella*, *S. cruzi* and *H. heydorni*. In the cat, *I. felis* becomes *C. felis*, *I. rivolta* becomes *S. rivolta* and *I. bigemina* could apply to so many species that it is impossible to identify the original descriptions with any certainty.

The majority of this group of parasites cause little harm to their hosts, the most important exception being *Toxoplasma gondii* (Fig. 1.9). *Toxoplasma gondii* is a parasite of felids with a very wide range of intermediate hosts, including humans. In cats, and other felids, the life cycle is

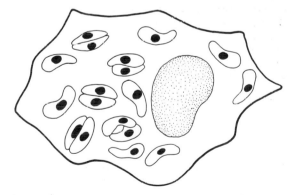

Fig. 1.9 *Toxoplasma gondii* zoites in a macrophage of an experimentally infected mouse. Drawn from a Giemsa stained slide. (The large stippled structure is the host cell nucleus.)

Table 1.6 Isosporan coccidia of humans and domesticated animals*

Definitive host	None	Humans	Cattle	Sheep	Horses	Pigs	Rodents
Man	I. belli		S. hominis S. bovihominis			S. suihominis	
Cats		T. gondii	S. hirsuta (S. bovifelis) B. besnoiti	T. gondii S. gigantea (S. ovifelis)		S. porcifelis	T. gondii H. hammondi S. muris B. wallacei C. felis C. rivolta
Dogs			S. cruzi (S. bovicanis) H. heydorni	S. tenella (S. ovicanis)	S. bertrami	S. miescheriana (S. suicanis)	C. canis C. ohioensis H. pardalis

* This list is not complete and the nomenclature of the group is still being clarified. Some alternative names are given in parentheses.

a normal eimerian one (Fig. 1.10). If, however, the oocysts are ingested by other warm-blooded animals, multiplication occurs in various cells of the body and eventually cysts are formed. If the intermediate host is eaten by a definitive host, the parasite enters the cells of the gut and reverts to a normal eimerian life cycle. If, on the other hand, the intermediate host is eaten by another potential intermediate host disseminated infections occur as before. The infection may cause no symptoms or it may kill the intermediate host. Most humans acquire their infections from undercooked meat or from cats. Infections are normally symptomless but in the unborn fetus or immunosuppressed patients they may be very serious and occasionally in healthy individuals they may cause ocular damage. Toxoplasmosis can also cause serious infections in puppies and lambs.

1.7.3 *Cryptosporidium*

Coccidians belonging to the genus *Cryptosporidium* are relatively common parasites in the intestinal and respiratory tracts of mammals, birds and reptiles. Two species occur in mammals, *C. muris* and *C. parvum*, the latter causing gastrointestinal disorders in cattle, sheep and humans.

The life cycle is typically coccidian and is confined to a single host. The oocysts, which contain four sporozoites, are long-lived and are extremely resistant to normal water purification procedures so contaminated drinking water constitutes the main source of infection. In recent years there has been an increasing number of epidemics of cryptosporidiosis. The infection is unpleasant but not normally dangerous except in immunocompromised individuals, such as AIDS sufferers, in which it can be fatal. There is no effective drug against this organism.

1.8 MALARIA PARASITES

The malaria parasites belong to the same phylum as the coccidians but to a different order, the Haemosporidida, members of which, as the name implies, are parasitic in the blood of vertebrates. All use dipteran insects as their vectors. The malaria parasites of mammals all belong to the genus *Plasmodium*, the life cycle of which is shown in Fig. 1.11, and are transmitted by female mosquitoes belonging to the genus *Anopheles*.

1.8.1 Malaria parasites of humans

Human malaria is one of the most important diseases in the world with over 500 million

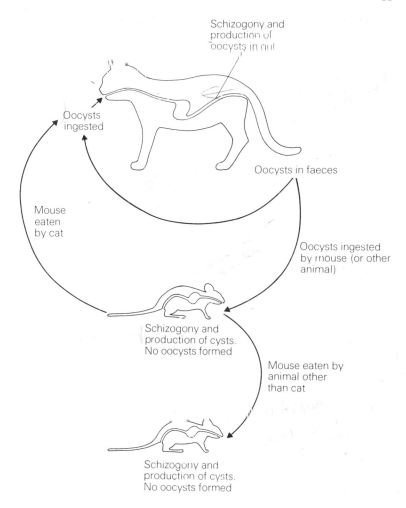

Fig. 1.10 The life cycle of *Toxoplasma*. The life cycle begins when oocysts are swallowed by a cat and the sporozoites are released in the gut. A typical eimerian life cycle with schizogony and the formation of gametocytes occurs and eventually oocysts are formed and the cycle is repeated if these are swallowed by cats. If the oocysts are swallowed by mice, schizogony occurs in various parts of the body but no oocysts are formed. Instead cysts are produced that lie dormant until the host is eaten. If eaten by a cat the normal eimerian life cycle follows. If eaten by another host an infection similar to that in mice occurs. This may occur in a variety of mammals including herbivores. Man becomes infected either by oocysts from cats or by eating infected meat. Congenital transmission in this alternative cycle also occurs.

people at risk in tropical and subtropical parts of the world especially Africa. Malaria is caused by four species of *Plasmodium*: *P. falciparum*, *P. vivax*, *P. ovale* and *P. malariae* (Table 1.7). The disease is characterized by periodic fevers coinciding with the liberation of merozoites during the erythrocytic phase of the infection; these fevers occur every 72 hours in the case of *P. malariae* and every 48 hours in the other species. In all species, there is a single phase of exo-erythrocytic schizogony and in *P. falciparum* and *P. malariae* this phase lasts for 5–15 days. After this, the only parasites in the body are those in the blood and subsequent bouts of fever are caused by recrudescences of these blood forms. In

P. vivax and *P. ovale* some of the parasites in the liver lie dormant for several years and subsequent infections due to the maturation of these forms are called relapses.

Plasmodium falciparum (Fig. 1.12a) causes malignant tertian malaria and is the most common and serious of all the forms of malaria. The infection is acute and the parasites tend to stick to endothelial cells causing blockage and cerebral damage, often resulting in death. *Plasmodium vivax* (Fig. 1.12b) causes benign tertian malaria and is the second most serious infection. *Plasmodium ovale* (Fig. 1.12c) causes ovale tertian malaria and is concentrated in West Africa. *Plasmodium malariae* (Fig. 1.12d) causes quartan

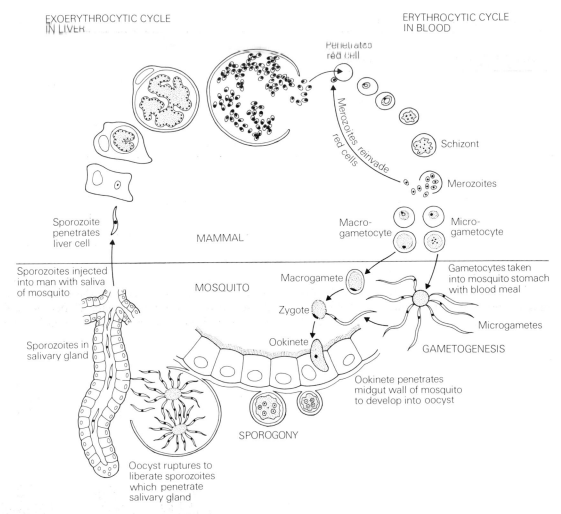

EXOERYTHROCYTIC CYCLE IN LIVER

ERYTHROCYTIC CYCLE IN BLOOD

Penetrated red cell

Merozoites reinvade red cells

Schizont

Merozoites

Sporozoite penetrates liver cell

Macro-gametocyte

Micro-gametocyte

MAMMAL

Sporozoites injected into man with saliva of mosquito

Macrogamete

Gametocytes taken into mosquito stomach with blood meal

MOSQUITO

Zygote

Microgametes

Sporozoites in salivary gland

Ookinete

GAMETOGENESIS

Ookinete penetrates midgut wall of mosquito to develop into oocyst

SPOROGONY

Oocyst ruptures to liberate sporozoites which penetrate salivary gland

Fig. 1.11 The life cycle of *Plasmodium* spp. in mammals. The infection begins when sporozoites are injected directly into the bloodstream from the salivary glands of a mosquito. The sporozoites enter liver cells where they begin a phase of multiplication called 'exoerythrocytic schizogony', during which thousands of uninucleate merozoites are formed. These enter red blood cells in which they undergo a second phase of multiplication or erythrocytic schizogony, during which fewer than 24 merozoites are formed. These merozoites invade new red blood cells and the cycle may be repeated many times. Some of the merozoites are capable of developing into sexual stages or gametocytes. These are taken up by a mosquito. In the gut of the mosquito microgametes are produced and these fertilize the macrogametes and the resulting zygote or ookinete bores through the gut wall to come to lie on the outer surface where it forms an oocyst. Within the oocyst a third stage of multiplication occurs resulting in the formation of sporozoites that enter the salivary glands of the mosquito. (After K. Vickerman and F.E.G. Cox, 1967, *The Protozoa*, John Murray, London.)

malaria and infections may last 30 years or more. Infections with these last three parasites, although debilitating, are seldom fatal in themselves.

1.8.2 Other malaria parasites

The parasites of humans, with the exception of *P. malariae*, are not naturally transmissible to

Table 1.7 Malaria parasites affecting humans

Species	Disease	Periodicity (hours)	Distribution
P. vivax	Benign tertian	48	Cosmopolitan, between summer isotherms 16°N and 20°S
P. ovale	Ovale tertian	48	Mainly tropical W. Africa
P. falciparum	Malignant tertian	48	Cosmopolitan, mainly tropics and subtropics
P. malariae	Quartan	72	Cosmopolitan but patchy

other animals so the malaria parasites of primates and rodents have received considerable attention both in their own rights and as models for the human infections. There are about 20 species of *Plasmodium* in non-human primates of which *P. cynomolgi*, which resembles *P. vivax*, has been the most studied. Another species from macaques, *P. knowlesi*, is now widely used in laboratory studies despite the fact that it has a 24-hour periodicity and does not closely resemble any of the human species. *Plasmodium cynomolgi*, *P. brasilianum*, *P. inui* and *P. knowlesi* infect humans under experimental (or accidental) conditions.

The malaria parasites of rodents have been much more extensively studied than any others. These fall into two major groups, *P. berghei* and *P. yoelii* and their subspecies and *P. vinckei* and

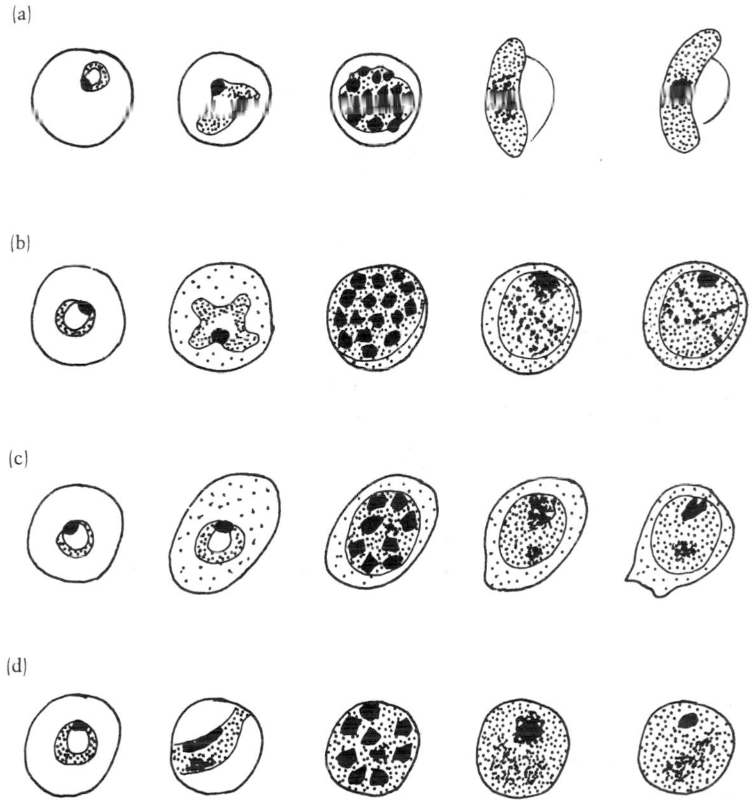

(a)

(b)

(c)

Fig. 1.12 Malaria parasites from human blood. (a) *Plasmodium falciparum*. (b) *P. vivax*. (c) *P. ovale*. (d) *P. malariae*. From left to right: ring stage, trophozoite, mature schizont, microgametocyte and macrogametocyte. In the case of *P. falciparum* only the ring stage and gametocytes appear in the peripheral blood. All drawn from Giemsa stained slides.

(d)

P chabaudi and their subspecies. The *bergei–youlii* group typically invade immature erythrocytes and the *vinckei–chabaudi* group invade mature cells. They all have 24-hour periodicities and only distantly resemble the human forms but, nevertheless, have provided a wealth of information on the biology of malaria parasites which would have been otherwise unobtainable.

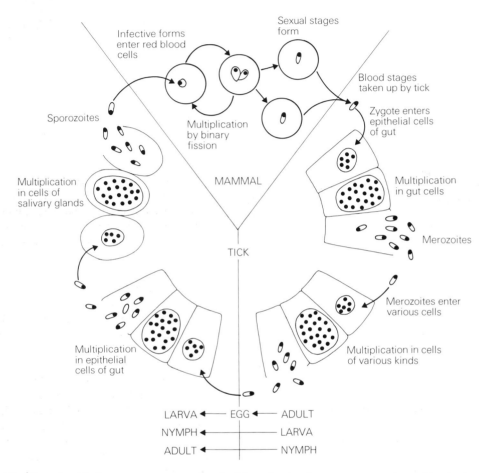

Fig. 1.13 The life cycle of *Babesia* spp. in mammals. The life cycle of *Babesia* spp. is complicated by the facts that any one of three stages in the life cycle of the tick vector may become infected and that ticks may have one, two or three hosts during their life cycles. These variations have marked effects on the life cycle of various *Babesia* species and the cycle described is a generalized one. The cycle in the mammal begins when infective forms are injected into the blood by a tick and the parasites enter red blood cells where they multiply by binary fission. Blood forms, probably gametocytes, are taken up by a tick when it feeds. These become gametes that fuse to form zygotes which enter epithelial cells of the gut where multiplication takes place. The resulting products are comparable with the merozoites in coccidian life cycles. These enter various cells in the body of the tick where further phases of multiplication occur. If the tick is a female, the merozoites may enter and multiply in the egg. When the larva hatches from the egg it is already infected and the parasites multiply once again in the epithelial cells of the gut. The uninucleate forms produced invade other tissues including the salivary glands from which small rounded infective bodies (sporozoites) are injected into the mammalian host. In some species, the nymph and not the larva harbours the infective parasites. The stage of the tick that is infected is never infective and usually the infection goes from larva to nymph, nymph to adult or adult to larva via the egg.

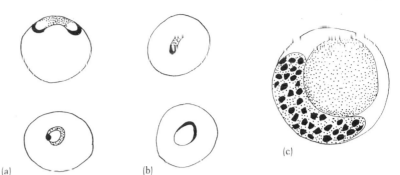

Fig. 1.14 Piroplasms from the blood of cattle. (a) *Babesia bigemina* and (b) *Theileria parva* in red blood cells. (c) *Theileria parva* schizont in a lymphocyte. All drawn from Giemsa stained slides.

(a)   (b)   (c)

1.9 PIROPLASMS

The piroplasms are parasites of the erythrocytes of vertebrates in which, if they multiply, they divide by simple binary fission. The parasites in the blood do not contain pigment and this, together with the absence of schizogony, distinguishes them from the malaria parasites. The vectors are ticks.

1.9.1 *Babesia*

Babesia species live in the blood of vertebrates and are transmitted by ticks. The life cycle is shown in Fig. 1.13. The form in the vertebrate is the trophozoite that lives in red blood cells in which it divides by binary fission to produce two merozoites, each of which invades a new cell (Fig. 1.14a). These blood stages cause serious disease in domesticated animals (see Table 1.8) and the general name given to these is babesiosis. The symptoms of babesiosis are fever, anaemia, jaundice and haemoglobinuria, and the infections are often fatal. The species are identified on the stages in the blood but as there are few morphological characters this is not altogether satisfactory. The main division is into 'large' and 'small' forms and in the past the number of genera have been created on this characteristic alone, but at present only one genus, *Babesia* is generally accepted. The distinction between 'large' and 'small' is a convenient starting point for the identification of any *Babesia* in a particular host (see Table 1.8).

There are a number of *Babesia* species in wild animals but whether or not these play any significant part in the transmission of these parasites to domesticated animals is not known. There are also two important species in rodents, *B. rodhaini* and *B. microti*, that have been widely used in laboratory studies. *Babesia microti* occasionally infects humans in America but the infections are not fatal. In Europe, splenectomized humans are sometimes infected with *B. bovis* and the infection frequently proves fatal.

1.9.2 *Theileria*

Theileria species are parasites of cattle, sheep and goats in which the majority of the life cycle occurs in the lymphoid tissues (Fig. 1.14c) and the stages infective to ticks occur in the red blood cells (Fig. 1.14b). The life cycle is shown in Fig. 1.15 and the various parasites that cause disease are listed in Table 1.9. The most serious diseases caused by *Theileria* occur in cattle in which *T. parva*, causing African theileriosis or East Coast Fever, is often lethal, and *T. annulata* is sometimes lethal but as the infected cattle remain carriers for long periods this infection is probably more important on a world scale than *T. parva*. *Theileria mutans* and *T. velifera* are not very pathogenic. *Theileria sergenti* is probably a synonym of *T. mutans*. In *T. parva* and *T. annulata* infections the damage is done during the phase of schizogony in the lymphoid tissues and the symptoms include hypertrophy of the lymphoid tissues, fever and loss of weight. In *T. mutans* (and *T. sergenti*) there may be anaemia and jaundice. In sheep and goats *T. hirci* is very pathogenic when newly introduced into an area but *T. ovis* which is morphologically indistinguishable from *T. hirci*, is hardly pathogenic at all.

Table 1.8 *Babesia* species causing disease in domesticated animals

Species	Size	Main hosts	Main vectors	Pathogenicity	Main distribution
B. bigemina	L	Cattle	Boophilus	+++	All continents
B. bovis	S	Cattle	Boophilus	++	All continents
B. divergens	S	Cattle	Ixodes	++	Europe
B. major	L	Cattle	Haemaphysalis	+	Europe, Africa
B. caballi	L	Horses	Dermacentor, Hyalomma, Rhipicephalus	++	All continents
B. equi	S	Horses	Dermacentor, Hyalomma, Rhipicephalus	+++	All continents
B. motasi	L	Sheep, goats	Dermacentor, Haemaphysalis, Rhipicephalus	+	Europe, Asia, Africa
B. ovis	S	Sheep, goats	Rhipicephalus	+	Europe, Asia, Africa
B. trautmanni	L	Pigs	Rhipicephalus	+	Europe, Asia, Africa
B. perroncitoi	S	Pigs	Not known		Europe, Africa
B. canis	L	Dogs	Rhipicephalus	+++	Europe, Asia, Africa, America
B. gibsoni	S	Dogs	Rhipicephalus	+++	Asia
B. herpailuri	L	Cats	?	+	Africa
B. felis	S	Cats	Haemaphysalis	++	Asia, Africa

+++, Most pathogenic, +, least pathogenic; L, Large; S, small.

Table 1.9 *Theileria* species causing disease in domesticated animals

Species	Main hosts	Main vectors	Pathogenicity	Main distribution
T. annulata	Cattle	Hyalomma	+++	Europe, Asia, Africa
T. mutans	Cattle	Amblyomma	+	Europe, Asia, Africa, Australia
T. parva	Cattle	Rhipicephalus	+++	Africa
T. sergenti	Cattle	Haemaphysalis	+	Asia
T. velifera	Cattle	Amblyomma	+	Africa
T. taurotragi	Cattle, sheep, goats	Rhipicephalus	+	Africa
T. hirci	Sheep, goats	Hyalomma	+	Africa
T. ovis	Sheep, goats	Rhipicephalus	+	Africa
T. separata	Sheep, goats	Rhipicephalus	+	Africa

+++, Most pathogenic; +, least pathogenic.

1.10 MICROSPORIDIA

There are some 700 species of microsporidians and these are found in nearly all groups of vertebrates and invertebrates, particularly fish and arthropods. Microsporidians are characterized by the possession of a thick-walled spore containing an infective body or sporoplasm surrounded by a coiled hollow tube, the polar filament, through which the sporoplasm is injected into its host. The infection begins when a spore is ingested and the sporoplasm enters a cell of the gut. The actual details of what follows vary from species to species, but multiplication by binary fission and the spread of the infection to other tissues occurs, and eventually spores are formed and released when the host dies or is eaten. *Encephalitozoon cuniculi* occurs in rodents, rabbits, carnivores and primates and there are a few records from humans. This is a parasite of peritoneal

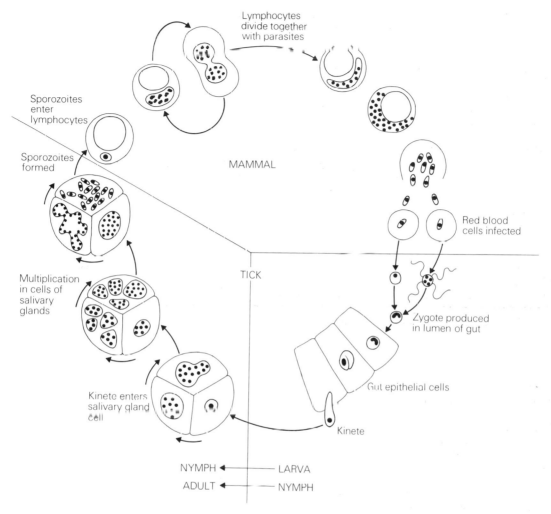

Fig. 1.15 The life cycle of *Theileria* spp. The cycle in a mammal begins when the infective stages, sporozoites, are injected by a tick. These enter lymphocytes in which they divide repeatedly causing the lymphocytes themselves to divide and the parasites to be distributed between the two daughter host cells. After a number of divisions, merozoites are formed and these invade red blood cells and are taken up by a tick when it feeds. Within the lumen of the gut of the tick, fertilization occurs and the zygote enters the epithelial cells of the gut where it develops into a motile kinete which subsequently enters the salivary glands where another phase of multiplication results in the formation of sporozoites. These are injected into the mammalian host when the tick feeds. The stage of the life cycle of the tick that is infected is never infective and there is no transovarian transmission. Thus, if the nymph is infected, multiplication and sporozoite formation occur in the salivary glands of the adult.

macrophages but may also spread to other parts of the body including the brain where spores can be found. Another microsporidian, *Enterocytozoon bieneusi*, is increasingly being reported in AIDS patients.

1.11 CILIOPHORA

The Ciliophora is a distinct phylum, considered by some to be a subkingdom, containing 4700 free-living and 2500 parasitic species. Parasitic

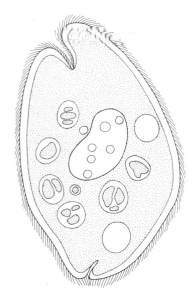

Fig. 1.16 *Balantidium coli* from *in vitro* culture.
Drawn from a haematoxylin stained slide.

ciliates occur in most groups of vertebrates and
invertebrates and those in amphibians and earth-
worms are frequently encountered in elementary
biology classes. Few of the parasitic ciliates are of
any economic importance.

1.11.1 *Balantidium coli*

Balantidium coli (Fig. 1.16) is a common parasite
of pigs in all parts of the world and has also
been recorded in rats, dogs, monkeys, apes and
humans. It is difficult to know how many human cases there have been but about 1000 have
been recorded, mainly in the tropics. The ciliate
lives in the lumen of the large intestine and may
invade the gut wall where it produces ulcers re-
sembling those caused by *Entamoeba histolytica*
although the majority of cases are asymptomatic.
Transmission is by cysts and epidemiological evidence suggests that most human infections are
acquired from pigs.

1.12 *PNEUMOCYSTIS*

The taxonomic position of *Pneumocystis* is uncertain but it is almost certainly a fungus. However, as it has been long regarded as a protozoan,
it is included here. *Pneumocystis carinii* is a
natural parasite occurring in the lungs of about
80% of humans in which it exists as a harmless
amoeba forming cysts which are the transmission
stages. In immunocompromised individuals parasite numbers increase until they fill the alveolar
spaces, causing pneumonia which may be fatal.
Pneumocystosis is a common complication in
people suffering from AIDS.

REFERENCES
AND FURTHER READING

Bruce-Chwatt, L.J. (1985) *Essential Malariology*, 2nd
edn. London, Heinemann.

Canning, E.U. & Lom, J. (1986) *The Microsporidia of
Vertebrates*. London, Academic Press.

Chang, K-P. & Bray, R.S. (eds) (1985) *Leishmaniasis*.
Amsterdam, Elsevier.

Coombs, G.H. & North, M. (eds) (1991) *Biochemical
Protozoology*. London, Taylor and Francis.

Cox, F.E.G. (1992) Systematics of parasitic protozoa. In
J.P. Kreier & J.R. Baker (eds) *Parasitic Protozoa*, 2nd
edn., Vol. 1, pp. 55–80. San Diego, Academic Press.

Dubey, J.P. & Beattie, C.P. (1988) *Toxoplasmosis of
Animals and Man*. Boca Raton, CRC Press.

Dubey, J.P., Speer, C.A. & Fayer, R. (1989) *Sarcocystosis
of Animals and Man*. Boca Raton, CRC Press.

Gardiner, C.H., Fayer, R. & Dubey, J.P. (1988) *An Atlas
of Protozoan Parasites in Animal Tissues*. U.S. Department of Agriculture Handbook No. 651.

Garnham, P.C.C. (1966) *Malaria Parasites and
Other Haemosporidia*. Oxford, Blackwell Scientific
Publications.

Hoare, C.A. (1972) *The Trypanosomes of Mammals*.
London, Tindall & Cox.

Honigberg, B.M. (1989) *Trichomonads Parasitic in
Humans*. New York, Springer-Verlag.

Hopkin, J. (1991) *Pneumocystic carinii*. Oxford, Oxford
University Press.

Knell, A.J. (ed.) (1991) *Malaria*. Oxford, Oxford University Press.

Kreier, J.P. & Baker, J.R. (1987) *Parasitic Protozoa*. Bos-
ton, Allen & Unwin.

Kreier, J.P. & Baker, J.R. (eds) (1992–3) *Parasitic
Protozoa*, Vols 1–7. New York, Academic Press.

Lee, J.J., Hutner, S.H. & Bovee, E.C. (1985) *An Illustrated Guide to the Protozoa*. Lawrence, Kansas,
Society of Protozoologists.

Levine, N.D. (1973) *Protozoan Parasites of Domesticated Animals and of Man*, 2nd edn. Minneapolis,
Burgess Publishing Co.

Levine, N.D. (1988) *The Protozoan Phylum Apicomplexa*, Vols 1 and 2. Boca Raton, CRC Press.

Long, P. (ed.) (1982) *The Biology of the Coccidia*. London, Edward Arnold

McGregor, I.A. & Wernsdorfer, W. (eds) (1988) *Malaria. Principles and Practice of Malariology*. Vols 1 and 2. Edinburgh, Churchill Livingstone.

Martinez-Palomo, A. (1986) *Amoebiasis*. Amsterdam, Elsevier.

Matsumoto, Y. & Yoshida, Y. (1986) Advances in *Pneumocystis* biology. *Parasitology Today*, 2, 137–142.

Mehlhorn, H. (ed.) (1988) *Parasitology in Focus: Facts and Trends*. Berlin, Springer-Verlag.

Meyer, E.A. (ed.)(1990) *Giardiasis*. Amsterdam, Elsevier.

Molyneux, D.H. & Ashford, R.W. (1983) *The Biology of Trypanosoma and Leishmania, Parasites of Man and Domestic Animals*. London, Taylor and Francis.

Norval, R.A.I., Perry, B.D. & Young, A.S. (eds) (1992) *The Epidemiology of Theileriosis in Africa*. London, Academic Press.

Peters, W. & Killick-Kendrick, R. (eds) (1987) *The Leishmaniases in Biology and Medicine*, Vol. 1. *Biology and Epidemiology*, Vol. 2. *Clinical Aspects and Control*. London, Academic Press.

Ristic, M. (ed.) (1988) *Babesiosis of Domestic Animals and Man*. Boca Raton, CRC Press.

Rondanelli, E.G. (ed.) (1987) *Amoebiasis and Amoebiasis. Human Pathology, Infectious Diseases Color Atlas Monographs 1*. Padua, Piccin Nuova Libraria.

Sleigh, M.A. (1989) *Protozoa and Other Protists*. London, Edward Arnold.

Smith, H.V. & Rose, J.B. (1990) Waterborne crytosporidiosis. *Parasitology Today*, 6, 8–12.

Stephen, L.E. (1986) *Trypanosomiasis. A Veterinary Perspective*. Oxford, Pergamon Press.

Tzipori, S. (1988) Cryptosporidiosis in perspective. *Advances in Parasitology*, 27, 63–129.

World Health Organization (1984) *The Leishmaniases*. WHO Technical Report Series No. 701. Geneva, WHO.

World Health Organization (1986) *Epidemiology and Control of African Trypanosomiasis*. WHO Technical Report Series No. 739. Geneva, WHO.

World Health Organization (1987) *The Biology of Malaria Parasites*. WHO Technical Report Series No. 743. Geneva, WHO.

Young, A.S. & Morzaria, S.P. (1986) Biology of *Babesia*. *Parasitology Today*, 2, 211–219.

Chapter 2 / Parasitic Helminths

P. J. WHITFIELD

2.1 INTRODUCTION

The vast majority of metazoan parasites of vertebrates are representatives of two phyla – the acoelomate Platyhelminthes and the pseudocoelomate Nematoda. 'Helminth' is a practically useful, but imprecisely defined, term which includes all the cestodes and digeneans in the former group and all the parasitic members of the latter. It is as the causative agents of a terrible list of debilitating, deforming and killing diseases of humans and their agricultural animals that helminths are principally studied.

Some helminth infections are numbered among the major human infectious diseases. Schistosomiasis (bilharzia), caused by digeneans of the genus *Schistosoma* which inhabit blood vessels, is a most important cause of morbidity with over 200 million infected persons in Africa, South America and the Far East. Even more sufferers exist in tropical and subtropical zones with the major nematode diseases. It is quite conceivable that one billion are infected with both ascariasis caused by *Ascaris lumbricoides* and trichuriasis whose causative agent is *Trichuris trichiura*. Insect vector-transmitted filarial nematodes like *Wuchereria bancrofti*, *Brugia malayi* and *Onchocerca volvulus* cause highly pathogenic forms of filariasis that harm hundreds of millions of patients. Human cestode disease is a relatively nonpathogenic affliction although two types of larval cestodiasis – hydatid disease and cysticercosis – can both be highly injurious.

2.2 STRUCTURE AND FUNCTION OF HELMINTHS

Helminths are very diverse in their structure, physiology and behaviour. This diversity results partly from their varying taxonomic origins, and partly from their multiple adaptations for their particular one-, two- or three-host life cycles.

Digenean and cestode platyhelminths share a solid triploblastic acoelomate body plan, with complex reproductive organs embedded in mesenchymal tissue, and a gut where present, possessing only a single, oral opening. The reproductive system is almost always hermaphroditic. All these worms possess a living syncytial body wall whose outer surface and secretory activity is differently modified in the two groups.

Nematodes possess a pseudocoelomate body cavity which plays an important hydrostatic skeletal role in locomotion. Most nematodes are dioecious, with the male and female reproductive tracts located in the body cavities of the separate sexes. These worms have a body wall consisting of a syncytial hypodermis surmounted by an apparently non-living, mainly collagenous cuticle and underlain by groups of longitudinal muscles. In almost all species there is a functional gut with a mouth and anus.

Helminths use many different life cycle modes, and some like the nematode *Strongyloides stercoralis* and the cestode *Hymenolepis nana*, can use more than one mode. Direct life-cycle strategies, with only a single host species involved, are used by many gut-dwelling nematodes and also by *H. nana* in one of its life-cycle modes. Indirect life cycles are those in which more than one host species is used in order to complete one circuit of the parasite's life history. In such cycles, the final or definitive host, that is the host in which parasitic sexual reproduction takes place, is almost always a vertebrate. The intermediate hosts, those in which development, growth, encystment or asexual multiplication occurs, can be vertebrates or invertebrates. Indirect life cycles with two or three hosts are the rule among digeneans and the vast majority of cestodes. Among the nematodes, filarial worms and *Dracunculus medinensis* (the guinea worm) utilize two-host indirect cycles. In the dioecious schistosomes and nematodes, sperm transfer between the sexes is obligatory. In the hermaphroditic digeneans and cestodes self-fertilization is sometimes possible but unidirectional or reciprocal cross-fertilization is nearly always favoured. Parthenogenesis is rare, but is used by both the nematode *Strongyloides stercoralis* and the lung-inhabiting digenean *Paragonimus*.

Transmission of helminth infections from host to host is achieved by eggs or larvae. Eggs are usually directly ingested by a host. Larvae may be similarly ingested, consumed while attached to a plant or eaten while located in an intermediate host which acts as a prey item for the next host in the life cycle. Free-living mobile larval forms such as digenean miracidia and cercariae and L_3 larval nematodes are often able to find, recognize and invade new hosts. They possess impressive sensory and locomotory abilities which enable them to carry out these functions. Transmission is discussed in more detail in Chapter 7, Section 7.2.

Table 2.1 An outline classification of helminths parasitic in vertebrates

---

PHYLUM: PLATYHELMINTHES
Class 1 Monogenea
Class 2 Cestoda (*Diphyllobothrium, Taenia, Echinococcus*)*
Class 3 Aspidogastrea
Class 4 Digenea (*Schistosoma, Fasciolopsis, Fasciola, Paragonimus*)

PHYLUM: NEMATODA
Order 1 Rhabditida (*Strongyloides*)
Order 2 Strongylida (*Necator, Ancylostoma*, etc.)
Order 3 Ascaridida (*Ascaris, Toxocara*, etc.)
Order 4 Oxyurida (*Enterobius*)
Order 5 Spirurida (*Dracunculus, Wuchereria, Brugia, Loa, Onchocerca*)
Order 6 Enoplida (*Trichinella, Trichuris*)

---

* Important genera which infect man are listed after the appropriate taxa.

2.3 CLASSIFICATION OF PARASITIC HELMINTHS

Table 2.1 provides a working outline classification for the helminth groups considered in any detail in this book. More detailed classification systems will be provided for each of the major helminth groups as they are considered in turn. In using these classifications it must be remembered that there are still major disagreements about the relationships between the higher taxonomic levels. It is also true that most helminth species descriptions are still centred upon morphological and host specificity criteria. Increasingly though, molecular criteria which involve genomic and isoenzyme polymorphisms are also being utilized in species discriminations. It is to be hoped that these will open the way for the development of effective natural taxonomies.

2.4 PLATYHELMINTH PARASITES OF VERTEBRATES

Of the four classes of entirely parasitic platyhelminths, only two, the Cestoda and Digenea, cause important diseases in humans or agricultural animals, although monogeneans of fish can cause serious losses in stocks kept under high-density fish farming conditions.

2.4.1 Cestodes

Cestodes are a unique and, in many ways, aberrant group of platyhelminth parasites, representing the most extreme specializations of the basic flatworm body plan for endoparasitism. With very few exceptions they all share two remarkable attributes: (1) they possess no gut; and (2) they have a very elongated body, often hundreds of times longer than broad. One small subclass, the Cestodaria, have compact non-segmented bodies, but members of the principal subclass, the Eucestoda, have a characteristically segmented adult body made up of a string of proglottids each of which contains in time a complete set of reproductive organs (Fig. 2.1). A mature adult eucestode may consist of several thousand proglottids behind an anterior attachment organ – the scolex – which is

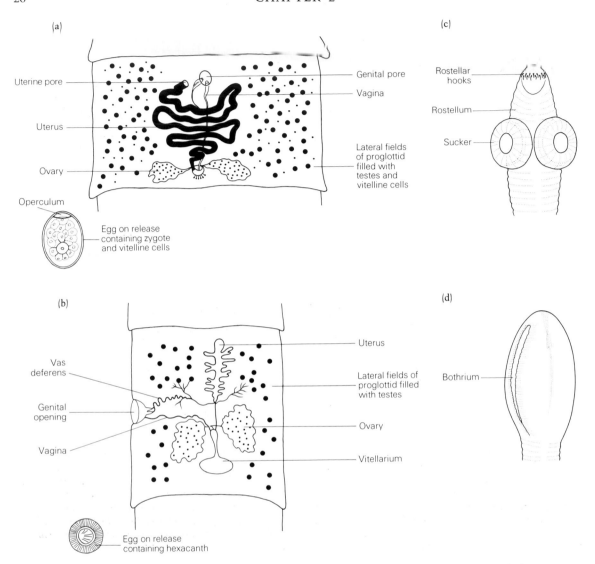

Fig. 2.1 Proglottids, eggs and scoleces of pseudophyllidean and cyclophyllidean cestodes. (a) Proglottid of *Diphyllobothrium latum*, a pseudophyllidean, demonstrating extensive areas of vitelline cells, genital openings on the ventral proglottid surface, and a uterine pore through which the gravid uterus can communicate with the outside world. The egg of *D. latum* when released is unembryonated. (b) Mature proglottid of *Taenia saginata* or *T. solium*, demonstrating a lateral genital opening and a uterus which does not directly communicate with the outside world. The egg of *T. saginata* or *T. solium* when released in human faeces contains a fully formed hexacanth larva. (c) The scolex of the cyclophyllidean *Echinococcus granulosus* from the dog gut, with four suckers and a hooked rostellum. (d) The unhooked scolex of the pseudophyllidean *D. latum* with two muscular grooves or bothria.

equipped with muscular grooves or suckers and sometimes also with hooks (Fig. 2.1). The serial multiplication of hermaphroditic reproductive organ sets which the segmented body represents, is an extraordinary modification for enhanced reproductive capacity. It avoids the production constraints of a single reproductive system and enables individual cestodes to sustain daily egg

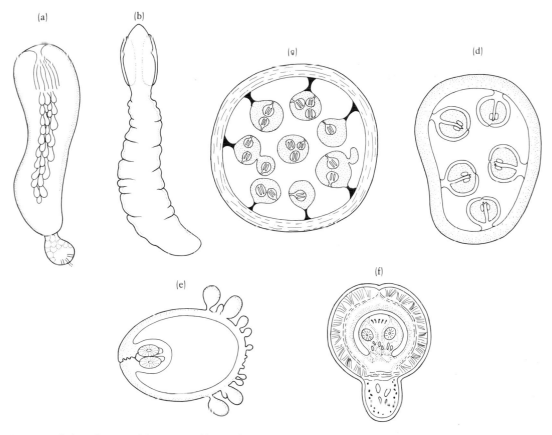

Fig. 2.2 Cestode larval stages. (a) Procercoid larva of *Diphyllobothrium latum* with a central core of gland cells and retained hooks from the hexacanth stage. (b) Plerocercoid larva of *D. latum* with anterior bothria present. (c) Hydatid cyst of *Echinococcus granulosus* with an outer laminated layer and proliferative brood capsules within it containing infective protoscoleces. (d) Coenurus larva of *Multiceps multiceps* with infective protoscoleces budding directly from a germinal membrane. (e) Cysticercus larva of *Taenia crassiceps* with scolex invaginated into wall of bladder. The posterior portions of the external bladder wall produce new cysticerci by exogenous budding. Such budding does not occur from the otherwise similar cysticerci of *T. saginata* and *T. solium*. (f) Cysticercoid larva of *Hymenolepis nana* with the invaginated scolex housed within an anterior vesicle.

outputs of hundreds of thousands or even millions of eggs.

The body wall of a cestode is a living syncytial tegument, the outer plasma membrane of which is thrown into a regular array of specialized microvilli termed microtriches, each surmounted by an electron-dense spine. The tegumentary plasma membrane is covered with a polyanionic glycocalyx. Microtriches produce a large amplification of surface area, operating in a digestive/absorptive manner in these gutless helminths to acquire external nutrients. Organic molecules of low molecular weight are absorbed by diffusion and active transport mechanisms across the plasma membrane. The latter also bears intrinsic phosphohydrolases which probably play a role in nutrient uptake. There is evidence that the tegument can also absorb macromolecules such as proteins by endocytosis. This repertoire of nutrient uptake techniques seems to restrict adult eucestodes to nutrient-rich internal locations within their vertebrate final hosts. The vast majority are found in the lumen of the small intestine attached to its mucosa by their scoleces. Except for *Hymenolepis nana*, all eucestodes exhibit only an indirect life cycle with larval (metacestode) development and/or asexual multiplication occurring in one or two intermediate hosts. Figure 2.2 illustrates a selection of cestode larvae.

2.4.2 Human cestodiasis

Within the Eucestoda there is considerable uncertainty about the appropriate subdivision of the taxon into orders. Conveniently however, the species that cause human disease all fall within two well-recognized and reasonably homogeneous orders the Pseudophyllidea and the Cyclophyllidea. Table 2.2 outlines the major differences between these two groups.

   Tapeworm-generated disease or cestodiasis in humans can be considered under two headings, namely larval cestodiasis in which larval cestodes in a variety of organ sites are the origin of disease symptoms and adult cestodiasis where the pathology is due to adult worms in the human gut. Table 2.3 lists the main examples of human cestodiasis and Fig. 2.3 describes the life cycles of a typical pseudophyllidean and cyclophyllidean – *Diphyllobothrium latum* and *Taenia saginata*.

Table 2.2 An outline classification of pseudophyllidean and cyclophyllidean cestodes

Class: Cestoda

Subclass: Eucestoda
Order 1: Pseudophyllidea. Scolex with two long superficial bothria. Mainly gut-dwelling parasites of non-elasmobranch fish, fish-eating mammals including man and birds. Proglottids dorso – ventrally flattened, usually bearing the uterine pore and genital apertures medially on the ventral surface. Each egg hatches in water to release a ciliated coracidium larva containing a hexacanth. Indirect life cycles including procercoid and plerocercoid larvae, e.g. *Spirometra*, *Diphyllobothrium*

Order 2: Cyclophyllidea. Scolex typically with four large suckers surmounted with a muscular rostellum normally armed with hooks. Gut-dwelling parasites of amphibians, reptiles, birds and mammals including man. Eggs contain non-ciliated hexacanth larvae. Genital apertures marginal on proglottids. Posterior gravid proglottids often shed containing eggs. Indirect life cycles include a variety of nonproliferative and proliferative larval forms in vertebrate and invertebrate intermediate hosts, e.g. *Taenia*, *Echinococcus*

Larval cestodiasis

There are three main types of human larval cestodiasis: sparganosis caused by plerocercoids (= spargana) of diphyllobothriid tapeworms in the

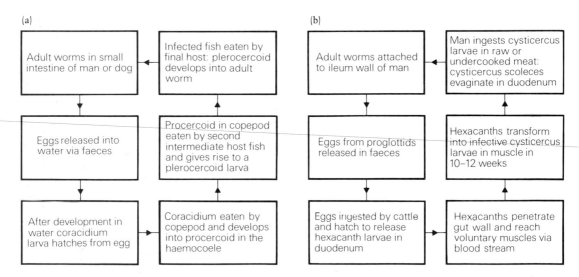

Fig. 2.3 Outline life cycles of cestodes. (a) A pseudophyllidean, *Diphyllobothrium latum*. (b) A cyclophyllidean, *Taenia saginata*.

Table 2.3 Human cestodiasis

Parasite*	Disease	Adult (A) or larval (L)	Location of worms in man	Geographical distribution	Typical pathology
PSEUDOPHYLLIDEANS					
Diphyllobothrium latum	Diphyllobothriasis	A	Lumen of small intestine	Finland, Central Europe, Italy, France, Ireland, Japan, Siberia, Argentina, Great Lakes area of USA and Canada	Very rarely pernicious megaloblastic anaemia
Spirometra spp.	Sparganosis	L	A variety of deep tissues	Tropical and subtropical regions of Africa, N. and S. America, Europe, Far East and Australasia	Mild, various
CYCLOPHYLLIDEANS					
Taenia saginata	Beef tapeworm infection	A	Lumen of small intestine	Cosmopolitan	Very infrequently gut obstruction or perforation
T. solium	Pork tapeworm infection	A	Lumen of small intestine	Cosmopolitan	Very infrequently gut obstruction or perforation; risk of cysticercosis
T. solium	Cysticercosis	L	A variety of deep tissues including the brain	Cosmopolitan, but particularly in S. and Central America, USSR, India and S. and E. Africa	A range of brain pathologies are produced by larvae in the CNS, including epilepsiform attacks
Hymenolepis nana	Dwarf tapeworm infection	A	Lumen of small intestine	Cosmopolitan especially in children	Usually none, diarrhoea and abdominal pain in heavy infections
Echinococcus granulosus	Echinococcosis: hydatid disease	L	A variety of deep tissue sites with liver and lungs predominating	Most sheep and cattle farming areas of the world	Various; depending on the site of the hydatid cysts

* A number of other cestodes occasionally infect man, namely *Bertiella studeri*, *Dipylidium caninum* and *Hymenolepis diminuta* (adult cestodiasis); *Echinococcus multilocularis*, *Mesocestoides* spp. and *Multiceps* spp. (larval cestodiasis).

genera *Diphyllobothrium* and *Spirometra*; cysticercosis caused by cysticercus stage larvae of *T. solium*; and hydatid disease resulting from the proliferative hydatid cyst larvae of *Echinococcus granulosus*. In none of these cases are humans obligate hosts for the parasites – they contribute very little to parasitic reproductive success being so rarely consumed by the relevant final hosts. (Some authorities refer to most non-adult cestode development stages as metacestodes.)

A variety of carnivores are the normal final hosts for sparganosis-producing *Diphyllobothrium* and *Spirometra* species. Eggs in their faeces hatch to produce coracidia which develop into procercoids in copepods of freshwater habitats. The crustaceans are eaten by a wide range of amphibians, reptiles and mammals, in which plerocercoids develop. Spargana are wandering plerocercoids in human tissues. Persons become infected in a number of ways. They may ingest copepods containing procercoids when drinking untreated water from streams and lakes. They may also acquire infections by eating raw poultry, frogs and snakes or by poulticing inflamed eyes or abscesses with fresh frog or snake skin. Such poultices enable the plerocercoids to migrate directly into human tissues. Spargana may wander slowly within the body causing short-lived lumps to appear. Ultimately the spargana die, are calcified and surrounded with a thin fibrous capsule. Most infections occur in Japan, Korea, China and Vietnam.

Adult worms of the pork tapeworm *T. solium* inhabit the human intestine while the larval stages normally develop in pig muscles. Human cysticercosis occurs when people eat *T. solium* eggs, and cysticercus larvae, each about 10 mm long, subsequently become lodged in a variety of tissues. Symptoms are particularly severe if the parasites occur and then die in the human central nervous system. The commonest mode of human infection is probably human faecal contamination of food, but it is possible that direct autoinfection occurs in patients harbouring adult worms, if segments containing eggs are regurgitated from the small intestine into the stomach so that egg activation begins.

Hydatid disease or echinococcosis is one of the most harmful cestodiases. It is a complex disease from the point of view of parasite taxonomy.

Most human infections are with the endogenously budding hydatid larvae of several strains of *E. granulosus* which has discrete spherical cysts. Humans can act as accidental intermediate hosts for *E. granulosus* in its domestic cycle (where it is transmitted between dogs as final hosts and sheep or cattle as intermediate hosts) or in one of its feral cycles utilizing wild carnivores and herbivores. Less frequently, humans become infected by strains of *E. multilocularis* whose irregularly expanding cysts are intensely pathogenic. The latter disease is always feral, with foxes acting as final hosts and rodents as intermediate hosts. In all instances, human infection occurs because of contact with faecally contaminated material from infected canine hosts. The highest community prevalences of hydatid disease around the world are always in cattle or sheep herding populations with high levels of human–dog contact.

Adult cestodiasis

The common adult cestodiases of human hosts are generally less pathogenic than cysticercosis or hydatid disease. Adult cestodes at low densities in the human gut often cause no symptoms. Only rarely, when parasite densities are high, do problems due to gut-wall damage or intestinal obstruction occur. The only other serious consequences of adult cestodiasis are the increased risk of cysticercosis in carriers of adult *T. solium*, and the geographically restricted risk of pernicious anaemia in those infected with *D. latum*.

In global terms only four adult cestodiases are at all common, namely those caused by *D. latum*, *T. saginata*, *T. solium* and *Hymenolepis nana*.

Diphyllobothrium latum (Figs 2.1, 2.2 & 2.3) is a pseudophyllidean tapeworm whose transmission is due to consumption of raw, undercooked or lightly smoked fish containing viable plerocercoids. Dogs can also act as final hosts. The pernicious megaloblastic anaemia caused by *D. latum* infections appears to be confined to a proportion of patients in Finland or of Finnish ancestry. It is caused by the competitive uptake of vitamin

B_{12} by the worms which restricts the amount of dietary B_{12} available for the human host

Taenia saginata (Figs 2.1, 2.2 & 2.3), the beef tapeworm can sometimes reach 20 m in length but 5 m is more usual. It has a cosmopolitan distribution due to the practice of eating raw or undercooked beef. In this way infective cysticercus larvae (about 8 mm in length) in the muscles of a cow are ingested. Larvae evert their unhooked scoleces in the duodenum and attachment and subsequent growth occurs in the duodenum. It is possible that over 60 million cases exist worldwide.

Taenia solium is very similar in morphology and life-cycle characteristics to *T. saginata*. It differs mainly in having a hooked rather than an

Table 2.4 Other cestodes

Name	Final hosts	Intermediate hosts	Comments
CYCLOPHYLLIDEA			
(a) **Taeniidae**			
Taenia crassiceps	Fox	Small mammals	Exogenously budding metacestode larvae in the peritoneal cavity and other tissue spaces of intermediate hosts. Can be maintained by syringe passage of larvae in mice and rats in the laboratory
T. ovis	Dog	Sheep	Larvae develop in sheep muscles. Economic importance in the meat industry
Multiceps multiceps	Dog, coyote, fox, jackel	Sheep, goats, cattle, horses, (occasionally humans)	Larvae form a budding coenurus developing in brain and spinal cord of sheep causing a disease called gid or staggers. Very dangerous larval cestodiasis in man
(b) **Hymenolepidae**			
Hymenolepis diminuta	Rats, mice, hamsters (occasionally humans)	Many insects including fleas and beetles (*Tenebrio* and *Tribolium* in laboratory)	The most commonly used laboratory model cestode. Life cycle easily maintained experimentally
(c) **Dilepidiidae**			
Dipylidium caninum	Dogs, cats (occasionally humans)	Fleas	Causes an uncommon and nonpathogenic human adult cestodiasis especially in children. Infections caused by ingesting infected fleas from pets
(d) **Anoplocephalidae**			
Moniezia expansa and *M. benedini*	Sheep, cattle	Soil inhabiting mites	Causes a common adult cestodiasis of low pathogenicity in sheep
(e) **Davaineidae**			
Raillietina cesticillus	Chickens	Many arthropod species, especially beetles	Causes a common adult cestodiasis in chickens

unarmed adult scolex and fewer lateral branches in the gravid proglottid uterus. Human consumption of undercooked or raw pork is the route of infection. Thorough cooking or prolonged deep freeze storage at −10°C or below kills the cysticercus larvae in the meat.

Hymenolepis nana is only 15–40 mm in length. It can occur at high densities in an infected person's small intestine. Children and young adults are particularly at risk especially those in institutions. Low density infections produce few or no symptoms, but at higher infection levels of more than 2000 worms, vomiting, diarrhoea, loss of appetite and abdominal pain have been recorded. Transmission of the parasite is mainly by a method which is extremely unusual among cestodes, that is the ingestion of eggs which complete larval then adult development in the same human host.

2.4.3 Other cestodes

A number of other cestode species can be considered of importance despite the fact that they are not normally human parasites. Their significance stems either from their use as laboratory models for studies on tapeworm physiology, development, biochemistry or chemotherapy, or from the parasites' utilization of agricultural animals as hosts. The main characteristics of some of these species are outlined in Table 2.4.

2.4.4 Digeneans

Adult digeneans are flattened or cylindrical platyhelminths that in humans and agricultural animals always inhabit endoparasitic locations. The vast majority of species live in the gut or its developmental offshoots such as the bile duct and lungs. Externally, adult digenean worms are char-

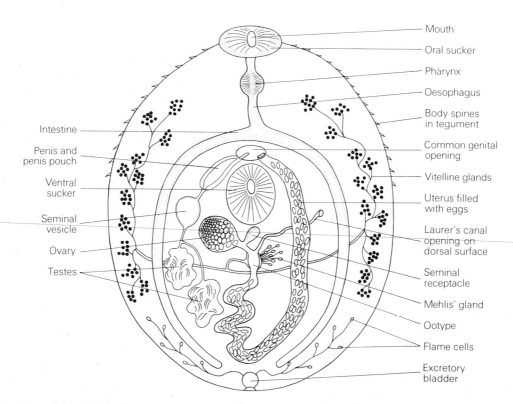

Fig. 2.4 A hypothetical adult digenean demonstrating the main morphological features.

acterized by an oral sucker around the anterior mouth and often an additional ventral sucker or acetabulum. The oral sucker is an important feeding organ while both sucker types are involved in attachment to internal host surfaces and locomotion. The outer surface of the body is a living syncytial tegument, the distal cytoplasm of which often contains spines. The internal organization of a typical hermaphroditic digenean (only schistosomes are dioecious) is described in Fig. 2.4. A basic taxonomy of digeneans is outlined in Table 2.5, with special emphasis on those groups that contain species that infect humans.

The life cycles of the great majority of digeneans display a remarkable and highly characteristic

Table 2.5 An outline classification of digeneans (several families excluded)

CLASS: DIGENEA
Superorder 1: Anepitheliocystida. In the cercaria the bladder wall of the excretory system is the retained wall of the primitive bladder formed from the fusion of the two main lateral excretory canals

Order 1: Strigeatida. Cercariae fork-tailed
Family (i) Bucephalidae (e.g. *Bucephalus*)
Family (ii) Strigeidae (e.g. *Alaria*, *Cotylurus*)
Family (iii) Schistosomatidae. The schistosomes: adults parasitic in the blood vessels of birds and mammals. Dioecious with males and females occurring in *in-copulo* pairs with the female held in the gynaecophoric canal of the male (e.g. *Schistosoma*)

Order 2: Echinostomida. Cercariae with cyst-producing gland cells. Encystation of cercariae occurs on vegetation or in molluscs
Family (i) Echinostomatidae. The echinostomes: adults parasitic in the intestine, bile ducts or ureters of reptiles, birds and mammals. Elongate forms with a raised collar behind the oral sucker carrying large backwards pointing spines. (e.g. *Echinostoma*)
Family (ii) Fasciolidae. Large, flattened, spinose leaf-shaped worms of mammals. (commonly herbivores) (e.g. *Fasciola*, *Fasciolopsis*)
Family (iii) Paramphistomatidae. The amphistomes: gut parasites of mammals (commonly herbivores). Large, thick-bodied digeneans with an anterior oral sucker and a ventral sucker at the extreme posterior end of the worms (e.g. *Paramphistomum*, *Gastrodiscoides*)

continued

Table 2.5 *Continued*

Superorder 2: Epitheliocystida. In the cercaria the bladder wall of the excretory system has a thick epithelial organization with mesodermal origins which replaces the original primitive bladder

Order 1: Plagiorchiida. Operculate eggs. Oral stylet usually present in oral sucker of cercaria
Family (i) Plagiorchiidae. Small flukes, mainly parasites of amphibians and birds, occasionally of fish, reptiles and mammals. Genital opening between well spaced oral and ventral suckers, tandem testes behind ovary. Y-shaped excretory bladder. Cercariae encyst in arthropods and possess oral stylet (e.g. *Plagiorchis*)
Family (ii) Dicrocoeliidae. Small flukes found in the intestine, liver, gall bladder and pancreas of most vertebrate groups. Oral sucker subterminal, vitelline glands do not extend in front of ventral sucker. Cercariae encyst in arthropods and possess oral stylet (e.g. *Dicrocoelium*)
Family (iii) Troglotrematidae. Flukes of birds and mammals, often in lungs or intestine but also in nasal cavities, frontal sinuses and subcutaneous tissues. Vitelline glands compact. Cercariae encyst in arthropods and possess oral stylet (e.g. *Paragonimus*)

Order 2: Opisthorchiida. Operculate eggs. Cercariae have no oral stylet
Family (i) Opisthorchiidae. Medium-sized flukes of the gall bladder and bile duct in mammals, birds and reptiles. Weakly developed suckers (e.g. *Opisthorchis*)
Family (ii) Heterophyidae. Small or minute flukes found in a variety of sites in mammals and birds. Genital sucker, closely associated with the ventral sucker (gonotyl) often present (e.g. *Heterophyes*, *Metagonimus*)

alternation of asexual and sexual reproductive phases, in molluscan and vertebrate hosts respectively. The basic life-cycle patterns employed by digeneans and examples of their larval stages are shown in Fig. 2.5. Hermaphroditic adults which normally cross-inseminate, utilize sexual reproduction in the final host, and tanned eggs are produced. These leave the host in faeces, urine or sputum and the zygote within the egg develops or has already developed by this stage into a ciliated larva – the miracidium. The miracidium then infects a gastropod mollusc either by direct penetration by a free-swimming, hatched miracidium

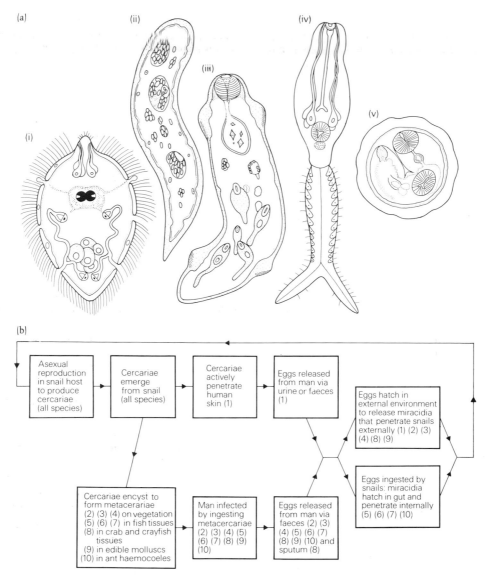

Fig. 2.5 (a) Diagrammatic representation of digenean larval stages. The larvae illustrated are completely schematic and do not represent an actual developmental sequence. (i) Miracidium: with tiers of epithelial ciliated cells, anterior gland cells, central 'eye spot' consisting of two pigment cup ocelli overlying a nervous cerebral ganglion, flame cells and posteriorly positioned germinal cells. (ii) Sporocyst: a sac shaped larval stage with no gut and a central cavity containing clusters of dividing germinal cells. (iii) Redia: with a large oral sucker and a simple gut, muscular processes in the body wall and a central cavity containing developing larval stages. (iv) Cercaria: a larva with a muscular propulsive tail and a head with gut, oral and ventral suckers and often penetration glands and cystogeneous glands. (v) Metacercaria: a transformed cercarial head surrounded by a secreted cyst wall. The cercarial head usually shows some development of somatic and reproductive characteristics towards the adult state. (b) Diagrammatic flow-chart of the basic organization of the life cycles of medically important digeneans that infect man. Species identities: 1: *Schistosoma*; 2: *Fasciolopsis*; 3: *Fasciola*; 4: *Gastrodiscoides*; 5: *Opisthorchis*; 6: *Heterophyes*; 7: *Metagonimus*; 8: *Paragonimus*; 9: *Echinostoma*; 10: *Dicrocoelium*.

or by the ingestion of the unhatched egg by a snail. Within the molluscan host the miracidium transforms into a sporocyst which is a tegument-covered, gutless germinal sac, containing germinal cells. These develop (in different species) into either a second generation of sporocysts or into a new larval type with a gut – the redia. Rediae can often produce further generations of rediae but eventually rediae or daughter sporocysts begin, again asexually, to produce cercariae, which are the larval stages that leave the molluscan host. A cercaria is a tailed, actively swimming larval form, the head of which will develop into an adult worm and often already possesses partially developed reproductive organs. Cercariae leave the snail, the first intermediate host, often with a marked circadian periodicity, and they or the metacercariae which develop from them, infect the final host.

The sequence of asexually reproducing larval stages in the mollusc is often able to sustain very high overall reproductive rates for this part of the life cycle. Ultimately many thousands of cercariae may develop from a single miracidial infection of a snail. Digeneans typically demonstrate greater host specificity with respect to their first intermediate hosts than they do to any subsequent hosts in the life cycle.

The majority of digenean life cycles involve metacercariae. These are cercariae that have shed their propulsive tails and have encysted either on external objects like snail shells or vegetation, or within second intermediate host species – vertebrate or invertebrate. Within these cysts, produced by cystogenous glands in the cercarial head, the cercaria carries out a partial development towards somatic and reproductive maturity. The remainder of this development is completed when the cyst is consumed by a final host. Only in a few digeneans is the free-swimming cercaria directly infective to the final host by an active penetration process that eliminates the need for a metacercarial phase. Although uncommon, this process has great epidemiological significance as it is the infection mode employed by schistosome cercariae.

Of the 6000 or so known digenean species only about a dozen are regular and important human parasites. Many more species, though, are spasmodically diagnosed as accidental human infections. The main non-schistosome digeneans which infect humans are listed in Table 2.3 along with some of their basic biological characteristics. Schistosomes, however, must be regarded as the most important human digeneans with a global total of over 200 million infected persons, a significant proportion of whom will experience moderate-to-severe levels of morbidity.

2.4.5 Human schistosomiasis

Four species of the digenean genus *Schistosoma* are important human parasites. Of these *S. mansoni*, *S. haematobium* and *S. japonicum* have widespread distributions, while *S. intercalatum* is restricted to specific foci in parts of Africa. *Schistosoma haematobium* causes urinary schistosomiasis whereas the other three species are the causative agents of the intestinal form of the disease.

Schistosomes are bizarre digeneans. Many aspects of their morphology, physiology and life-cycle tactics are unorthodox or unique when compared with the other members of the taxon. Adult worms live in the lumina of blood vessels and feed directly on the cellular and plasma fractions of blood. The species causing intestinal schistosomiasis live in the mesenteric veins of the gut while *S. haematobium* occupies the vesical veins of the bladder wall. As an adaptation for these unusual microhabitats, the worms are threadlike in shape, 10–30 mm in length but only 0.2–1.0 mm in width. They are also dioecious and the male and female worms show considerable sexual dimorphism (Fig. 2.6). Cylindrical elongate females live permanently in an extensive ventral groove, the gynaecophoric canal, of the shorter, stouter males. In most schistosomes, female sexual maturity only follows successful pairing with a male worm.

Living as they do in a host location in which immunological defences might be expected to be both rapidly engendered and effective, adult schistosomes possess a range of adaptations in the molecular organization of their outer surface which enable them to partially counter these defences (see Chapter 8, Section 8.6.3). In this way adult worms are able to live and reproduce

Table 2.6 Some features of non-schistosome digenean diseases of humans

Name	Site of adult worms in man	Route of egg emergence	Snail (first intermediate) host and mode of snail infection	Mode of human infection	Geographical distribution
Fasciolopsis buski	Small intestinal mucosa, rarely in stomach or colon	Faeces	*Segmentina* (external miracidial invasion)	Metacercarial cysts on water plants, such as water caltrop and water chestnut: eaten	China, Taiwan, India, Thailand, Laos, Kampuchea and Bangladesh
Heterophyes heterophyes	In crypts of jejunum and upper ileum	Faeces	*Pironella* and *Cerithidia* (egg ingestion)	Metacercarial cysts in fish, such as mullet and *Tilapia*: eaten	Egypt, Israel, Romania, Greece, Japan, China, Taiwan, Philippines
Metagonimus yokogawai	Mucosal folds of jejunum	Faeces	*Semisulcospira* (egg ingestion)	Metacercarial cysts in fish, such as carp and trout: eaten	Japan, Korea, China, Taiwan, Siberia
Gastrodiscoides hominis	Mucosal lining of caecum and ascending colon	Faeces	*Helicorbis* (external miracidial invasion)	Probably by ingestion of metacercarial cysts on water plants such as water caltrop	India, Bangladesh, Vietnam, Philippines
Opisthorchis sinensis	Bile duct	Faeces	*Bulinus, Parafossarulus, Alocima* (egg ingestion)	Metacercarial cysts in freshwater fish: eaten. Juvenile flukes migrate directly up bile duct from gut	China, Taiwan, Korea, Japan, Vietnam
Fasciola hepatica	Bile duct	Faeces	*Lymnaea* (external miracidial invasion)	Metacercarial cysts on watercress or lettuce: eaten. Juvenile flukes penetrate gut wall then liver from the perivisceral cavity	Central and S. America, Cuba, France, UK, N. Africa
Paragonimus westermani	In cysts in lungs, rarely in a variety of extra-pulmonary sites	Sputum and faeces	*Semisulcospira, Thiara* and *Oncomelania* (external miracidial invasion)	Metacercarial cysts in freshwater crabs and crayfish: eaten	China, Taiwan, Korea, Japan, Philippines, India, Malaysia, Indonesia

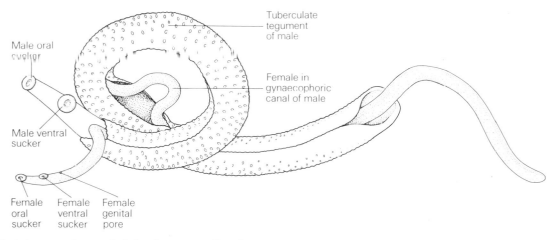

Fig. 2.6 *In copulo* pair of adult schistosomes (based on *S. mansoni*).

for many years in a host who is relatively resistant to new infections.

Pairs of adult worms produce eggs which are laid into the lumina of the venules in which they live. There is no direct, non-pathological route by which these eggs can reach the outside world from this location. In fact, most schistosome eggs possess a sharp shell spine which helps to provide a means of escape. Spines lodge in the intima of the venule and impede the movement of eggs by blood flow. Small blood vessels packed with eggs may rupture, enabling them to move into surrounding connective tissue. A proportion of these eggs eventually reach the outside world via the lumen of the gut or bladder in faeces or urine. This necessarily unusual exit route for schistosome eggs is at the heart of their considerable pathogenicity. Many eggs become lodged in the tissues all over the body. In these locations, living and then moribund or dead eggs become immobilized in spherical granulomatous lesions. It is these progressively accumulating lesions in many different organs which give rise to most chronic schistosome-induced pathology.

The eggs which do leave an infected person in urine or faeces hatch on contact with freshwater and the emergent miracidia infect a range of aquatic and amphibious snails in which infective cercariae are produced. All human digenean infections other than schistosomiasis are initiated when people eat metacercarial cysts. Schistosome transmission is quite different. Furco-

cercariae emerging from the snails survive on average for about a day at 20°C, swimming tail-first through the water. During this brief, free-swimming and non-feeding existence fuelled by endogenous glycogen reserves, they are directly infective to people entering the water in which they are swimming. Aided by small backward-pointing tegumental spines and cytolytic secretions a cercaria rapidly penetrates bare human skin, often down the side of a hair shaft and sheds its tail in the process. The penetrant cercarial head can now be considered as an immature adult schistosome or schistosomulum. Schistosomula enter the peripheral blood system and are carried eventually to the lungs, usually within a week of skin penetration. From the lungs the worms migrate to the liver where they pair up and mature. Pairs of worms then move to their final egg-producing sites. Figure 2.7 illustrates the general organization of human schistosome life cycles while Table 2.7 outlines specific information on the four major species.

Schistosomiasis is very largely a rural disease of the tropics. Contact with mud or water carrying infective cercariae comes about by children playing in water, water gathering or washing and very importantly in the course of work. Agricultural practices which involve direct human contact with irrigation water are especially hazardous. One reason for the increasing prevalence of all rural schistosomiases is the expanding programme of construction of dams for hydroelectric

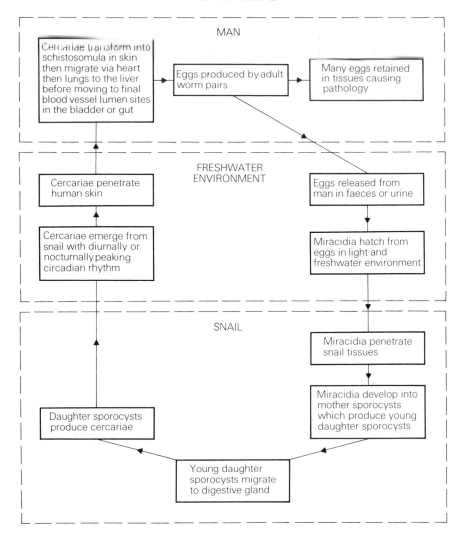

Fig. 2.7 Diagrammatic flow-chart of the main features of the life cycles of human schistosomes.

or irrigation purposes in developing countries. Such schemes inevitably produce new water-bodies for snail colonization.

The pathological effects of intestinal and urinary schistosomiasis are somewhat different. Those, for instance, of *S. mansoni* intestinal schistosomiasis are slowly progressive and complex. During the initial invasion stage there may be a transient dermatitis or 'swimmer's itch'. Thereafter most pathology is the direct or indirect result of the host's immunological responses to parasite eggs (see Chapter 8, Section 8.8). In heavy

infections eggs in the lining of the gut cause granulomatous reactions there which can extend into the gut lumen as pseudopapillomas, with or without egg calcification. Extensive change of this sort may cause colonic obstruction and blood loss. The hepatic portal system carries many eggs to the liver and eggs trapped there are the foci of new granulomatous lesions. General damage to the liver induced in this way produces liver enlargement (hepatomegaly) and following portal hypertension, splenomegaly also occurs. Damage to the portal circulation causes collateral

Table 2.7 Some features of human schistosomes

Name	Site in man	Route of egg emergence	Intermediate hosts	Geographical distribution	Reservoir hosts
Schistosoma mansoni	Mesenteric veins	Faeces	*Biomphalaria* spp.	Egypt, Middle East, W. Central and S.E. Africa, Malagasy, Brazil, Venezuela and some Caribbean islands	Probably not important except in S. America where some rodents are infected
Schistosoma haematobium	Vesical veins of the bladder	Urine	*Bulinus* spp.	Africa, Malagasy and Middle East	No important reservoir hosts
Schistosoma japonicum	Mesenteric veins	Faeces	*Oncomelania* spp.	China, Philippines, Japan, Vietnam, Thailand, Laos and Kampuchea	Important throughout range dogs, rats, pigs, cattle, etc
Schistosoma intercalatum	Mesenteric veins	Faeces	*Bulinus* spp.	Limited distribution in Zaire, Central African Republic, Cameroon and Gabon	No important reservoir hosts

circulatory shunts to be set up and these enable eggs in the bloodstream to bypass the filtering action of the liver. Consequently, eggs begin to reach the lungs and heart causing pathology there.

The pathology caused by *S. intercalatum* is similar to that of *S. mansoni* but less severe. *Schistosoma japonicum* infections are often characterized by colonic pathology associated with the presence of large numbers of calcified eggs.

The symptoms of urinary schistosomiasis caused by *S. haematobium* are closely linked with the bladder location of the adult worms and their eggs. Eggs are deposited in the walls of the bladder itself and the ureters. Granulomatous reactions, and later fibrosis and calcification, induce a wide range of pathological manifestations including haematuria and dysuria, hyperplasia of the bladder lining which can progress to bladder cancer, partial blockage of the ureters and secondary damage to the kidneys.

2.4.6 Other digeneans

Fasciola hepatica, the liver fluke, is primarily a parasite of sheep and cattle but it also infects humans. There is also a wide range of other digenean infections of agricultural stock that only very rarely cause human disease. Among the liver and bile-duct-inhabiting species are *F. gigantica*, living in a variety of agricultural and feral ruminants in Africa and the Far East; *F. magna* that infects deer, horses, cattle and sheep in North America; and *Dicrocoelium dendriticum* infecting sheep and other herbivores particularly in Europe and Asia. Other important parasites of ruminants include the paramphistomes, usually in the genus *Paramphistomum*, which are especially damaging to cattle.

Other schistosomes also warrant some attention. *Schistosomatium douthitti*, a parasite of rodents, has been used as an easily maintained laboratory analogue of human schistosomes. In many of the warmer parts of the world cattle and sheep are severely affected by a large number of *Schistosoma* species. *Schistosoma bovis* in Southern Europe, Africa and Asia, *S. mattheei* in Southern Africa and *S. spindale* in Africa and India all cause serious losses.

2.5 NEMATODE PARASITES OF VERTEBRATES

The nematodes are the most successful group among the pseudocoelomic, eutelic Aschelminthes. The relative lack of interspecific morphological variation among nematodes belies their ecological diversity. They are almost all unsegmented, spindle-shaped roundworms with bilateral symmetry (Fig. 2.8). Both free-living and parasitic species take this form and the vast majority have a highly conservative life-cycle developmental sequence (Fig. 2.9). This starts when dioecious adults sexually produce eggs that hatch to release L_1 larvae, which with intervening cuticular moults and size increases, develop through three more larval phases (L_2, L_3 and L_4) before attaining full sexual maturity after the final post-L_4 moult.

The cuticle which is repeatedly moulted, secreted anew and rearranged during development is a central element in the structural organization of all nematodes. It is a multi-layered covering of the nematode body consisting of collagen and other components. A number of surface invaginations of the body wall (buccal cavity, pharynx, excretory pore, vulva, chemosensory pits called amphids and phasmids, rectum, cloaca and spicule pouches) are also cuticle-lined. Stylets, teeth and cutting blades in the buccal cavity and the copulatory spicules are essentially cuticular structures.

The body plan and behavioural repertoire of nematodes appear to have provided a degree of preadaptation for endoparasitic lifestyles within this group. The chemically and physically resistant cuticle presumably enabled originally free-living species to exist in the damaging conditions in vertebrate guts into which they could be accidentally transferred by host feeding. Similarly the substrate burrowing abilities of many nematodes must have made the penetration of host integuments and internal tissues a distinct possibility.

Table 2.8 outlines a classification of nematodes that are parasitic in vertebrates itemizing species that infect humans. The horrifyingly long inventory of human parasitic diseases can be split

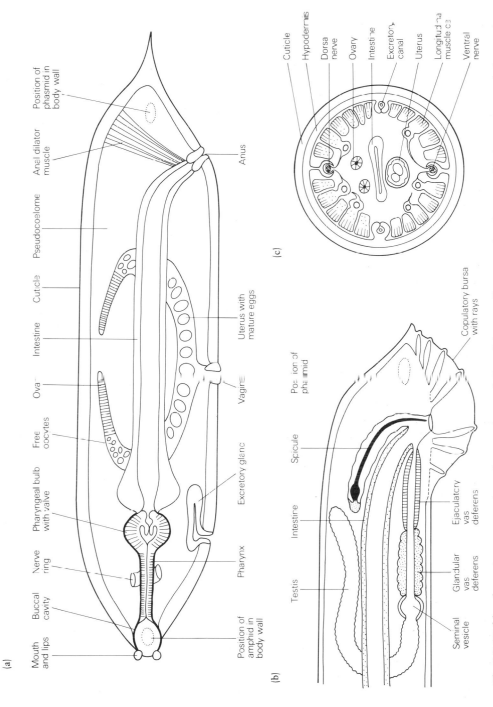

Fig. 2.8 Highly schematic representations of nematode morphology. (a) A sexually mature female worm. (b) The posterior end of a sexually mature male worm. (c) A transverse section through a female worm.

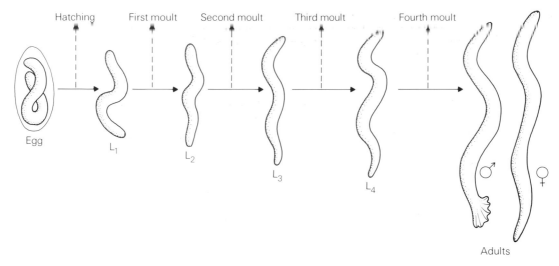

Fig. 2.9 A typical nematode developmental sequence.

Table 2.8 Outline classification of the nematode parasites of vertebrates with particular emphasis on orders and superfamilies containing important human parasites

PHYLUM: NEMATODA
Subclass 1: Secernentea (Posterior phasmid chemoreceptive organs present)
   Order 1: Rhabditida. Parasitic female adults parthenogenetic, pharynx often possesses prominent posterior muscular bulb (e.g. *Strongyloides*)
   Order 2: Strongylida. Adult males possess copulatory bursa with supporting rays. L_1 and L_2 larvae often free living
      Superfamily 1: Ancylostomatoidea. Prominent buccal capsule often with cuticular teeth or cutting plates – 'hookworms' (e.g. *Ancylostoma, Necator*)
      Superfamily 2: Strongyloidea
      Superfamily 3: Trichostrongyloidea
      Superfamily 4: Metastrongyloidea
   Order 3: Ascaridida. Large intestinal nematodes; three-lipped mouth; simple pharynx (e.g. *Ascaris, Toxocara*)
   Order 4: Oxyurida. Pharynx possesses posterior bulb with valves. Adult female has long pointed post-anal tail (e.g. *Enterobius*)
   Order 5: Spirurida. Pharynx divided into two sections, a shorter anterior muscular portion and a longer glandular posterior one
      Superfamily 1: Filarioidea. Very elongate adults; usually viviparous; intermediate hosts blood sucking arthropods – 'filarial worms' (e.g.

continued

Table 2.8 *Continued*

      Wuchereria, Brugia, Loa, Onchocerca, Dipetalonema, Mansonella)
      Superfamily 2: Dracunculoidea. Very elongate adults with extreme sexual dimorphism – females much longer than males. Mature female vulva nonfunctional; viviparous; intermediate hosts copepods (e.g. *Dracunculus*)
      Superfamily 3: Gnathostomatoidea
      Superfamily 4: Thelazoidea
      Superfamily 5: Habronematoidea
      Superfamily 6: Physalopteroidea

Subclass 2: Adenophorea (Phasmids absent; pharynx usually forms a stichosome)
   Order 1: Enoplida
      Superfamily 1: Trichuroidea. Adult body divided into slim anterior region and broader posterior section. Female possesses only a single ovary and uterus; males have single or absent spicule (e.g. *Trichinella, Trichuris*).
      Superfamily 2: Dioctophymatoidea

epidemiologically into three subsets (Table 2.9). Direct life cycle intestinal forms and insect vector-transmitted filarial species are two homogeneous groupings. The remaining species are a heterogeneous mixture of larval and adult infections that are neither insect transmitted nor have

Table 2.9 Nematode species causing important disease in humans

Name	Geographical distribution
INTESTINAL PARASITES	
Strongyloides stercoralis	Worldwide in tropical zone
Ancylostoma duodenale	Worldwide in tropics and subtropics plus Mediterranean fringes
Necator americanus	Worldwide in tropics and subtropics including S.E. states of USA
Enterobius vermicularis	Cosmopolitan
Ascaris lumbricoides	Cosmopolitan
Trichuris trichiura	Cosmopolitan
FILARIAL PARASITES	
Wuchereria bancrofti	Asia, Africa, S. America, Pacific Islands
Brugia malayi	S.E. Asia
Onchocerca volvulus	Yemen, Africa, S. and Central America
Loa loa	W. and Central Africa
Dipetalonema perstans	Africa, S. America
Dipetalonema streptocerca	W. and Central Africa
Mansonella ozzardi	S. America, Caribbean
OTHER NEMATODES	
Toxocara spp. (only larvae in humans)	Cosmopolitan
Dracunculus medinensis	Africa, Middle East, Pakistan, India
Trichinella spiralis (adults intestinal)	Cosmopolitan

simple direct life cycles with free-living eggs or larvae as infective stages.

2.5.1 Human intestinal nematodes

These are the most important human intestinal helminth parasites in terms of both their overall prevalences and their potential for causing serious clinical harm. All the six important species (Table 2.9) have direct life cycles and, apart from *Enterobius vermicularis*, they are all soil-transmitted in the sense that the eggs or larvae responsible for transmission normally become infective during a period of development in the soil. They are sometimes termed geohelminths because of this char-

acteristic. Similar modes of transmission and the extensive and overlapping geographical distributions of these helminths, together ensure that many individuals suffer from concurrent infections with two or more species of these intestinal parasites. Such multiple infections are often particularly harmful.

Ascaris lumbricoides (Fig. 2.10), is the largest human intestinal nematode and probably one of the most prevalent with between 800 million and a billion persons infected globally. Although most cases occur in the Far East and tropical Africa it is a truly cosmopolitan parasite. Transmission potential is determined largely by human

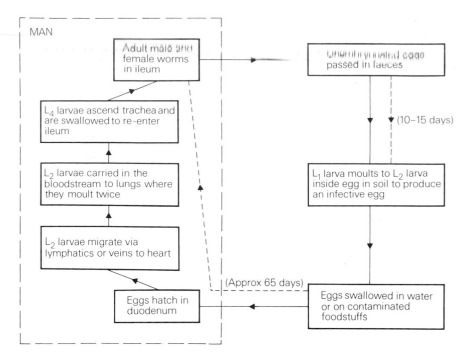

Fig. 2.10 The life cycle of *Ascaris lumbricoides*.

faecal disposal practices and is highest where soil around dwellings is heavily contaminated with faeces or where faecal slurry (night soil) is used as an agricultural fertilizer. The very long-lived eggs (maximum survival around 7 years) are typically ingested via uncooked, unwashed vegetables or in contaminated water.

Migrating *Ascaris* larvae cause allergic broncho-pneumonia in previously infected and sensitized patients along with bronchitis, bronchospasm and urticaria. Adult worms in the small intestine in high density can cause obstruction and worms can migrate out of the gut into the bile duct, pancreatic duct, oesophagus, mouth and occasionally the liver. Worms may also perforate the intestine inducing peritonitis.

Trichuris trichiura (Fig. 2.11), the causative agent of trichuriasis or whipworm infection, is an unusually shaped worm about 4 cm long with an extremely long, thin pharyngeal region and a wider posterior section containing the rest of the gut and the reproductive organs. The buccal cavity has a small stylet with which the worms

disrupt the caecal mucosa. The thin anterior region lies embedded shallowly in the surface of the mucosa with the rear portion free in the gut lumen. Infections are most common in humid tropical countries with global infections probably totalling well over 500 million cases. Low density *Trichuris* infections can be without symptoms but higher density infections, especially in undernourished children can be highly pathogenic and result in chronic bloody diarrhoea, colic, anaemia and rectal prolapse.

Ancylostoma duodenale and *Necator americanus* (Fig. 2.12), two morphologically and developmentally similar blood-feeding nematodes, cause human hookworm disease (ancylostomiasis). Both forms are extensively distributed in the tropical and subtropical zones of the world with a probable total prevalence of 700–900 million cases. Hookworms are transmitted to humans by skin-penetrating L_3 larvae that develop in faecally contaminated soil. Adult worms attach themselves to gut villi and use cutting blades or teeth in their buccal capsules to abrade the mucosal surface.

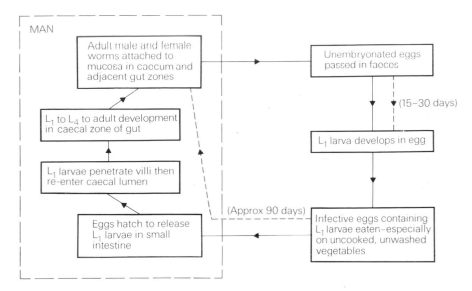

Fig. 2.11 The life cycle of *Trichuris trichiura*.

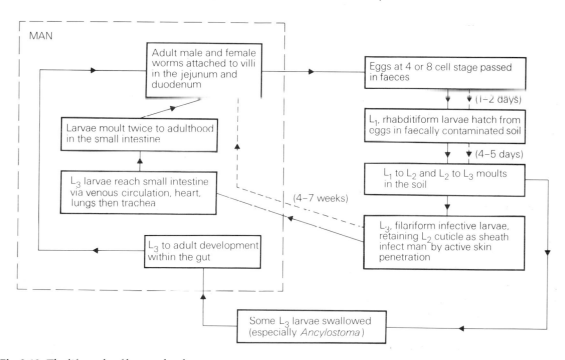

Fig. 2.12 The life cycle of human hookworms.

They then feed on blood so that anaemia is a prominent symptom in many cases of hookworm disease.

The disease induces three quite distinct types of pathology. Invading L_3 larvae cause an allergic dermatitis with a papular and sometimes vesicular rash that is termed ground itch. Lung pathology – focal haemorrhages and allergic

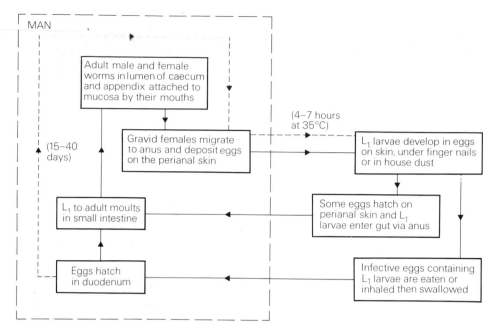

Fig. 2.13 The life cycle of *Enterobius vermicularis*.

pneumonia – can be provoked by migrating larvae. The most important damage, though, is caused by the blood-feeding adults. Up to 200 ml of blood per day may be lost due to this cause by a patient with a heavy infection although some of the iron in this deficit is reabsorbed. The final degree of anaemia and associated protein loss in a particular patient is a complex resultant of hookworm species, worm load and host nutritional status.

L_3 larvae of a number of non-human hookworm species can penetrate human skin and cause irritating 'cutaneous larval migrans' as they move laterally in the skin. Such larvae never develop to adulthood so the infections are self-limiting.

Enterobius vermicularis (Fig. 2.13), the causative agent of enterobiasis or pinworm disease affects some 500 million individuals, mainly children, globally and is more common in temperate than tropical countries. It is of only minor pathogenic significance and the mode of infection can be considered contaminative as the eggs are infective almost immediately after release onto the perianal skin. Infective eggs are found in clothing and bedding and children can reinfect themselves

by transferring eggs from anus to mouth with the fingers. This is facilitated by the pruritus ani which is the commonest symptom of infections to become apparent. Probably most infections are symptomless and self-limiting. Ectopic worms may be found in some girls in the vagina and uterus.

Strongyloides stercoralis (Fig. 2.14) infections (strongyloidiasis) occur either when L_3 larvae penetrate the skin or develop directly within the gut. The life-cycle organization of this parasite is further complicated by the existence in some circumstances of an entirely free-living cycle of sexually reproducing adults and their larvae in the soil. This alternative route can produce L_3 larvae that are infective. The pathology associated with this disease can be divided into larval and adult-generated phases. The larvae give rise to dermatitis at their sites of invasion. They then migrate to the lungs where they cause allergic pneumonia. High densities of egg-laying adult females in the gut cause mucosal inflammation and malabsorption. Some larvae can re-enter the gut mucosa or penetrate the perianal skin to

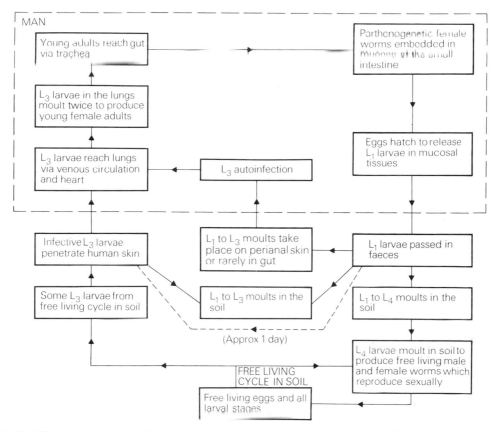

Fig. 2.14 The life cycle of *Strongyloides stercoralis*.

maintain a chronic infection status. In immuno-suppressed hosts this autoinfection route can be amplified to produce a considerable and some-times life-threatening larval invasion of many organ systems.

2.5.2 Human filariasis

Filarial worms are long-lived nematodes that all require a period of larval development (L_1 to L_3) in a blood-sucking insect host. These biting vectors initiate new human infections. Table 2.9 lists the seven major filarial species and Fig. 2.15 outlines the basic features of their essentially similar life cycles. Of the seven, *Wuchereria bancrofti* which causes lymphatic filariasis and *Onchocerca volvulus* which induces river blindness

and onchodermatitis are widespread and extre-mely damaging infections. *Brugia malayi* and *Loa loa*, although pathogenic, are much more restricted in distribution while *Dipetalonema perstans*, *D. streptocerca* and *Mansonella ozzardi* cause relatively few serious infections.

Adult filarial worms are rarely seen diagnos-tically. Diagnosis depends on collection and microscopical identification of L_1 larvae, termed microfilariae, in either the peripheral blood or skin. Most, if not all, filarial species are divided into geographical subspecies or strains which may utilize different vector species, display differing circadian periodicity of microfilarial density in the blood and be associated with distinctive hu-man pathological syndromes (see Chapter 7, Sec-tion 7.2.6).

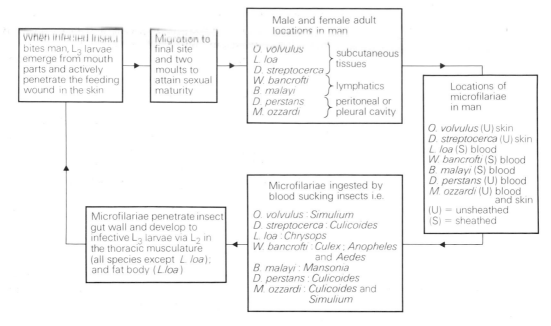

Fig. 2.15 The life-cycle organization of the seven most important human filarial parasites.

Lymphatic filariasis

Two filarial species *Wuchereria bancrofti* and *Brugia malayi*, which are very similar morphologically, live as adults in the afferent lymphatic vessels of the human host and produce blood-dwelling sheathed microfilariae. Diagnosis of lymphatic filariasis depends largely on finding and identifying these larvae in the peripheral blood. The characteristic pathologies associated with these infections, however, are hardly ever related to the larvae but instead to the adult worms in the lymphatics. The clinical course of pathogenesis is strongly linked with inflammatory and immunological host responses to the adults and commonly passes through three phases. The first or incubation phase – the time between infection and the first appearance of blood microfilariae – is rarely associated with symptoms although there may be some lymphatic inflammation and fever. The second phase of indeterminate duration is associated with adult worms producing microfilariae and is termed the acute inflammatory phase. It is marked by intermittent bouts of lymphatic inflammation and pain along with fever. In

Wuchereria infections there may also be episodes of inflamed lymph nodes in the inguinal region and scrotal changes including orchitis, hydrocoele and epididymitis. When it occurs, the final or obstructive phase displays irreversible lymphatic-centred damage with both filarial species. The most disfiguring aspect of these changes is the condition termed elephantiasis. This can occur to the legs, arms, scrotum, vulva and breasts. It is a chronic lymphoedema associated with very extensive fibrous infiltration and thickening of the skin.

Taken together, Bancroftian filariasis and *Brugia* infections occur in some 250 million people at any one time. *Wuchereria* cases, transmitted by *Culex, Anopheles* and *Aedes* mosquitoes, are predominantly found in India, the Far East, Polynesia and East Africa. The urban cycle of this disease which typically occurs in shanty town conditions is gradually assuming an increasing importance. *Brugia* infection, transmitted by *Mansonia* vectors, is an almost entirely rural disease restricted to South-East Asia. Some forms of *Brugia*-induced lymphatic filariasis possess forest-dwelling animal reservoir hosts such as monkeys and forest cats. Bancroftian filariasis has no such zoonotic complication.

Onchocerciasis

Onchocerca volvulus causes human infections and considerable morbidity across Africa and in minor pockets in South and Central America and in Yemen. Within this geographical range about seven distinct parasite strains exist, each transmitted with its own species of blackfly in the genus Simulium, and each associated with its own particular pathological pattern. Most blackfly species have to breed in well-oxygenated, fast-flowing freshwaters and the detailed spatial distribution of onchocerciasis infections largely reflects this riverine pattern.

The disease causes two types of severe pathology – ocular changes including blindness (river blindness) and onchodermatitis. Both are the result of immune responses to microfilariae, in contrast to the damage caused in lymphatic filariasis where adult worms are responsible. The unsheathed microfilariae of Onchocerca are found in the skin rather than the blood system and arise from groups of copulating adult worms located in fibrous nodules in subcutaneous tissues. These nodules, which may be several centimetres in diameter, can often be seen externally as rounded elevations of the skin. Other, deeper nodules are, however, invisible externally. Diagnosis of infections is based on microfilarial identification in skin snips, each 2–3 mm in diameter.

Microfilariae induce a wide range of skin changes in infected patients. Early, light infections may produce an itchy rash. In more chronic infections patchy depigmentation or increased pigmentation can occur and the skin becomes thickened and coarse while losing its elasticity. These changes in the integument predispose to hernias, 'hanging groins' or pendulous pouches under the eyes. Microfilariae moving through the skin reach the eyes and here they produce the most striking pathology linked with Onchocerca. The final stage of this is blindness, the 'river blindness' which is especially common in the savannah areas of West and Central Africa. In heavily infected villages close to rivers with high Simulium densities, as many as 30% of the population may become blind. Total blindness, though, is only the distressing endpoint of a progressive and accumulative series of changes that over a number of years involve all portions of the eyes.

Other human filarial infections

Loa loa causes loiasis in the rain-forest zones of West and Central Africa and is transmitted by large day-feeding tabanid flies in the genus Chrysops. Adult worms live subcutaneously but do not form permanent nodules. They move around inducing transient 'calabar swellings' and sometimes visibly cross the front of the eye under the conjunctiva.

Other filarial species, Dipetalonema streptocerca from Africa, D. perstans in Africa, Central and South America and Mansonella ozzardi from the Caribbean and South America are not of great pathological significance.

2.5.3 Other human nematode diseases

Dracunculiasis

This disease which is also called guinea worm disease, is caused by Dracunculus medinensis, the mature females of which may reach almost a metre in length. Dracunculiasis is widespread in much of India as well as parts of West and Central Africa, with more restricted foci in Pakistan, Iraq, Iran and Saudi Arabia. The parasites have an indirect life cycle and human infections occur when people drink water contaminated with copepods containing infective L_3 larvae of the nematode. Ingested larvae penetrate the human gut wall and spend about 12 weeks growing, moulting to adulthood and mating in the subcutaneous tissues. After copulating, female worms move down through the body reaching a lower extremity like the ankle or foot about 8–10 months after the original infection. Here the mature female, with its uterus filled with about one million L_1 larvae, induces a blister in the host's skin which subsequently bursts. This enables large numbers of actively swimming L_1 larvae to leave the lesion each time it is immersed in water, over a period of 3–4 weeks. Copepods then ingest larvae which develop into L_3 larvae in these intermediate hosts.

This indirect life cycle means that the disease is usually focally transmitted. Sites of infection are often small discrete water sources such as pools with gently shelving sides in Africa, or constructed step wells in India. In both such circumstances persons washing, drinking or collecting water are required to stand in the water. This contact enables L_1 larvae from infected persons to enter the water and infect copepods. Blister formation can be an extremely painful process and the open wound of the blister often becomes secondarily infected. Adult female worm infections closely associated with bony joints can lead to arthritis.

Trichinosis

Trichinosis, caused by *Trichinella spiralis*, is a cosmopolitan disease which demonstrates a very low vertebrate host specificity. Short-lived adult infections in the guts of a wide range of carnivorous and omnivorous mammals give rise to large numbers of invasive L_1 larvae which migrate to voluntary muscles throughout the bodies of these same hosts. Here they encyst. The cysts are the infective stages which can be transmitted to any new host when infected flesh is eaten. In domestic cycles throughout the world, human hosts become infected by eating cyst-containing

Table 2.10 Nematodes of domesticated animals (including species utilized as laboratory models)

Name	Final hosts	Comments
ORDER STRONGYLIDA		
Dictyocaulus viviparus	Cattle	Adult worms parasitic in lungs
Haemonchus contortus	Sheep	'Stomach worm'; blood-feeding injurious parasite, particularly pathogenic in young hosts
Nippostrongylus brasiliensis	Rats	Adults parasitic in rat intestine; much used laboratory nematode in immunological, developmental and physiological studies
Ostertagia ostertagi	Sheep	Stomach-inhabiting trichostrongylid
Stephanurus dentatus	Pigs	Adults in kidney of pigs; eggs passed out in urine
ORDER ASCARIDIDA		
Ascaridia galli	Chickens	Adults parasitic in small intestine
Ascaris suum	Pigs	Adults parasitic in the intestine of pigs; closely related to *Ascaris lumbricoides*
Heterakis gallinarum	Chickens and turkeys	Large numbers found in the intestinal caecae of turkeys; important as the carrier of the infective agent of the protozoan disease 'black spot', *Histomonas meleagridis*
ORDER OXYURIDA		
Syphacia obvelata	Rodents	Adults parasitic in the gut of rats and mice; used in laboratory studies
ORDER SPIRURIDA		
Dirofilaria spp.	Dogs	Adult filarial worms in the heart and pulmonary artery of dogs; immature worms occasionally found in lungs or pulmonary artery of humans
Onchocerca gutturosa	Cattle	Adult filarial worms in connective tissue on the surface of the nuchal ligament

pork from pigs. In feral cycles meat from wild boar, bears, bushpigs or warthogs can give rise to infections.

Although adult worms in the small intestine do give rise to some gut damage, the main pathogenic phase of the infection in humans is the population of migrating and encysting larvae. Usually no symptoms occur until the larvae reach the muscles and then, in heavy infections, a confusingly diverse range of serious symptoms can arise. Usually, a marked eosinophilia is apparent, with vomiting and diarrhoea followed by a high fever, muscle pain and in severe cases, evidence of cardiac and central nervous system pathological damage.

Toxocariasis

In humans this is a zoonotic infection brought about by the inadvertent ingestion, usually by children, of infective eggs of the ascarid species *Toxocara canis* and *T. cati*. These are normally found as adults in dogs and cats. The L_2 larvae which hatch from the eggs in children migrate round the body causing widespread small granulomas in the sites where they eventually die. The disease is a form of visceral larval migrans. Lesions are commonest in the liver and lungs but also occur in muscles, brain, skin and eyes; ocular pathological damage being the most serious consequence of human toxocariasis. Such changes occur in only a small proportion of infected children in which the larvae cause granulomatous endophthalmitis with retinal degeneration occurring.

2.5.4 Other nematodes

In addition to those species which are the causative agents of medically important human diseases, a number of other nematodes are of significance either because they infect agricultural animals or because they have been utilized extensively as laboratory models. Table 2.10 lists such species.

REFERENCES AND FURTHER READING

Anderson, R.C., Chabaud, A.G. & Willmot, S. (1976–1985) *C.I.H. Keys to the Nematode Parasites of Vertebrates*. Farnham Royal, Commonwealth Agricultural Bureaux.

Arai, H.P. (ed) (1980) *Biology of the Tapeworm Hymenolepis diminuta*. London, Academic Press.

Arme, C. & Pappas, P.W. (eds) (1983) *The Biology of the Eucestoda*, Vols 1 and 2. London, Academic Press.

Basch, P.F. (1991) *Schistosome Biology*. Oxford, Oxford University Press.

Bird, A.F. & Bird, J. (1991) *The Structure of Nematodes*, 2nd edn. New York, Academic Press.

Bundy, D.A.P. & Cooper, E.S. (1989) *Trichuris* and trichuriasis in humans. *Advances in Parasitology*, **28**, 107–173.

Campbell, W.C. (ed.) (1983) *Trichinella and Trichinosis*. New York, Plenum Press.

CIBA Foundation (1987) *Symposium 127 – Filariasis*. Chichester, John Wiley.

Coombs, I. & Crompton, D.W.T. (1991) *A Guide to Human Helminths*. London, Taylor and Francis.

Craig, P.S. & Macpherson, C. (eds) (1990) *Parasitic Helminths and Zoonoses in Africa*. London, Unwin Hyman.

Crompton, D.W.T., Nesheim, M.C. & Pawlowski, Z.S. (eds) (1989) *Ascariasis and its Prevention and Control*. London, Taylor and Francis.

Farley, J. (1991) *Bilharzia. A History of Imperial Tropical Medicine*. Cambridge, Cambridge University Press.

Frenkel, J.K., Taraschewski, H. & Voigt, W.P. (1988) Important pathological effects of parasitic infections of man. In H. Mehlhorn (ed.) *Parasitology in Focus*, pp. 538–590. Berlin, Springer-Verlag.

Gilles, H.M. & Ball, P.A.J. (eds) (1991) *Hookworm Infections*. Amsterdam, Elsevier.

Grove, D.I. (1989) *Strongyloidiasis: A Major Roundworm Infection of Man*. London, Taylor and Francis.

Grove, D.I. (1990) *A History of Human Helminthology*. Wallingford, C.A.B. International.

Kim, C.W. (ed.) (1985) *Trichinellosis*. Albany N.Y., State University of New York Press.

Lee, D.L. & Atkinson, H.J. (1977) *Physiology of Nematodes*, 2nd edn. New York, Columbia University Press.

Mahmoud, A. (1987) *Schistosomiasis*. New York, Ballière Tindall/Saunders.

Mehlhorn, H. & Walldorf, V. (1988) Life Cycles: Metazoa. In H. Mehlhorn (ed.) *Parasitology in Focus*, pp. 1–148. Berlin, Springer-Verlag.

Miyazaki, I. (1991) *An Illustrated Book of Helminthic Zoonoses*. International Medical Foundation of Japan.

Muller, R. & Baker, J.R. (1990) *Medical Parasitology*. London, Gower Medical Publishing.

Peters, W. & Gilles, H.M. (1989) *A Colour Atlas of Tropical Medicine and Parasitology*, 3rd edn. London, Wolfe.

Rollinson, D. & Simpson, A.J. (eds) (1987) *The Biology*

of Schistosomes – From Genes to Latrines. London, Academic Press

Schad, G.A. & Warren, K.S. (eds) (1990) Hookworm Disease: Current Status and New Directives. London, Taylor and Francis.

Schmidt, G.D. & Roberts, L.S. (1989) Foundations of Parasitology, 4th edn. St. Louis, Times Mirror Mosby.

Smyth, J.D. & Halton, D.W. (1983) The Physiology of Trematodes, 2nd edn. Cambridge, Cambridge University Press.

Smyth, J.D. & McManus, D.P. (1989) The Physiology and Biochemistry of Cestodes. Cambridge, Cambridge University Press.

Thompson, R.C.A. (ed.) (1986) The Biology of Echinococcus and Hydatid Disease. London, George Allen and Unwin.

Thompson, R.C.A. & Allsopp, C.E. (1988) Hydatosis: Veterinary Perspectives and Annotated Bibliography. Wallingford, C.A.B. International.

Warren, K.S. (1989) Selective primary health care and parasitic diseases. In K.P.W.J. McAdam (ed.) New Strategies in Parasitology, pp. 217–231 Edinburgh, Churchill Livingstone.

World Health Organization (1987) Onchocerciasis. WHO Technical Report Series No. 752. Geneva, WHO.

Chapter 3 / Vectors

D.H. MOLYNEUX

3.1 INTRODUCTION TO VECTORS

The association of insects with disease and pestilence has been recognized since the biblical plagues. It was only just over 100 years ago that a haematophagous arthropod was implicated in human disease transmission by Manson's discovery of the developmental stages of *Wuchereria bancrofti* in mosquitoes. Since then, the life cycles and transmission of numerous arthropod-borne infections have been studied in detail both in the field and laboratory. Natural vectors have been incriminated and laboratory colonies established to aid experimental studies in parasite–vector associations and vector biology itself. In parallel with laboratory studies, the understanding of vector biology in the field is a prerequisite for the interpretation of epidemiology and the development of control strategies. However, ecological changes which take place associated either with natural or artificially induced changes in the habitat have a profound influence on the biology of vectors and hence parasitic disease incidence and prevalence. These changes are often unpredictable, emphasizing the need for continuing awareness of potential problems where environmental changes are occurring. Such changes pose the risk of epidemics of vector-borne diseases in hitherto immunologically naive populations, primarily due to the successful adaptation of both vectors and parasites. Recent examples include malaria epidemics in Madagascar; sleeping sickness in Busoga, Uganda, due to *Lantana* (a toxic shrub) growth; leishmaniasis in Amazonia, associated with development projects involving penetration into the forest, and increased urbanization in the Middle East; and schistosomiasis, associated with the Aswan Dam in Egypt, the Volta Lake in Ghana and the Senegal River barrage in Senegal.

3.2 FLIES: ORDER DIPTERA

3.2.1 Mosquitoes: Family Culicidae
Subfamily Anophelinae: *Anopheles* (vectors of *Plasmodium* and *Wuchereria*)
Subfamily Culicinae: *Culex, Aedes, Mansonia* (vectors of *Wuchereria* and *Brugia*)

Mosquitoes are the most abundant widely distributed and important vectors of human and animal diseases. Table 3.1 lists the parasitic diseases transmitted by mosquitoes of the above genera. Besides parasites, mosquitoes also transmit viruses to man, e.g. *Aedes aegypti* transmits the viruses which cause yellow fever and dengue. *Anopheles* are vectors of *Plasmodium* and various filarial worms. *Aedes, Culex* and *Mansonia* also transmit filariae.

The basic anatomy of all mosquitoes is similar (Fig. 3.1), but genera differ in important details of egg morphology and disposition, larval position in relation to the water surface and in the resting position of the adult. These differences are summarized in Fig. 3.2 and Table 3.1.

Adult female mosquitoes (around 5 mm in length) have a single pair of wings with scales on the wing veins of various colours and, like all Diptera, a pair of halteres on the metathorax; on the head the proboscis emerges from between conspicuous compound eyes positioned anteriorly and containing the mouthparts modified for sucking blood; antennae (the male antennae are plumose) and maxillary palps are also present on the head and the relative size of the palps enables the different families to be easily identified. The legs of mosquitoes are delicate and also have scales.

There are ten abdominal segments. The abdomen can vary in colour significantly but lacks scales. At the posterior end, however, there are

Table 3.1 Biology of the most important mosquito vectors and differences between them

Anopheles gambiae	Aedes aegypti	Culex quinquefasciatus	Mansonia uniformis
Eggs			
Eggs laid singly with floats. Not resistant to desiccation	Eggs laid singly. No floats. Resitant to desiccation	Eggs in rafts/masses. Never have floats. Not resistant to desiccation	Eggs in masses under leaves of aquatic plants
Larvae			
Larval habitat typically temporary, e.g. hoof marks, tyre tracks. Larva horizontal to water surface and have palmate or float hairs	Larval habitat in water pots, tins, tyres; non-polluted water. At angle to surface. No palmate hairs	Organically polluted water, e.g. cesspits ditches, ponds. At angle to surface. No palmate hairs	Larger bodies of water containing swamp plants. Larvae obtain oxygen by penetrating root system of water hyacinth (*Eichornia* or *Pistia*). No palmate hairs. Respiratory siphon
Pupae			
Comma shaped. Breathing trumpet broad apically	Comma shaped. Opening not broad	Comma shaped. Opening not broad	Comma shaped. Respiratory tube to plants under water
Adults			
Rest at acute angle to surface. Dark and pale scales on wing. Wing veins arranged in blocks.	Parallel to surface at rest. Wings uniform brown or black	Parallel to surface at rest. Wings uniform in coloration	Heavily marked with boat-shaped terminal part to abdomen
Palps in female are long as proboscis. Male palp as above but swollen at end	Palps of female much shorter than proboscis. Male and female palps never swollen	Palps of females much shorter than proboscis. Male and female palps never swollen	Palps of females much shorter than proboscis. Male and female palps never swollen

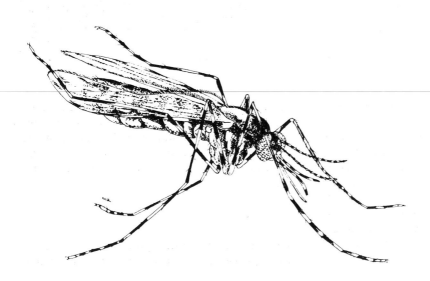

Fig. 3.1 Female *Anopheles* mosquito.

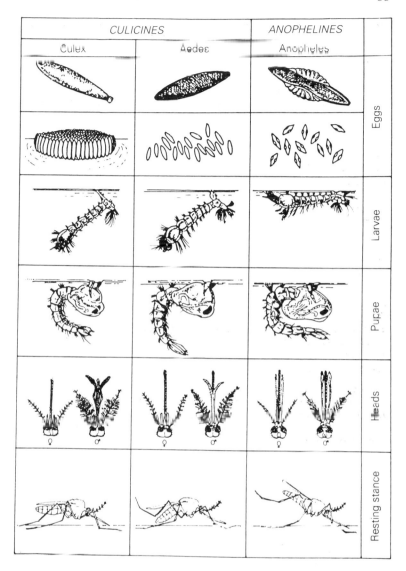

	CULICINES		ANOPHELINES	
	Culex	Aedes	Anopheles	
				Eggs
				Larvae
				Pupae
				Heads
				Resting stance

Fig. 3.2 Differences between Culicine and Anopheline mosquitoes.

cerci in females and, in males, two claspers on the genitalia.

Larvae vary in length depending on instar. The head contains a complex arrangement of mouth brushes for collecting particulate matter, together with a pair of antennae and a compound eye. There is a conspicuous thorax which is the widest part of the body and a long segmented abdomen. Terminally the structure is complicated by the siphon or breathing tube, anal gills and dorsal brush.

The pupa is a comma-shaped non-feeding structure with a breathing trumpet which emerges from the cephalothorax. On the abdomen are terminal paddles which are used to enable the pupa to move, when disturbed, by characteristic jerks.

Life cycle

The life cycle of all mosquito species is similar, with egg, larval and pupal stages. The duration of the cycle is dependent on temperature, but at temperatures around 25°C *A. aogypti* takes 10 days from egg to adult. There are four larval instars. Eggs are deposited in selected habitats (see Table 3.1). However, any water containing structures, either natural or artificial, can serve as a larval breeding site. Eggs of *Aedes* can survive desiccation; development commencing when the rains arrive. Larval habitats are usually highly specific and are due to selection of oviposition sites by the female. Natural breeding sites range from temporary foot prints to rice fields for *Anopheles gambiae*, plant axils (particularly banana) for *Aedes simpsoni*, bromeliads for *Anopheles cruzi*; tree holes and artificial structures such as cans, water tanks or tyres for *Aedes*. The pupae have a duration of around 3 days at tropical temperatures before the adult emerges, although in temperate zones this may be over a week.

On the emergence of the adult female, a blood meal is sought and, following the acquisition of blood, the gonotrophic cycle begins; this cycle lasts between 2 and 3 days in the tropics before the production of eggs. This cycle is divided into four stages: unfed, bloodfed, half-gravid and gravid. When the blood is fully digested the gravid female lays its eggs in a suitable habitat and seeks a further blood meal. The number of eggs laid varies from tens to hundreds in any one batch.

Both sexes of adult mosquito may search out sources of sugar; in the case of females to supplement their blood meal; males take sugar as their only source of nutrients.

3.2.2 Blackflies: Family Simuliidae
Simulium (vectors of *Onchocerca volvulus* and *Leucocytozoon*)

Blackflies of the genus *Simulium* are small dark flies with a characteristic thoracic hump (Fig. 3.3). They are active during the day and only the females suck blood. The sexes are recognized in adults by the comparative size and position of the eyes which, in females, are well separated above the antennae, whereas in males the eyes join above the antennae. The ommatidia which make up the compound eye are smaller in females than in males. The antennae of blackflies are small with a characteristic beaded appearance. The other head appendages are five-segmented pendulous palps which are longer than the proboscis. The wings are short and broad and have a well-defined anal lobe. Despite the small size of these flies they are efficient and strong fliers.

Species identification in blackflies is based on the structure of the male and female terminalia, the respiratory organ of the pupae, the larval head structure and thoracic scutal patterns in adults. However, *S. damnosum*, and other important vectors, are made up of species complexes; isomorphic (morphologically identical) adult females cannot be distinguished; cytotaxonomy (identification of polytene chromosomes) in salivary glands of larvae is the usual method of identification; additional methods are isoenzyme electrophoresis, DNA probes or analysis of cuticular hydrocarbons by gas–liquid chromatography.

Life cycle

Female blackflies lay eggs in batches of 200–500 on vegetation or other objects in or adjacent to fast-flowing well-oxygenated rivers, usually at dusk. Oviposition behaviour is communal, egg batches being deposited in close proximity to each other. The eggs are 100–400 μm in length, oval or triangular in shape, initially smooth and are embedded in a gelatinous substance derived from the egg itself. The eggs are sensitive to desiccation and if water levels are lowered they can be easily destroyed. The duration of the egg is dependent on temperature but at tropical temperatures hatching to the larval stages takes 4–6 days. They have a prominent sclerotized head with cephalic fans which filter the water flowing over them. The larvae attach to substrates by their posterior ends, usually within 30 cm of the water surface. The current is directed over the fans by a ventral proleg, and the larvae feed on particles 10–100 μm in size, such as diatoms and bacteria. Larvae are extended by the current in the direction of flow (Fig. 3.3).

Larval salivary glands secrete a silk-like material which is spun on to the substrate, enabling them to drift downstream to change substrate if

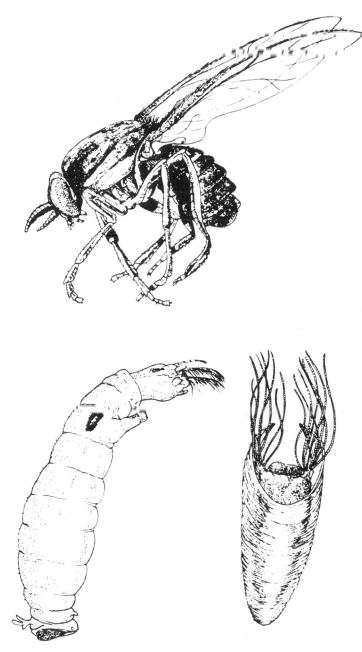

Fig. 3.3 Blackfly, *Simulium*, adult female, larva (bottom left) and pupal case (bottom right).

water oxygen tension is not optimal. Blackfly larvae also move via a 'looper-like' movement. Where the river is very large and water is flowing very fast, larvae can be found several metres deep. Larvae are also frequently found at outfalls of large bodies of water, such as lakes, dam spillways and causeways.

The larvae of *S. neavei* have a phoretic association with fresh-water crabs of the genera *Potamon* and *Potamoneutes*; the larvae and pupae live on the carapace. Eggs are not laid on the crab so larvae must locate their site.

There are four larval instars before pupae are formed. Blackfly pupae are conical, firmly attached to the substrate and have characteristic anterior respiratory threads. They do not feed and have a cephalothorax and a segmented abdomen with spines which engage the threads of the cocoon and keep the pupae in position. The duration of the pupal stage varies from 3 to 7 days. The adult emerges in a bubble and either flies immediately or first crawls on to a suitable substrate to harden its wings. Emergence usually takes place during daylight hours. Mating on the wing occurs near breeding sites. Females also supplement blood meals by feeding on plant sugars. Blackflies have a range of hosts, but medically important species are strongly anthropophilic and bite man outside (exophagy) and during the daytime (diurnal) with two peaks of biting activity, around 09.00 and 17.00 hours, although, in more shaded areas, biting can be spread throughout the day. The gonotrophic cycle at tropical temperatures can be 24–48 hours, blood meals being taken every 3–4 days after oviposition.

Simulium damnosum are monitored by human landing catches to assess parameters such as annual biting rate (ABR) and from this annual transmission potential (ATP) (risk of onchocerciasis as measured by the number of L_3 larvae in flies caught over a period) as measures of the effectiveness of control. The methods of catching blackflies are light traps, sticky aluminium plaques to catch ovipositing flies, car mounted traps or cattle mimicking traps. However, human landing catches are the most effective way of collecting anthropophilic species.

Blackflies feed specifically at certain sites on the body; *S. damnosum* on lower limbs, particularly the ankles and *S. ochraceum* on upper limbs.

In West Africa, *S. damnosum* can disperse up to 400 km. This is associated with the seasonal movements of the Inter Tropical Convergence Zone.

Unlike many vector species, *Simulium* are not easily reared in self-sustaining laboratory populations, although larvae/pupae brought into the laboratory can be reared into adults.

3.2.3 Sandflies: Family Psychodidae
Subfamily Phlebotominae: *Phlebotomus*, *Lutzomyia* (vectors of *Leishmania*)

Sandflies are small (2–4 mm in length), pale brown flies covered with delicate hairs (Fig. 3.4). They hold their wings above the body and are not strong fliers but move in a series of wing-assisted hops. They have 16-segmented antennae, 5-segmented pendulous palps and their stilt-like legs elevate the body above the substrate. The females are haematophagous but also take sugar meals from plant sources or aphid honeydew. Males may also feed on sugar sources but do not take blood. Males are recognized by the conspicuous claspers on the terminal abdominal segment. The female mouthparts are long with functional mandibles for piercing the skin. Sandfly salivary glands contain a potent vasodilator to increase the ability of the fly to feed.

The features of sandflies used for species identification are the structure of the spermatheca (sperm storage organ) in females; the genitalia of males; the shape and number of ridges in the pharynx and teeth in the cibarium; the length of antennal segments and the sensory ascoids on the antennae. These characters separate the genus *Phlebotomus* into subgenera (see Table 3.2).

Sandflies have a widespread distribution in the tropics and subtropics, extending into southern European climates. They are usually seasonally abundant during the hotter months of the year in the Old World. Habitats of Old World vectors are usually associated with more arid environments but within such habitats the sandflies, which are nocturnal insects, seek out humid cooler microclimates (cracks in walls, animal burrows). Larval habitats of sandflies are poorly known, although it is usually considered that larvae develop in soil with a high level of organic matter (such as may be found in cellars, animal shelters, burrows and caves).

Life cycle

Adult females lay eggs in batches of 50–70 in laboratory colonies, although fewer may be laid in the wild. The eggs are white (300–400 μm long and 90–150 μm wide) when laid but gradually darken to a brown-black colour. Different

Table 3.2 Main vectors of important human and animal parasitic diseases

Parasite	Vectors	Geographical zone
PLASMODIUM		
P. falciparum	Anopheles gambiae complex	
P. vivax	A. funestus	Sub-Saharan Africa
P. malariae		
P. ovale	A. culicifacies	
	A. stephensi	Indian subcontinent
	A. minimus	
	A. dirus	S.E. Asia
	A. balabacensis	
	A. farauti	
	A. punctulatus	Australasia
	A. koliniensis	
	A. albimanus	Latin America
	A. darlingi	South America
	A. saccharovi	Middle East
TRYPANOSOMA		
T. cruzi	Triatoma infestans	Brazil, Argentina, Bolivia
	T. dimidiata	Central America
	Rhodnius prolixus	Venezuela
	Panstrongylus megistus	Brazil
T. brucei brucei	Glossina spp.	Sub-Saharan Africa
T. b. rhodesiense	G. morsitans group	E. and S. Africa
	G. fuscipes	Uganda
T.b. gambiense	G. palpalis	W. and Central Africa
	G. tachinoides	
T. vivax	Glossina spp.	Sub-Saharan Africa
T. congolense		
LEISHMANIA		
Old world vectors		
L. donovani	Phlebotomus argentipes	India
	P. orientalis	Sudan
	P. martini	Kenya
L. infantum	Phlebotomus spp.	France
	P. perniciosus	Italy
	P. perfiliewi	Italy
	P. major	Greece
L. donovani	P. orientalis	Sudan
L. aethiopica	P. longipes	Ethiopia, Kenya
	P. martini	Kenya

continued on p. 60

Table 3.2 *Continued*

Parasite	Vectors	Geographical zone
L. major	*P. papatasi* *P. dubosqi*	N. Africa Middle East Sub-Saharan Africa
L. arabica	*P. sergenti*	Middle East
New world vectors		
L. chagasi	*Lutzomyia longipalpis*	S. America
L. mexicana	*L. olmeca*	Central America
L. amazonensis	*L. flaviscutellata*	Brazil
L. braziliensis	*L. wellcomei* *L. whitmani*	Brazil
L. guyanensis	*L. umbratilis*	Brazil, Guyana
L. panamensis	*L. trapidoi*	Central America
L. peruviana	*L. peruensis*	Peru
ONCHOCERCA *O. volvulus*	*Simulium damnosum* complex *S. neavei* complex *S. metallicum* *S. ochraceum* *S. roraimense*	W. Africa E. Africa Central America Brazil, Venezuela
WUCHERERIA *Wuchereria*	*Culex quinquefasciatus* *Anopheles*	S. America Africa Asia
BRUGIA *B. malayi*	*Mansonia* spp. *Aedes* spp.	Asia
LOA *L. loa*	*Chrysops dimiata* *C. silacea*	W. and Central Africa
MANSONELLA *M. ozzardi*	*Culicoides* spp.	Central and S. America
M. perstans	*Culicoides* spp.	Africa, S. America
M. streptocerca	*Culicoides* spp.	Africa
DIROFILARIA *D. immitis*	*Aedes* spp.	Worldwide

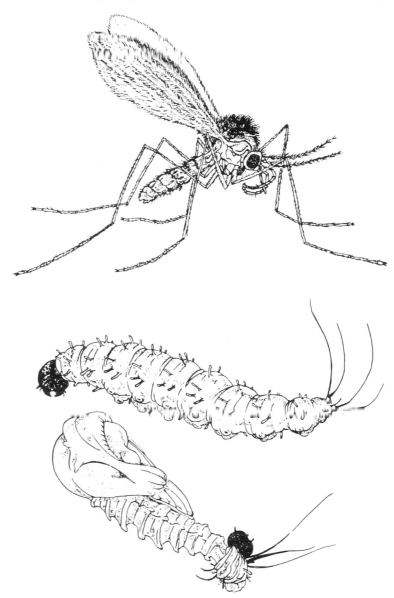

Fig. 3.4 Sandfly, *Phlebotomus*, adult female, larva (middle) and pupa (bottom).

species have a characteristic sculptured chorion which assists survival if immersed in water. Larvae emerge from eggs after 3 days at temperatures around 25°C; there are four larval instars. Duration of larval life is dependent on temperature but in the Old World where a defined winter occurs larvae undergo diapause (arrested development).

Larvae are greyish white in colour with a dark head. The thorax is not differentiated from the abdomen, and abdominal segments have ventral pseudopodia. Larval sandflies have characteristic matchstick hairs and one or two pairs of caudal setae (Fig. 3.4). The head has a pair of chewing mouthparts for feeding on organic debris. Larvae in laboratory colonies are fed on a diet of autoclaved rabbit faeces, powdered leaf mould, yeast, liver powder, dried *Daphnia* or soil.

Pupae assume an upright position secured to the substrate by the exuvium of the larva which is retained at the posterior of the abdomen. Larval setae are visible through the pupae. Pupae are described as exarate, with developing legs and wings free from the body (Fig. 3.4).

Some sandflies are autogenous; they produce eggs without a bloodmeal. Mating takes place in flight or near hosts to which males are also attracted. Recently a male sex pheromone has been identified in *Lutzomyia* and oviposition pheromones in *Phlebotomus*.

Populations of adult sandflies are sampled in a variety of ways. Some of the most commonly used methods are castor-oil coated cards, light traps, animal baited traps, aspiration from resting sites – trees, walls of habitations or animal shelters – or in funnel traps placed at the entrances to animal burrows.

3.2.4 Tsetse flies: Family Glossinidae

Glossina (vectors of African trypanosomes, *Trypanosoma*)

The genus *Glossina* is restricted to sub-Saharan Africa. It is the only genus in the family Glossinidae. Twenty-three species are known and some of these species have well-defined but geographically restricted subspecies. Both sexes are haematophagous and are the vectors of pathogenic trypanosomes of humans and domestic animals. Trypanosomes undergo cycles of development in *Glossina* but may occasionally be transmitted mechanically not only by *Glossina* but by other biting flies (Tabanids).

Glossina are brown in colour and are approximately the size of houseflies. They are characterized at rest by the position of the proboscis which extends in front of the fly, and by the configuration of the wings, which are folded over the dorsal surface of the abdomen. *Glossina* also possess characteristics which distinguish them from all other flies, a hatchet cell on the wing defined by outlines of the wing veins and an arista with branched hairs on both antennae. Species of *Glossina* are identified by the structure of the male genitalia, colour and patterns on the dorsal surface of the abdomen and colour of the tarsi of the legs (Fig. 3.5).

The genus is divided into three groups or subgenera. These are the subgenus *Glossina* (the *G. morsitans* group), subgenus *Nemorhina* (the *G. palpalis* group) and the subgenus *Austenina* (the *G. fusca* group). The *G. morsitans* group is typically found in savanna and thicket habitats. The *G. palpalis* group are riverine flies requiring a higher humidity; this group, however, is also capable of adapting to peridomestic habitats in West Africa or *Lantana* thickets in Uganda. The *G. fusca* group is found in high forest habitats or

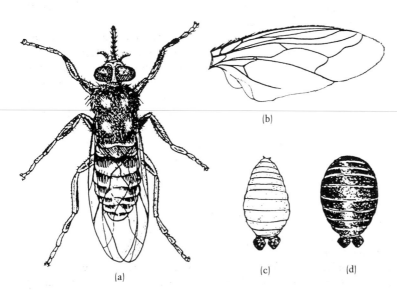

Fig. 3.5 (a) Tsetse fly, *Glossina*, adult. (b) Venation of wing showing hatchet cell. (c) Larva. (d) Puparium with polypneustic lobes.

in more humid relict forests on the periphery of the rain forest belts. All species can transmit trypanosomes and infection of *Glossina* is dependent on a variety of factors.

Analysis of blood meals has demonstrated that species of *Glossina* have different food preferences, *G. morsitans* and *G. fusca* groups feed mainly on Bovidae and Suidae, whereas members of the *G. palpalis* group have less well-defined host preferences and are capable of feeding on most available hosts. Usually the most abundant host is the one from which most blood meals are derived.

Glossina take a blood meal approximately every 3 days. Host seeking behaviour is dependent on visual and olfactory stimuli; having obtained a blood meal, flies rest in well-defined resting sites and refuges providing shade and humidity. Flies can be assessed for the stage in the hunger cycle by the amount of blood in the gut. Stress in a *Glossina* population can be assessed by the amount of fat and haematin present in flies. *Glossina* trapping has played an increasingly important part in our knowledge of the ecology of the fly. *Glossina* can be caught by a variety of trapping devices which have superseded the use of fly rounds in which humans or cattle act as bait. Odour-baited (acetone, octenol, phenols or cattle urine) box traps are used for *G. morsitans* and biconical or pyramidal traps for the *G. palpalis* group. *Glossina* are active during the day. Most activity is confined to the periods around dawn and dusk; such activity is based on intrinsic internally driven rhythms.

Mating occurs on the host; females only mate once and store the sperm transferred by the males in spermathecae.

Life cycle

Female *Glossina* are larviparous; after mating an egg passes into the uterus of the female, is fertilized *en route*, and develops over a period of 12–14 days being nourished by milk glands within the uterus. The fully developed larva weighs more than the female herself. The female chooses a shady humid larviposition site in which to deposit her offspring, which rapidly burrows into the moist soil to a depth of 2–4 cm

where it hardens to form a puparium. The puparia are black and barrel-shaped with two characteristic polypneustic lobes (Fig. 4.8). Imperial slits are underneath leaf litter and fallen trees or branches. The duration of pupal life varies between 30 and 60 days depending on temperature. The adult emerges from its puparial case by forcing it open by pressure from haemolymph pumped into the head capsule which has an expandable sac, the ptilinum, anteriorly. The adult then inflates its wings and waits for its cuticle to harden before flying off to search for its first blood meal. At this stage, the young fly is referred to as a teneral; characterized by a soapy feel, an eversible ptilinum and the absence of blood in the gut. The fly is more susceptible to infection with trypanosomes, particularly *T. brucei*, ingested with the first blood meal than in later blood meals.

Female flies live longer than males, and during their life produce up to 12 offspring. However, the mean numbers produced by a population of females will depend on many factors. More offspring are produced during the rainy season when conditions are less harsh than in the dry season when fly numbers are lower. *Glossina* also disperse more widely during the rainy season, whilst in the dry season they seek out the shade and humidity of dense vegetation or refuges. Under harsh dry season conditions flies are also likely to require food more frequently to survive, which may entail a higher proportion of time devoted to host-seeking behaviour.

It is important to be able to ascertain the mean age of *Glossina* populations. Analysis of wing fray of male and female gives a rough indication of the mean age of the population, but a more precise estimate of the age of females can be obtained from dissection of the reproductive system. Recently a spectrofluorometric technique of analysing residual pteridine in the eye has permitted a more precise estimate of the age of both males and females.

3.2.5 Biting midges: Family Ceratopogonidae
Culicoides (vectors of *Mansonella* and *Onchocerca* of horses)

Culicoides are amongst the smallest of the vectors (2–5 mm) and are also a severe biting

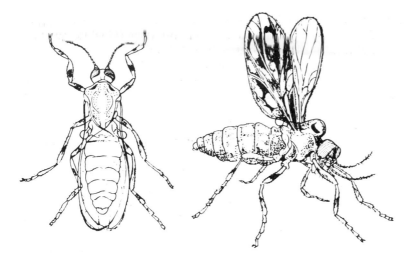

Fig. 3.6 Midge, *Culicoides*.

nuisance. The wings, which are folded over the abdomen like scissors, have a series of well-defined patterns of pale and dark markings, although wing venation is not distinct (Fig. 3.6). The thorax also has characteristic dark markings. Only the females take blood and, as in mosquitoes, females and males have different antennae (male plumose; female non-plumose). Males take only sugar meals.

Life cycle

Culicoides eggs are dark and approximately 0.5 mm in length. The number laid varies between 30 and 450 depending on the species. They are laid on damp surfaces in habitats such as swamps, edges of lakes, ponds and rivers, damp leaf litter or decaying organic materials (compost, rotting vegetation). Temperate species overwinter in the egg stages. The larvae, which emerge after approximately 7 days (temperature- and season-dependent), are 5 mm long, whitish with relatively few characteristics except 'gill-like' structures on the terminal abdominal segment. Larvae have a small triangular head and a segmented body (three thoracic segments and nine abdominal ones). Larvae feed on organic material and move to damper parts as the habitat dries out. There are four larval instars and the duration of larval life is 15–30 days in the subtropics and tropics. The pupae are 2–4 mm in length and

have a characteristic appearance with a pair of respiratory trumpets on the cephalothorax and tubercles with terminal hairs on abdominal segments. The pupal period varies from 3 to 11 days.

Adults emerge in large numbers, are exophilic and feed on a range of hosts; biting activity tends to be at its peak in the evening. Although midges do not disperse actively by flight, long distance movement on wind has been responsible for bringing animal virus diseases to Europe from Africa.

3.2.6 Deer flies: Family Tabanidae
Chrysops (vectors of *Loa*), *Tabanus* (mechanical transmitters of *Franciscella tularensis*, *Trypanosoma evansi* and *T. vivax*)

Tabanid genera of medical and veterinary importance are *Chrysops*, *Tabanus* and *Haematopota*. They are also referred to as horseflies or clegs. The most important genus is *Chrysops*, which are large stoutly built flies 10–15 mm in length with spotted and coloured eyes, banded wings and well-developed wing venation. The abdomen is dark with orange or yellow markings. The genera are distinguished by the structure of the three segmented antennae, particularly the type of annulation on the distal segment. Only female *Chrysops* feed on blood although both sexes also feed on plant sugars and water. *Chrysops* are of medical importance as vectors of *Loa loa*, a filarial parasite of humans in West and Central Africa. The two important species of *Chrysops* – *C. dimidiata* and *C. silacea* – are diurnal feeders (coinciding with diurnal periodicity of *Loa* microfilaraemia in human blood).

Chrysops are common in low-lying humid habitats although some species are associated with savanna/grassland habitats. Most species feed outdoors although *C. silacea* can enter houses. *Chrysops silacea* feeds in the forest canopy and descends to the forest floor only in clearings or when attracted by smoke which increases the biting of *C. silacea* at ground level up to tenfold. Whilst a wide range of hosts provide blood meals for tabanids, *C. dimidiata* and *C. silacea* feed on monkeys and humans.

Tabanus species transmit non-pathogenic trypanosomes of the subgenus *Megatrypanum*, e.g. *T. theileri* to cattle; tabanids transmit mechanically *T. evansi* of horses and camels in North Africa and Asia and *T. vivax* in S. America and also the agent of tularaemia, *Franciscella tularensis*. When these biting flies are abundant they are also important as a nuisance to livestock.

Life cycle

Eggs are laid on vegetation or rocks around water; up to a thousand eggs may be laid in a mass.

Chrysops eggs are laid as a single layer attached to a substrate. Initially they are white but they darken in about 48h. Larvae hatch in 4 days at optimal temperature and drop into aquatic habitats (swamps, ditches, slow flowing streams) or adjacent mud. The larvae are pointed at both anterior and posterior ends; *Chrysops* larvae feed on plant debris whilst other genera are cannibalistic. The number of larval instars is variable and the duration of the different instars can vary up to several months hence delaying pupation. Pupae are found at the edge of the larval habitats where an upright pupal cell of mud is formed by the larva. After 5–20 days the adult emerges from the pupal cell by breaking a protective cap. Females mate before the first blood meal by entering swarms of males. The adult longevity is of the order of 3–4 weeks during which time several batches of eggs will be produced. Control of tabanids is difficult due to the habitats of both larvae and adults.

3.3 BUGS: ORDER HEMIPTERA
Family Reduviidae (biting bugs), Subfamily Triatominae (vectors of South American trypanosomes, *T. cruzi*)

Triatomines are blood-sucking bugs sometimes called kissing or assassin bugs. Three major genera act as vectors of *T. cruzi*: *Triatoma*, *Rhodnius*

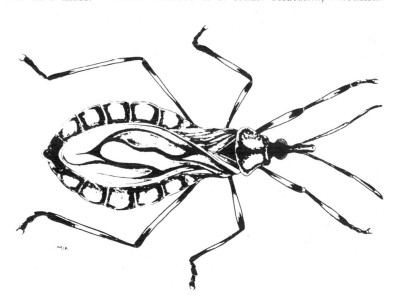

Fig. 3.7 Bug, *Triatoma*, female.

and *Panstrongylus*, and it is these species which have become adapted to live in close contact with human habitation. However, several other genera of bugs are responsible for transmission of *T. cruzi* amongst a wide range of sylvatic mammalian reservoirs. *Trypanosoma cruzi* and the Chagas Disease which it causes in humans are restricted to Central and South America.

Bugs are hemimetabolous insects; there are five nymphal instars before the adult; each instar takes a blood meal and ecdysis to the next stage occurs after varying periods of time. Both males and females take blood.

Triatominae (Fig. 3.7) have piercing and sucking mouthparts. Adults are winged and although the degree of dispersal undertaken by flight is not well documented, bug flight is probably an important method for distribution of bugs to new habitats or previously uninfested areas. Adults vary from 1 to 4 cm in length depending on the species.

The head of a bug is elongated with a pair of lateral compound eyes and a pair of ocelli. The head is divided by a transverse sulcus. Four jointed antennae are present. The forewings (hemielytra) are thickened whereas the hind wings are entirely membranous and are folded beneath the forewings when at rest.

Life cycle

Eggs, which are oval in shape and pale in colour (1.5–2.5 mm in length), are laid in crevices, cracks or behind wall-hangings in houses or in the roofs of houses thatched with palm leaves, straw or other vegetation. Eggs are also laid in burrows or nests of mammalian reservoir hosts. Eggs are sticky and adhere to the substrate on which they are laid. Females begin to lay a few days after the final moult, but the number of eggs laid is variable depending on the nutritional state of the bug. Eggs hatch in 7–14 days. The five wingless nymphal instars take 4–48 months between hatching to the moult to an adult bug, depending on temperature and nutritional state. During blood meals, which are taken at night, nymphs take up to ten times their weight and adults three to four times. Adult bugs can take 100–300 mg of blood, and blood meals take around 30 minutes to

obtain. Blood meals are essential for viable egg production. Defaecation on the host follows feeding; faeces containing *T. cruzi* are then rubbed into the wound. Females mate 1–3 days and males 5–9 days post moult.

Bugs can live for many months and all stages can live without feeding for several months. Houses in rural areas of South America can harbour several thousand bugs. Sampling of bugs can be undertaken by pyrethrum spraying, by systematic destruction of homes or by construction of artificial refuges (boxes of folded paper nailed to walls). Bugs associated with sylvatic mammals have been traced by trapping mammals and attaching thin fishing lines to them and allowing them to return to the nest. The location of the nest can then be found and the bugs examined.

3.4 ECTOPARASITIC INSECTS

Some insects and arachnids are classified as ectoparasites. These are usually macroscopic, visible to the naked eye, and haematophagous (e.g. fleas) or microarthropods which dwell within the skin (e.g. scabies mites). The life history varies in the length of time the ectoparasite spends in association with the host. Some ectoparasites such as the scabies mite *Sarcoptes scabei*, and human head lice, *Pediculus humanus capitis*, spend all their life on the host, whereas most fleas (Siphonaptera) have a temporary association with their hosts as adults and leave the host to lay eggs, with the consequence that the preimaginal stages are free-living in the nest of the host. However, some fleas, such as the adult female

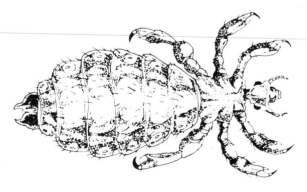

Fig. 3.8 Human louse, *Pediculus humanus*.

jigger flea *Tunga penetrans*, live embedded in the skin of humans (under toenails).

As the ability to act as a vector requires the facility to move from one host to another during the period of adult life only those ectoparasites which have a transitory association with a host, or can transfer from one host to another via contact, can act as vectors.

3.4.1 Lice: Order Anoplura
Vectors of *Borrelia* (relapsing fever) and *Rickettsia* (typhus)

Human body and head lice, *Pediculus humanus corporis* and *P.h. capitis* respectively, are common ectoparasites of humans throughout the world. Human lice are dorsoventrally flattened greyish insects about 2.5–4.5 mm in length. The male is smaller than the female but both are haematophagous. Lice cling to hairs and clothing by claws on the distal segment (tarsus) of the legs. They have a leathery cuticle with the lateral area of the abdomen black in colour (Fig. 3.8).

The females glue the operculated eggs (nits) to hair or clothing; about 250 eggs are laid by a female during a month, which is the mean longevity of lice. Eggs hatch to nymphs in 7–9 days, but the duration of nymphal stages (particularly of body lice) is dependent on whether clothes are removed at night, hence exposing them to reduced temperatures away from the body. De-

velopment of the three nymphal instars can take up to 15 days before adults are produced.

Body lice are commonly found on individuals who live in unsanitary conditions. Heavy louse infestations are associated with refugee camps, crowded conditions during warfare and in individuals living at high altitude who rarely remove their clothes. Head lice may be prevalent in children living in crowded environments. Both body and head lice are transmitted from one individual to another by close contact.

Body lice *P.h. corporis* are vectors of *Borrelia recurrentis*, a spirochaete which causes louse-borne relapsing fever; infestations are transmitted to humans by crushing the spirochaetes from the haemolymph of the louse into the skin; spirochaetes enter the body through abrasions or lesions produced as a result of the irritation from the bites of the louse.

Lice also transmit *Rickettsia prowazeki* which develops in the epithelial cells of their midguts. The infestation is passed out in the faeces, humans becoming infected by contamination of abrasions with infective faeces.

The pubic louse *Phthirus pubis* is smaller than the genus *Pediculus* and is usually restricted to the pubic and perianal hairs. Their characteristic shape resembles a crab as the thoracic segments are reduced. The two hind pairs of legs are large and have well-developed claws; because of their appearance they are described as crab lice.

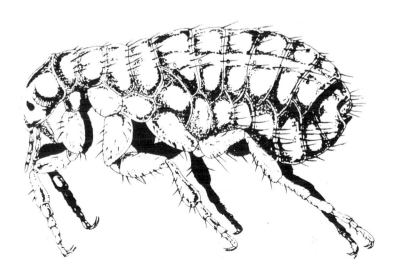

Fig. 3.9 Flea, *Pulex irritans*, female.

The life cycle of *P. pubis* is similar to that of *Pthi... ulius*. Transmission of *P. pubis* is usually via sexual contact, and the infestation is frequent in patients present at venereal disease clinics.

3.4.2 Fleas: Order Siphonaptera

Vectors of *Yersinia* (plague), *Rickettsia* and
    cestodes *Hymenolepis diminuta* and
    Dipylidium caninum

Adult fleas are laterally compressed, small (2–7 mm in length), wingless insects which are obligatory haematophagous (both male and female), and as adults have a variable duration of association with their mammalian or avian hosts. They have a holometabolous life cycle, eggs are laid in the nest of the host and the larvae are free-living. Adult fleas are usually a rich brown in colour, move rapidly through the pelage of their hosts, or in the nest, and are capable of jumping via powerful hind legs. The head of the flea is triangular in outline with conspicuous eyes. Groups of fleas are identified by the presence or absence of combs on the head (genal) or thorax (pronotal). The important genera, *Xenopsylla* and *Tunga*, are combless; the presence of combs together with the presence of a meral rod on the middle segment of the thorax are important features for identification of other genera. The abdomen has eight segments and the dorsal region of the last segment is a conspicuous sensilliary structure, the pygidium. The male genitalia are more conspicuous at the terminal end of the abdomen than those of the female, whose abdomen is more rounded (Fig. 3.9).

Life cycle

The adult female leaves the host to lay eggs (except in the case of the jigger flea, *T. penetrans*) in the nest of the host. The small oval pale eggs have an adhesive coat, which becomes covered with debris; eggs hatch after approximately 7 days. The larvae are pale with hairs and a black head capsule with mandibles. There are 13 similar segments and the terminal segment has a pair of ventral struts. Larvae are negatively phototactic and feed on organic debris in the nest and deposited blood from the adults. Some species take

blood direct from adult fleas. There are three larval instars over a 3-week period, although at lower temperatures this may be prolonged. The larva spins a silk cocoon within which the adult flea develops; again the cocoon is sticky and debris attaches to this surface. Providing a stimulus is provided – vibration – adults emerge from the cocoon after about 10 days. However, adults can remain quiescent within the cocoons for long periods (12 months) with the result that, in nests which have been abandoned, reoccupation by a host will result in simultaneous emergence of many fleas due to mechanical or chemical (carbon dioxide) stimuli. Fleas feed frequently and can survive for up to a year as adults.

Although some species of fleas have a degree of host specificity on animals (e.g. cats, dogs) they will also feed on humans. For example, the initiation of plague epidemics arises from rat fleas, *Xenopsylla cheopis*, leaving dead rats and feeding on humans, having acquired *Yersinia pestis* the causative organism of plague from the rats. Like many ectoparasites, fleas abandon dead hosts as soon as they detect a drop in temperature.

The female jigger flea, *T. penetrans*, has a different life cycle. The female, on reaching a human host, penetrates the soft skin beneath the toenails and swells to become a bag of eggs which are released into the ground. Infestations of *Tunga* can be highly irritating and can cause serious secondary infections.

Fleas, particularly human fleas, *Pulex*, and cat and dog fleas, *Ctenocephalides*, represent nuisance problems although the latter can act as intermediate hosts of tapeworms *Hymenolepis diminuta* and *Dipylidium caninum*. Eggs of these cestodes are passed in the faeces of host rodents (*H. diminuta*) or dogs (*D. caninum*) and are taken in by the larval stages. Larval cestodes migrate through the gut wall and form cysticercoids in the haemocoele. They remain in this stage until adult, whereupon, if they are ingested by the mammalian host during grooming, the adult cestode develops. Human infestation can result from children swallowing fleas as result of intimate contact.

Fleas also transmit murine typhus (*Rickettsia mooseri*). Reservoir hosts are rats. *Rickettsia* remains infective for several years in dried flea faeces. Infection of humans results from contamination of abrasions or mucous membranes with infected flea faeces.

3.5 TICKS: CLASS ARACHNIDA

The soft ticks (Argasidae) are distinguished from the hard ticks (Ixodidae) by the absence of a scutum (hard shield) which is prominent in all Ixodidae. In the Argasidae, the capitulum (false head) is in a ventral position and cannot be seen from a dorsal view. In Ixodidae the nymphs and adults have a forward projecting capitulum which is easily seen. The capitulum structure itself differs between the two groups; in Ixodidae the palps are club-shaped; they are leg-shaped in Argasidae. The basal segment of the palp is tiny in Ixodidae. The hypostome is toothed in both groups but the cheliceral sheaths have denticles or small teeth only in the Ixodidae which have terminal cutting teeth for puncturing the skin. Only soft ticks have coxal organs.

Significant differences also occur in the life cycle of the two groups. Soft ticks lay fewer eggs and have a series of nymphal stages, each stage of which may take a blood meal. In the soft tick *Ornithodoros* the larvae are retained in the egg shells. Blood meals are taken more rapidly by Argasidae than by Ixodidae – hours compared with days.

3.5.1 Soft ticks: Family Argasidae
Vectors of *Babesia* and *Borrelia*

The soft ticks, sometimes called tampans, are common in arid areas. Unfed adults are dorsoventrally flattened with a leathery cuticle embellished, depending on the species, with tubercles or granulations, radially arranged discs or polygons. Occasionally simple eyes are present. The capitulum is ventral and has pedipalps, chelicerae and centrally a toothed hypostome. Four pairs of legs are present, each with terminal claws. The anus is ventrally positioned, extending about two-thirds along the length of the body. Respiratory openings (spiracles) open at the base of the hind legs and coxal organs at the base of the first two pairs of legs. The genital opening is more prominent in male ticks. Both sexes are haematopha-

gous; blood is taken into a branching diverticulum (midgut or 'stomach') which enables ticks to engorge several times their own weight of blood. The coxal organ of *Ornithodoros* acts as a filter and excess fluid derived from blood meals passes out via the openings of these organs. Ticks infested with *Borrelia*, the spirochaete which causes relapsing fever, can have spirochaetes in the coxal fluid.

Life cycle

Eggs are deposited following a blood meal in batches of between 20 and 100. Soft ticks can live for several years so many thousands of eggs can be laid in a life-time. Eggs are laid in resting places – cracks in walls, mud floors or in animal holes or burrows, and have a waxy protective coating which enables them to remain viable for several months if conditions are not favourable, although they normally hatch within weeks.

A six-legged larva emerges from the eggs. Larvae are highly active and take a blood meal as rapidly as possible. Fully engorged larvae drop off the host and within 2 days moult to an eight-legged nymph. There are several nymphal instars (4–8) in Argasid ticks, and a blood meal is taken by each instar. Feeding usually takes place at night. The duration of the nymphal stages can be up to 12 months, and thereafter adult ticks can survive for many years; all stages can withstand starvation in the absence of hosts.

The most important genus is *Ornithodoros*. *Ornithodoros moubata*, which transmits the spirochaete *Borrelia duttoni* to humans, and is common in huts, particularly in dry areas where people sleep on earthen floors. *Borrelia* are present in the haemocoele and salivary glands of infected ticks. Female ticks are infected in the reproductive organs and transovarial transmission through several generations can occur. Ticks thus act as reservoirs of infection.

3.5.2 Hard ticks: Family Ixodidae
Vectors of *Borrelia*, *Rickettsia*, *Babesia* and *Theileria*

There are several hundred species of hard tick. Several genera are of considerable economic importance, particularly in the transmission of the diseases of livestock, *Ixodes, Rhipicephalus, Amblyomma, Hyalomma, Boophilus*, but also of human disease caused by rickettsiae, e.g. *Dermacentor*.

Adults are usually dark brown, sometimes coloured. When not engorged they are flattened dorsoventrally and are oval in outline. Size varies considerably depending on the state of engorgement, but females (which are larger than males) can be up to 2 cm in length. The capitulum is visible beyond the outline of the body. The palps are club-shaped and both hypostome and chelicerae are toothed. Some genera have indentations called festoons on the abdomen. A scutum is present which is smaller in females than in males and enables ticks to be sexed and identified. The adult females can be distinguished from nymphs by the location of the genital opening near the base of the second pair of legs. The anus is located at the posterior of the body. There are no coxal organs. Larval *Ixodes* have three pairs of legs, nymphs and adults four (Fig. 3.10).

Life cycle

All stages (larvae, nymphs and adults) and both sexes are haematophagous. A blood meal is necessary for the completion of the moults from larvae to nymphs and nymphs to adults. On emergence from the egg, the six-legged larval tick (seed ticks) which are small and resemble mites, become active after a quiescent phase and climb to the tips of vegatation responding to host odours, heat and vibration by questing behaviour ('searching' by front legs). On finding a host they attach to favoured sites such as ears or genitals when the mouthparts are inserted deep into the skin. Following engorgement the larvae 'drop' to the ground and seek shelter to digest their blood meal. Moulting follows after a few days to the eight-legged nymph which follows a similar pattern of behaviour in host searching, feeding and quiescence following engorgement. Only a single nymphal stage is present in *Ixodes* ticks. Adult females take a larger blood meal than males. Mating takes place on the host, males copulating with engorging females.

In temperate climates, ticks overwinter as adults and can survive for several years. Because Ixodid ticks remain attached for relatively long periods, ticks can be dispersed by hosts (for example migrating birds). The life cycle described is that of a three host tick – a species that feeds at each stage, larva, nymph and adult. Two host ticks feed as larvae but remain on the same host to moult to an adult. One host tick (*Boophilus*) remains on the host from larva to adult, with two

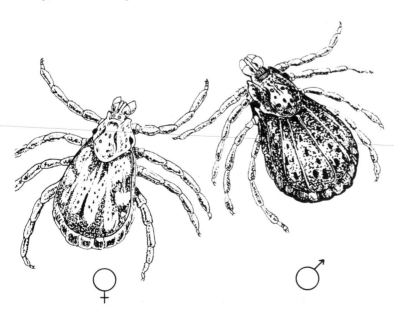

Fig. 3.10 Hard ticks, *Ixodes*, female (left) and male (right).

moults on the host. Only following adult engorgement does the tick drop. This life cycle will only permit the tick to act as a vector if transovarial transmission of the pathogen occurs.

Hard ticks are vectors of several pathogens. The genera *Rhipicephalus*, *Hyalomma* and *Amblyomma* transmit parasites of the genus *Theileria*, the most important being the causative agent of East Coast Fever in cattle, *T. parva*. *Babesia* infections in *Boophilus* are transmitted from one generation to the next transovarially. Ticks of the genus *Ixodes* transmit *Babesia* amongst small mammals and *B. microti*, a human infection in the eastern part of the USA, is derived from voles (*Microtus*). The larval and nymphal stages of the vector tick *I. damini* normally feed on voles with adult ticks feeding on larger animals such as deer. *Ixodes* have also been incriminated as vectors of the spirochaete responsible for Lyme disease in the USA and Europe, *Borrelia burgdorferi*. Hard ticks also transmit various viruses responsible for encephalitis which have a zoonotic origin; rickettsiae (*Rickettsia*

rickettsii, the cause of Rocky-Mountain spotted fever (transmitted by *Dermacentor*)) and tularaemia (*Francisella tularense*). The toxins in hard tick saliva can cause motor paralysis, a condition referred to as tick paralysis.

3.6 MOLLUSCS: INTERMEDIATE HOSTS OF DIGENEANS

Families of the Pulmonata, the Gastropoda, the Planorbidae and Lymnaeidae are intermediate hosts of Trematodes. The term vector is not used in this section as snails are not active transmitters of trematode infections, merely releasing cercariae rather than bringing infective forms directly onto or inoculating them into the mammal, as do arthropod vectors. Snails are essential but passive participants in trematode life cycles.

Those snails which act as intermediate hosts of trematodes are listed in Table 3.3. Hosts are freshwater dwellers but can also be amphibious (e.g. *Oncomelania* or *Lymnaea*). Typical habitats

Table 3.3 Snail vectors of digeneans

Species	Snail Vector	Distribution
SCHISTOSOMA		
S. mansoni	Biomphalaraia pfeifferi	Africa
	B. glabrata	W. India, Brazil
S. haematobium	Bulinus truncatus	Middle East
	B. rohlfsi	W. Africa
S. japonicum	Onchomelania hapensis	China
FASCIOLOIDEA		
Fasciola hepatica	Lymnaea truncatula	Europe, Asia, Africa
Fasciolopsis buski	Segmentina hemisphaerula	Far East
OPISTHORCHIOIDEA		
Opisthorchis sinensis	Bithynia manchocericus	Far East
O. belineus	Bithynia leachi	Far East
Heterophyes heterphyes	Pironella conica	Egypt, Far East
Metagonimus yokogawi	Semisulcospira libertina	Far East
PLAGIORCHOIDEA		
Paragonimus westermani	Thiara	
	Semisulcospira	

include ponds, lakes, dam impoundments, slow moving parts of streams and rivers.

Life cycle

Snails are hermaphrodite but cross-fertilization is usual. Eggs are laid in masses in water (Oncomelania is an exception, being dioecious and laying eggs out of water). Egg masses are up to 1 cm in diameter and contain up to 30 eggs. Eggs hatch at tropical temperatures in 1–2 weeks, and growth to maturity takes 3–6 months. Initially growth is between 0.5–0.8 mm per week, but as the snail reaches maturity the growth rate is retarded. Snails have a big reproductive potential and egg-laying is usually most frequently recorded at the beginning of the rainy season, the stimulus being either, or a combination of, change in temperature, the addition of nutrients, dilution of substances in solution and changes in size of habitat.

The ecological requirements of snails are difficult to define in view of the combination of biological and abiotic variables to which they may be exposed. It is believed that snails require water at temperatures between 22 and 24°C, which contains suspended solids, and is calcium-rich (for shell growth) and hence alkaline. Other factors which are measured in studies in snail ecology are water conductivity (to measure salinity), light intensity, pH (variable due to carbon dioxide fluctuation in water), geology of the area providing different ions in solution and rate of flow. Temperature reduction will reduce breeding, as will extreme heat. Snail populations benefit from the presence of a rich algal microflora and this can support snails in the absence of higher aquatic plants. However, specific snail–plant interactions are found, for example the snail Bulinus rolfsi and the plants Ceratophyllum and Polygonium in the Volta Lake, the roots of the latter providing shelter and oviposition sites whilst rotting Polygonium provides a feeding substrate.

The identification of snails requires advanced techniques in addition to morphology which is often inadequate for species determination. Biochemical, genetic and immunological analyses have assisted in clarifying relationships of different groups of snails, but no method is available which can be linked to susceptibility to trematode infection.

In the tropics Bulinus species aestivate during the dry season, as does Lymnaea, the host of Fasciola. Oncomelania is an amphibious snail which hibernates in cold weather and it has an operculum to close the shell. During aestivation development of platyhelminths stops. Lymnaea lives in muddy margins of water courses or poorly drained land in clay soils. Lymnaea remains active down to 0°C, although breeding is restricted to Spring and Summer in temperate regions when temperatures rise above 10°C, a critical temperature below which Fasciola does not develop.

Development of trematodes in snails follows penetration of miracidia which are attracted to particular snails by chemicals produced by the snails themselves. Trematode development in snails results in changes in snail physiology — pigment changes and increased fragility of the

Table 3.4 Examples of species complexes of vectors

Vector	Distribution
ANOPHELES GAMBIAE COMPLEX	Africa
A. gambiae sensu strictu	
A. arabiensis	
A. quadrimaculatus	
A. bwambae	
A. melas	
A. merus	
ANOPHELES MACULIPENNIS COMPLEX	
A. maculipennis sensu strictu	Europe
A. melanoon	N. America
A. atroparvus	
A. labranchiae	
A. messeae	
L. sacharovi	
SIMULIUM DAMNOSUM COMPLEX	Africa
S. damnosum sensu strictu	Savanna
S. sirbanum	Savanna
SIMULIUM SOUBRENSE SUBCOMPLEX	
S. sanctipauli subcomplex	Forest cytospecies
S. yahense	
S. squamosum	Forest cytospecies

shell, although usually there is also external evidence of infections.

3.7 VECTOR IDENTIFICATION AND SPECIES COMPLEXES

The identification of vectors is usually undertaken by detailed morphological examination of appropriate structures – genitalia, patterns, structures or colours on wings, antennae or legs. However, valid as these characters are, in many situations there has been an increasing recognition that many important vectors are in fact species complexes (Table 3.4). The term sibling species is also sometimes applied to these groups. A species complex is a group of insects which are morphologically identical or difficult to distinguish by reliable morphological criteria, containing subgroups which do not interbreed and hence are genetically isolated. Some species complexes of vectors may be distinguishable in one sex (males) but not in the females which transmit the disease, e.g. *Lutzomyia* sandflies which transmit *Leishmania braziliensis*. The recognition of the existence of such species complexes has led to the search for methods of identification to replace the original method which was the inability of crosses (of mosquitoes of the *Anopheles gambiae* complex) to produce fertile offspring. At present, the most widely used technique is the examination of giant polytene chromosomes of *Anopheles* mosquitoes and blackflies. The *A. gambiae* complex is differentiated by diagnostic chromosome inversion polymorphisms when chromosomes from ovarian nurse cells are examined. Blackflies (*Simulium damnosum* complex) are differentiated by examination of larval salivary gland chromosomes. This vector complex can, in the first instance, be separated into savanna and forest cytospecies by the colour of the wing tufts (pale in savanna; dark in forest). Adult identification, however, of blackflies, continues to pose problems although morphometric methods are becoming increasingly reliable.

The need to differentiate members of species complexes relates to the differences in behaviour (proportion of meals taken from human or animals; differences in resting sites; vectorial capacity) and hence the differences in approach or requirement for control and implications for epidemiology. Recently, other techniques have been developed for differentiation of species complexes, particularly for the *A. gambiae* and *S. damnosum* complexes. These techniques are isoenzyme differentiation, using stable marker enzymes which enable differentiation of *Anopheles* and forest cytospecies of *Simulium*, molecular methods such as DNA probes and the polymerase chain reaction (PCR) (see Chapter 6, Section 6.2.5) and analysis by gas–liquid chromatography of cuticular hydrocarbons.

There is an increasing recognition, not only that there is considerable intraspecific variation within vector insect species, but also that true species complexes are widespread, particularly in the genera *Anopheles* and *Simulium* as well as in other groups of vectors, e.g. sandflies.

REFERENCES AND FURTHER READING

Brown, D.S. (1979) *Freshwater Snails of Africa and Their Medical Importance*. London, Taylor and Francis.

Bruce-Chwatt, L.J. (1986) *Essential Malariology*, 2nd edn. London, William Heinemann.

Buxton, D.A. (1955) *The Natural History of Tsetse Flies*. London, H.K. Lewis.

Crosskey, R.W. (1990) *The Natural History of Blackflies*. Chichester, John Wiley.

Curtis, C.F. (ed.) (1991) *Control of Disease Vectors in the Community*. London, Wolfe.

Ford, J. (1971) *The Role of the Trypanosomiases in African Ecology*. Oxford, Oxford University Press.

Gale, K.R. & Crampton, J.N. (1987) DNA probes for species identification of Mosquitoes of the *Anopheles gambiae* complex. *Medical and Veterinary Entomology*, **1**, 127–136.

Jordan, A.M. (1986) *Trypanosomiasis Control and African Rural Development*. London and New York, Longman.

Ke Chung Kim & Merritt, R.W. (eds) (1987) *Blackflies: Ecology, Population Management and World List*. University Park and London, Pennsylvania State University.

Kettle, D.S. (1984) *Medical and Veterinary Entomology*. London and Sydney, Croomhelm.

Laird, M. (ed.) (1981) *Blackflies: The Future for Biological Methods in Integrated Control*. London, Academic Press.

Lewis, D.J. & Ward, R.D. (1987) Transmission and Vectors. Chapter 5. In W. Peters & R. Killick-Kendrick (eds) *The Leishmanias in Biology and Medicine*, Vol. I, pp. 235–262. London, Academic Press.

McKelvey, J.J., Eldridge, B.I. & Maramorosch, K. (1981) *Vectors of Disease Agents: Interactions with Plants, Animals and Man.* New York, Praeger.

Mattingley, P.F. (1969) *The Biology of Mosquito-borne Diseases.* London, George Allen and Unwin.

Molyneux, D.H. & Ashford, R.W. (1983) *The Biology of Trypanosoma and Leishmania Parasites of Man and Domestic Animals.* London, Taylor and Francis.

Muirhead Thomson, R.C. (1982) *Behaviour Patterns of Blood Sucking Flies.* Oxford, Pergamon Press.

Mulligan, H.W. (ed.) (1970) *The African Trypanosomiases.* London, George Allen and Unwin, Ministry of Overseas Development.

Peters, W. (1992) *A Colour Atlas of Arthropods in Clinical Medicine.* London, Wolfe.

Service, M.W. (1980) *A Guide to Medical Entomology.* London, Macmillan.

Service, M.W. (1986) *Lecture Notes on Medical Entomology.* Oxford, Blackwell Scientific Publications.

Service, M.W. (ed.) (1988) *Biosystematics of Haematophagous Insecta.* Oxford, Clarendon Press.

Service, M.W. (1989) *Demography and Vector-borne Diseases.* Boca Raton, CRC Press.

Traub, R. & Starke, H. (eds) (1980) *Fleas.* Proceedings of the International Conference on Fleas. Rotterdam, Balkema.

Vajime, C.G. & Dunbar, R.W. (1975) Chromosomal identification of eight species of the subgenus *Edwardsellum* and including *Simulium (Edwardsellum) damnosum* Theobald (Diptera: Simulidae). *Zeitschrift für Tropenmedizin und Parasitologie*, **26**, 111–138.

Ward, R.D. (1985) Vector Biology and Control. Chapter 12. In K.P. Chang & R.S. Bray (eds) *Leishmaniasis*, pp. 199–212. Amsterdam, Elsevier.

Willmott, S. (ed.) (1978) *Medical Entomology.* Centenary Symposium. London, Royal Society of Tropical Medicine and Hygiene.

Chapter 4 / Epidemiology

R. M. ANDERSON

4.1 INTRODUCTION

Epidemiology is the study of disease behaviour within populations of hosts and, as such, it is a discipline with many links to a broader area of scientific investigation – namely, that of ecology.

Our knowledge of the epidemiology of infectious disease agents has expanded rapidly in the past few decades, drawing on research from many separate areas of scientific study. It is a subject of great significance in the world today with respect to human diseases and those of crops and livestock.

Epidemiology is a quantitative science which relies on statistical methods for the accurate measurement of disease parameters, and mathematical techniques for the provision of a theoretical framework to aid in the interpretation and integration of field and experimental observations. A sound and detailed knowledge of the biology of the organisms under study, however, is an essential prerequisite for the successful application of these methods. When interpreting the pattern of infection within a community, the course of infection within an individual host is as important as the rate of transmission between hosts.

This chapter concentrates on outlining general principles that may be applied successfully to epidemiological studies of disease agents as diverse as the measles virus and the schistosome fluke. Before proceeding to describe these, however, some brief comments on epidemiological measures and units of study are necessary. Among the many possible measures of the patterns of infection and disease within a population we must clearly aim to choose the most objective, the clearest and least ambiguous and the most practically useful. What this measure will be depends on many factors, including the type of disease agent, but care must be devoted to its choice.

4.2 UNITS OF STUDY

4.2.1 The population as a unit of study

The definition and description of the host and parasite populations is clearly important in the study of disease epidemiology. A population is an assemblage of organisms (hosts or parasites) of the same species which occupy a defined point in the plane created by the dimensions of space and time. The basic unit of such populations is the individual organism.

Populations may be divided into a series of categories or classes, the members of which possess a unifying character (or characters) such as age, sex or their stage of development (egg, larvae or adult). Such subdivisions may be made on spatial criteria where distinctions are made between local populations within a larger assemblage or organisms.

Host–parasite associations are often characterized by the existence of a number of host populations (final and intermediate hosts) and several distinct parasite populations formed by the various developmental stages within the parasite life cycle (eggs, larvae and adults). We will use the term *subpopulation* to describe the number of parasites of a defined developmental stage within an individual host. The host here provides a convenient sampling unit. The total population of a given parasite stage is formed from the sum of all the subpopulations within the total host population. In the case of parasite transmission stages, their population is defined as the total number of organisms in the habitat of the host population.

The boundaries in space and time between these various host and parasite populations are often vague, but it is important to define them as clearly as possible.

4.2.2 Parasitic infection as a unit of study

The basic unit of study for the parasite population varies according to the type of infectious disease agent.

Microparasites (viruses, bascteria and protozoa) are small/and possess the ability to multiply directly and rapidly within the host population. The measurement of the number of parasites within an individual host is usually difficult, if not impossible, and as a direct consequence the infected host provides the most convenient unit for the study of these parasites. The choice of this unit leads to the division of the host population into a series of classes such as susceptibles, infecteds and immunes, categories which are based on the current or past infection status of the host.

Macroparasites (helminths and arthropods) are much larger than microparasites and in general do not multiply directly within their hosts. Moreover, the pathology induced by infection is often positively correlated with the burden of parasites harboured by the host. The *individual parasite* provides the basic unit of study for these infections. It is therefore desirable to measure the number of parasites within the host by direct (e.g. worm expulsion by chemotherapeutic agents) or indirect (e.g. egg output in the faeces of the host) measures. In certain helminth life cycles, such as those of the schistosomes, the parasite multiplies directly within the intermediate host. In these cases, the infected host forms the unit of study for the intermediate host segment of the life cycle and the parasite acts as the basic unit for the phase involving humans.

The duration of viral and bacterial infections in vertebrate hosts is typically short in relation to the expected lifespan of the host, and therefore is of a transient nature. This is a direct consequence of the ability of such directly multiplying parasites to elicit a strong immunological response from the host which tends to confer immunity to reinfection in those hosts that overcome the initial onslaught. There are of course some exceptions of which the herpes and human immunodeficiency viruses are particularly remarkable examples.

The immune responses elicited by macroparasites and many protozoan microparasites generally depend on the number of parasites present in a given host, but they are often unable to eliminate the parasite from the host (see Chapter 8). This may be due to antigenic variation within the parasite subpopulation in a given host, the immunosuppresive action of the parasite or the evasive mechanisms adopted by the infectious agent. Such infections therefore tend to be of a persistent nature with hosts being continually reinfected.

4.2.3 Measurement of infection within the host population

The most widely used epidemiological statistic is the prevalence of infection which records the proportion or percentage of the host population that is infected with a specific parasite (Fig. 4.1a–d). This measure is particularly convenient for the study of microparasitic infections where the infected host forms the basic unit of study but it is also widely used for helminth infections.

Prevalence may be measured in different ways: by direct observation of parasites within or on the host – the examination of blood films for malarial parasites; on the basis of serological evidence – the detection of *Trichinella* infections; and by the emission of infective stages – the examination of host faeces for helminth eggs. Immunological evidence provides the basis for a further epidemiological measure, namely the proportion of serologically positive individuals within the host population. This approach is widely adopted for human viral diseases, such as measles and yellow fever, which induce lifelong immunity to reinfection. In such circumstances, the measure records the proportion of the population that have experienced infection at some time in their life (Fig. 4.1e & f).

The severity of disease symptoms shown by a host is often related to the number of parasites harboured. This number is referred to as the

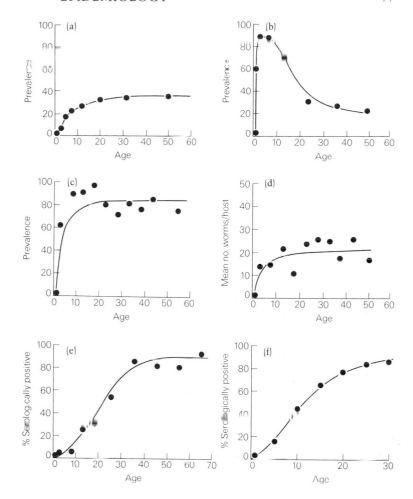

Fig. 4.1 Some examples of horizontal cross-sectional surveys of the prevalence and intensity of parasitic infection in human communities (age–prevalence and age–intensity curves). (a) *Entamoeba histolytica* in West Africa. (After Bray & Harris, 1977.) (b) Malaria in Nigeria. (After Fleming *et al.*, 1979.) (c) & (d) *Ascaris* in Iran. (After Arfaa & Ghadirian, 1977.) (e) Yellow fever in Brazil. (After Muench, 1959.) (f) Measles in New York. (After Muench, 1959.)

intensity of infection while the mean number of parasites per host is denoted as the *mean intensity of infection* (note that this mean should be calculated for the total host population including the uninfected individuals) (Fig. 4.1d).

The measurement of disease intensity is rarely straightforward. It may be estimated by direct counting (e.g. ectoparasites such as ticks and lice), from samples of host tissues or fluids (e.g. the use of blood films to count malaria parasites), or by the use of chemotherapeutic agents to expel gut helminths in the faeces of the host (e.g. *Ascaris* and hookworms). Indirect measures may often be necessary, such as the recording of helminth egg and protozoan cyst numbers in the faeces of the host, or the measurement of the

levels of host immunological responses to parasite antigens. It is usually difficult, however, to relate these indirect measures to parasite numbers.

The prevalence and intensity of infection are often recorded within different age classes of human populations in order to ascertain the sections of the community most at risk. Data of this form are often plotted as *age–prevalence* or *age–intensity* curves (Fig. 4.1). The examination of alternative and finer stratifications of the population may be necessary for certain infections. Sex, religion, age and occupation, for example, are important factors in the epidemiology of many directly transmitted helminths of humans.

The examination of various classes within a

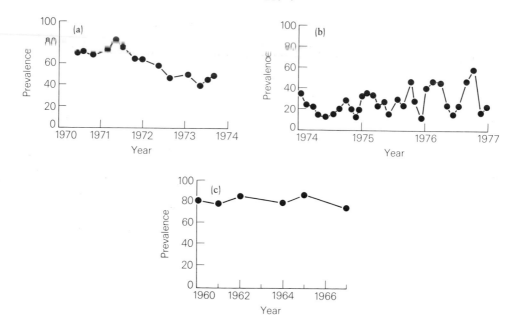

Fig. 4.2 Some examples of longitudinal surveys of parasitic prevalence. (a) *Trypanosoma vivax* in tsetse flies in Nigeria. (After Riordan, 1977.) (b) *Entamoeba histolytica* in West Africa. (After Bray & Harris, 1977.) (c) *Schistosoma haematobium* in Iran. (After Rosenfield *et al.*, 1977.)

community at one point in time, or over a short time interval, is referred to as a *horizontal cross-sectional epidemiological survey* (Fig. 4.1). The monitoring of these classes over successive points in time is described as a *longitudinal cross-sectional study* (Fig. 4.2). Horizontal surveys of prevalence and intensity within different age classes of a community can provide valuable information on the rate at which hosts acquire infection through time (age = time), provided that the host and parasite populations have remained approximately stable for a period of time (stable endemic disease).

4.3 FREQUENCY DISTRIBUTION OF PARASITE NUMBERS PER HOST

Measures such as parasite prevalence and mean intensity are statistics of the probability or frequency distribution of parasite numbers per host. The form of this distribution determines the relationship between prevalence and intensity (Fig. 4.3). Dispersion or distribution patterns of parasites within the host population can be broadly divided into three categories: (1) underdispersion (regular, homogeneous, variance < mean); (2) random (variance = mean); and (3) overdispersion (contagious, aggregated, heterogeneous; variance > mean).

These patterns are well-described empirically by three well-known probability distributions: (1) the positive binomial (underdispersion); (2) Poisson (random); and (3) negative binomial (overdispersion).

Parasites are almost invariably overdispersed or aggregated within their host populations, where many hosts harbour a few parasites and a few hosts harbour large numbers of parasites (Fig. 4.4). The causes of such heterogeneity are many and varied but they are usually associated with variability in 'susceptibility' to infection within the host population. Such variability may be due to differences in host behaviour, spatial aggregation of infective stages or differences in the ability of individual hosts to mount effective immunological responses to parasite invasion (due either to past experiences of infection, other parasitic species within the host or genetic con-

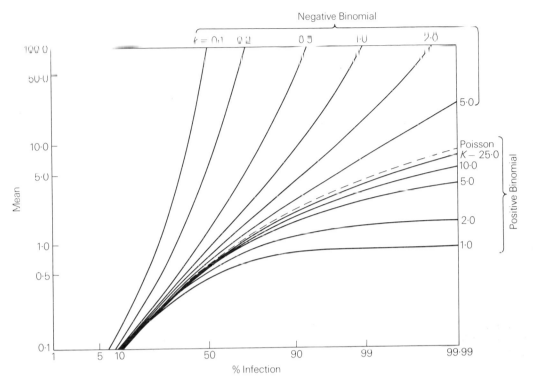

Fig. 4.3 The relationship between the prevalence and mean intensity of infection for the positive binomial (*K*), Poisson and negative binomial distributions (*k*) of parasite numbers per host.

stitution) (Table 4.1). Those hosts with heavy infections are often predisposed to this state by as yet undermined factors involving genetic, behavioural and environmental components. Good examples of predisposition are provided by the intestinal helminth infections of humans, where, following chemotherapeutic treatment to remove current parasite burdens, those with heavy burdens prior to treatment tend on average to acquire heavy burdens following a period of reinfection (Fig. 4.5).

The aggregation of parasites within the host population has many implications for epidemiological study. The most important concerns the sampling of the host population to measure infection intensity. A high degree of variability in parasite numbers per host necessitates the examination of large numbers of hosts (large sample size) if an accurate picture is to be obtained of parasite abundance within the host population.

Very misleading results may be obtained from small samples.

Parasite aggregation also has important implications for the regulation of parasite numbers within the host population and these are discussed in a latter section.

4.4 TRANSMISSION BETWEEN HOSTS

Parasites may complete their life cycles by passing from one host to the next either directly or indirectly via one or more intermediate host species (see Chapter 7). *Direct transmission* may be by contact between hosts (for example venereal diseases) or by specialized or unspecialized transmission stages of the parasite that are picked up by inhalation (respiratory viruses), ingestion (such as pinworm) or penetration of the skin (such as hookworm). *Indirect transmission* can involve biting by vectors (flies, mosquitoes, ticks

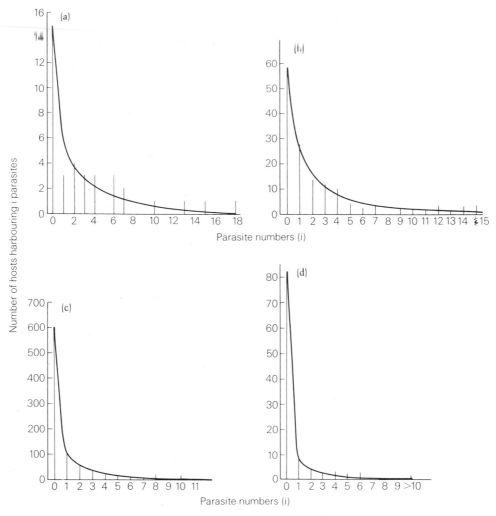

Fig. 4.4 Some examples of aggregated distributions of parasite numbers per host for which the negative binomial is a good empirical model (vertical bars: observed frequencies; solid line: fit of negative binomial). (a) A gut nematode (*Toxocara canis*) of foxes (Watkins & Harvey, 1942). (b) Nematode microfilariae (*Chandlerella quiscoli*) in mosquitoes (*Culicoides crepuscularis*). (After Schmid & Robinson, 1972.) (c) Lice (*Pediculus humanis capitis*) on humans. (After Williams, 1944.) (d) A (*Diplostomum gastrostei*) of fish (*Gastrosteus aculeatus*). (After Pennycuick, 1971.)

and others) that serve as intermediate hosts (the parasite undergoing obligatory development within the vector). In other cases, the parasite is ingested when an infected intermediate host is eaten by the predatory or scavenging final host. A special case of direct transmission arises when the infection is conveyed by a parent to its unborn offspring (egg or embryo), as can occur in rubella virus or HIV transmission. This process has been termed *vertical transmission* in contrast to the variety of *horizontal transmission* processes discussed above and in Chapter 7.

The prevalence and intensity of a parasitic infection within the host population is in part determined by the rate of transmission between hosts. Many factors influence this rate including

Table 4.1 Parasite dispersion in host populations

←————————————— Dispersion spectrum —————————— ———————————→

All hosts harbour the same number of parasites	Parasites randomly (independently) distributed (variance = mean)	All parasites in one host, all other hosts uninfected
Underdispersion (variance < mean) (regularity, homogeneity)		Over dispersion (variance > mean) (aggregation, heterogeneity)
Some factors which generate underdispersion Parasite mortality Host immunological processes		*Some factors which generate overdispersion* Heterogeneity in host behaviour Heterogeneity in effective immunity within the host population
Density-dependent processes Parasite induced host mortality when the death rate of the host is positively correlated with parasite burden		Direct reproduction within the host Spatial heterogeneity in the distribution of infective stages

climatic conditions, host and parasite behaviour, and the densities of both host and infective stage, plus their respective spatial distributions. The rate of transmission (sometimes referred to as the rate or force of infection) is measured in a variety of ways, the method depending on the type of parasite and its mode of transmission.

4.4.1 Transmission by contact between hosts

For many directly transmitted viral and protozoan diseases, where infection results either from physical contact between hosts or by means of a very short-lived infective agent, the net rate of transmission is directly proportional to the frequency of encounters between susceptible (uninfected) and infected hosts. In these cases no attempt is made to measure the number of parasites transmitted between hosts (due to their small size and ability to multiply rapidly within the host); attention is simply focused on the rate at which hosts become infected.

If we define the parameter β as the rate of contact between hosts which result in transmission (where $1/\beta$ is directly proportional to the average time interval between contacts), then during an interval of time Δt, in a population of N hosts consisting of X susceptibles and Y infecteds ($N = X + Y$), the

number of new cases of an infection is often approximated by the quantity $\beta XY\Delta t$. The parameter β is called the *per capita* rate of infection and is the product of two components, namely the average frequency of contact between hosts, multiplied by the probability that an encounter between susceptible and infected will result in the transference of infection.

If the time interval Δt becomes very small we arrive at a differential equation representing the rate of change of the number of infected hosts through time:

$$dY/dt = \beta XY = \beta(N - Y)Y. \qquad (1)$$

The equation has the solution

$$Y_t = NY_0/[Y_0 + (N - Y_0)\exp(-\beta Nt)], \qquad (2)$$

where Y_0 represents the number of infecteds introduced into the population at time $t = 0$. In practice we are often more interested in the rate at which new cases of infection arise (or are reported) (i.e. dY/dt). Calling this rate ω and substituting equation (2) into equation (1) we find that

$$\omega = \frac{\beta Y_0(N - Y_0)N^2 \exp(\beta Nt)}{[(N - Y_0)\exp(-\beta Nt)]^2} \qquad (3)$$

Equation (3) describes an epidemic curve which is roughly bell shaped in form (Fig 4.6b). Note that

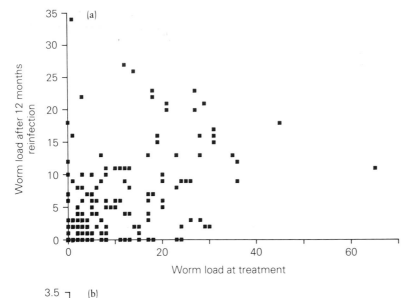

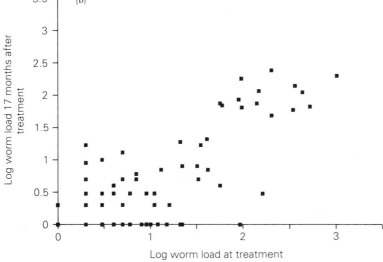

Fig. 4.5 Two examples of predisposition to light and heavy infection in humans infected with the intestinal nematodes *Ascaris lumbricoides* and *Trichuris trichuria*. Worm burdens were measured in samples of the human population, at the point of initial chemotherapeutic treatment and after several months of host exposure to reinfection. (From Haswell-Elkins *et al.*, 1987; Bundy, 1988.)

as the density of the host population (*N*) increases, the development of the epidemic occurs more rapidly. The net rate of parasite transmission is always greater in dense populations than sparse ones.

Equations (2) and (3) accurately mirror the growth and decay of various types of disease epidemics, including infections within laboratory populations of animals and viral diseases within human communities (Fig. 4.6a). The agreement between observation and theory supports the

assumption that transmission of many direct life cycle microparasites is directly proportional to the rate of encounter between hosts. It is important to note, however, that within certain human communities the rate of transmission of some viral and bacterial diseases appears to be non-linearly related to population density. This is thought to arise as a consequence of non-homogeneous mixing within such communities.

The value of the transmission rate β in equation (1) often varies seasonally due to the influence of

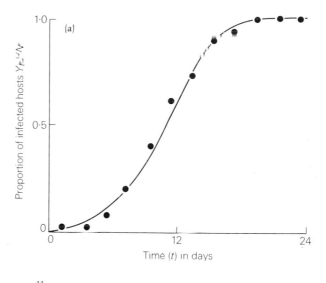

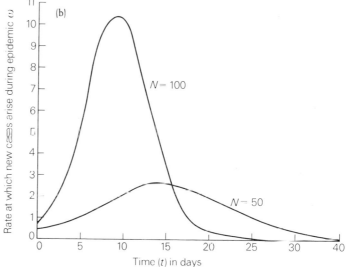

Fig. 4.6 (a) An epidemic of the protozoan *Hydramoeba hydroxena* in a population of the coelenterate *Chlorohydra viridissima*. (Stiven, 1964.) (b) Two examples of epidemic curves for populations of different sizes ($N = 100$ and $N = 50$, see text).

climatic factors on either the frequency of host contact or the life expectancy of the infective stages which determines the probability of parasite transference during host encounters. Seasonal contact rates are important in the epidemiology of many common viral infections of humans such as measles, mumps and chickenpox.

Within large communities where there exists a continual inflow of new susceptibles (births), a disease may exhibit recurrent epidemic behaviour or may even persist in a stable endemic manner through time. In the case of human viral infections such as measles, where children who have recovered from infection are immune from further attack, the persistence of the parasite is very dependent on the rate of inflow of susceptibles. The number of new births per unit of time is related to community size and thus such infections are more likely to persist from year to year in large communities. Interestingly, seasonal contact rates can, in conjunction with other factors (such as the rate of input of susceptibles and

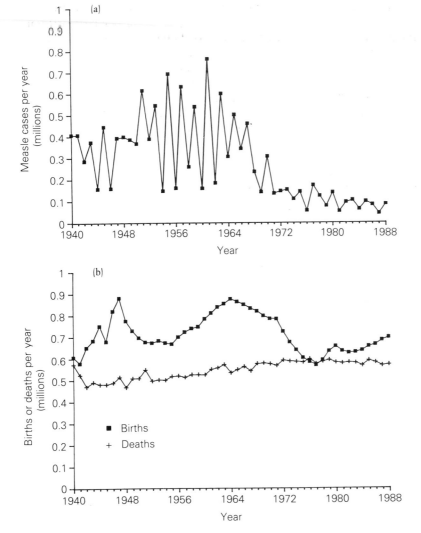

Fig. 4.7 (a) Reported cases of measles in England and Wales over the period 1940–1988. Mass vaccination was initiated in 1967. The bottom graph (b) records net births and deaths over the same period.

incubation periods during which hosts are infected but not infectious), produce complicated non-seasonal cycles (perhaps chaotic) in the prevalence of viral infections. Figure 4.7 shows a biannual cycle of measles incidence in England and Wales over the period 1940–1988. Note the introduction of mass vaccination in 1967 and also that the high birth rate in 1946–48, post the Second World War, disrupted the two-year cycle. The two-year cycle is superimposed over a regular seasonal cycle induced by changes in the infection rate β.

Aside from the influence of climatic factors on β, it is important to note that human *behaviour patterns* have a major effect on disease transmission. 'Who mixes with whom' is an important determinant of the pattern of infection observed for directly transmitted infectious agents.

4.4.2 Transmission by an infective agent

Many directly and indirectly transmitted parasites produce transmission stages with a not insignificant lifespan outside of the host. Examples are the miracidia and cercariae of schistosomes, the infective larvae of hookworms and the eggs of *Ascaris*. In the case of helminth parasites, we are usually interested in the number of parasites a host acquires, since parasite burden is invariably related to the severity of disease symptoms induced by parasitic infection. The measurement of transmission must therefore be based on the rate of acquisition of individual parasites rather than simply the rate at which hosts become infected (as in equations 1–3).

The rate of acquisition is often directly proportional to the frequency of contact between hosts and infective stages. If we define the parameter $\hat{\beta}$ as the rate of contact between hosts (N) and infective stages (I) which results in successful infection, the number of infections that occur in a small interval of time Δt is $\hat{\beta}IN\Delta t$ (the

mean number per host is $\hat{\beta}I\Delta t$). The parameter β again represents the product of two components, namely, the rate of contact times the probability that a contact results in the establishment of the infective stage within the host.

The properties of this form of transmission are well-illustrated by the following simple example. If N uninfected hosts are introduced into a habitat with I_0 infective stages, whose life expectancy in the free-living habitat is $1/\mu$, the rates of change through time of I_t and the mean number of parasites acquired by each host, M_t, are described by the following differential equations:

$$dI/dt = -(\hat{\beta}N + \mu)I \tag{4}$$
$$dM/dt = \hat{\beta}I. \tag{5}$$

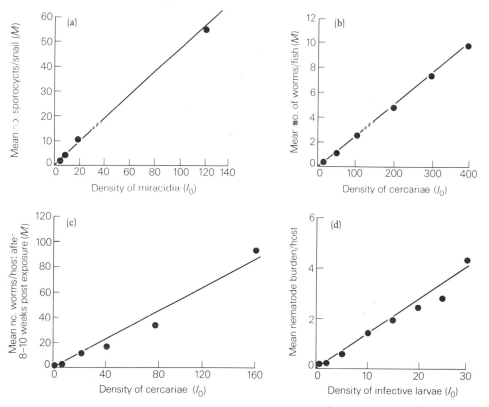

Fig. 4.8 Some examples of the linear relationship between the number of parasites that establish on or in a host when exposed to varying numbers of infective stages. (a) *Biomphalaria* exposed to miracidia of *Echinostoma lindoense*. (After Lie *et al.*, 1975.) (b) Fish exposed to cercariae of *Transversotrema patialense*. (After Anderson *et al.*, 1978.) (c) Hamsters exposed to cercariae of *Schistosoma mansoni*. (After Grove & Warren, 1976.) (d) Mosquitos exposed to infective larvae of the nematode *Romanomermis culicovorax*. (After Tingley & Anderson, 1986.)

These equations have the following solutions:

$$I_t = I_0 \exp[(-\hat{\beta}N + \mu)t] \tag{6}$$

$$M_t = [\hat{\beta}I_0/(\hat{\beta}N + \mu)][1 - \exp[-(\hat{\beta}N + \mu)t]]. \tag{7}$$

In an interval of time $\bar{t}$ the relationship between the mean number of successful infections per host (M) and the initial exposure density of infective stages (I_0) is linear where

$$M = I_0 A. \tag{8}$$

The constant A is defined for notational convenience as

$$A = \hat{\beta}[1 - \exp[(-\hat{\beta}N + \mu)\bar{t}]]/(\hat{\beta}N + \mu). \tag{9}$$

The linear relationship predicted by equation (8) is observed in many experimental situations, various examples of which are shown in Fig. 4.8. Such experiments enable the parameters $\hat{\beta}$ and $1/\mu$ to be estimated under defined laboratory conditions.

This simple model (equations 4 and 5) illustrates a number of important points concerning the dynamics of parasite transmission.

1 The existence of a *linear* relationship between infective stage density and the average number of parasites acquired per host (Fig. 4.8) indicates that the net rate of infection is *directly* proportional to the density of infective stages times the density of hosts.

2 The number of parasites acquired during an interval of time ($\bar{t}$) is very dependent on the expected lifespan of the infective stage ($1/\mu$) (Fig. 4.9). Very high infection rates ($\hat{\beta}$) may not necessarily lead to the rapid accumulation of parasites within the host population if the ex-

pected lifespan of the infective stage is short (e.g. the infection of molluscs by the miracidia of digeneans). Conversely, high parasite burdens may accumulate even when the contact rate is low, provided the infective stage is long-lived (e.g. *Ascaris* eggs and the acquisition of infection by humans). Overall transmission success (measured by the accumulation of parasites within the host population) is therefore dependent on a number of rate determining processes (influencing various developmental stages in the parasite life cycle), and not simply on infection or contact rates ($\hat{\beta}$).

3 The total number of parasites that manage to establish within the host population (MN) during a fixed time interval ($\bar{t}$) increases as host density (N) rises (Fig. 4.9a). This relationship is non-linear, the total approaching an asymptote whose size is determined by both the number of infective stages available within the habitat (I_0) and their expected lifespan ($1/\mu$).

4 A relationship exists between the mean number of parasites acquired (M) and the prevalence of infection (the proportion of the population infected, p) within the exposed host population. The precise form of the relationship depends on the statistical distribution of parasite numbers per host (Fig. 4.3). If the parasites are randomly distributed (Poisson distribution), this relationship is defined by

$$p = [1 - e^{-m}]. \tag{10}$$

The prevalence rapidly rises to unity, where all hosts carry infection, as the exposure density of infective stages (I_0) increases (Fig. 4.10). Random patterns are, however, rarely observed even under laboratory conditions. More usually the distri-

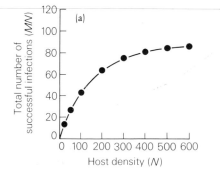

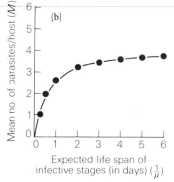

Fig. 4.9 (a) The relationship between the total number of parasites that establish within the host population and host density, N (see text). (b) The relationship between the mean number of parasites established per host and the expected lifespan of the infective stage, $1/\mu$ (see text).

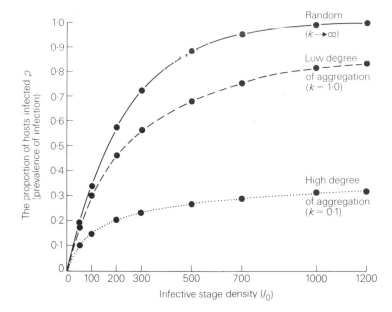

Fig. 4.10 The relationship between the prevalence of infection and infective stage density for varying distribution patterns of parasite numbers per host (see text).

bution is aggregated or contagious, where a high degree of variability exists between the worm burdens of individual hosts (the statistical variance is greater than the average burden). This variability is, as discussed earlier, principally due to heterogeneity in susceptibility to infection within the host population.

The negative binomial distribution is a good empirical model of aggregated patterns of parasite numbers per host. It is defined by two parameters, the mean (M) and a parameter k, which varies inversely with the degree of parasite aggregation. The relationship between prevalence and mean parasite burden predicted by this distribution is given by

$$p = 1 - [1 + M/k]^{-k}. \tag{11}$$

If the host population is highly heterogeneous with respect to susceptibility (k small), the prevalence of infection will not rise rapidly to unity as the density of infective stages rise. A few hosts will acquire the majority of parasites, always leaving a proportion of the host population uninfected (Fig. 4.10). The results of an experimental investigation of the relationship between infective stage density, host density and the resultant prevalence of infection are recorded in Fig. 4.11.

A good illustration of the type of mechanism that generates heterogeneity in the distribution of parasite numbers per host, is provided by the influence of various spatial patterns of infective stages on the acquisition of *Hymenolepis diminuta* infections by the beetle intermediate host *Tribolium confusum*. For a fixed density of eggs in experimental infection arenas, increasing degrees of aggregation in the spatial distribution of these infective stages resulted in a concomitant increase in the level of aggregation of worm numbers per beetle (Fig. 4.12). Interestingly, the spatial pattern of eggs does not affect the average rate of parasite acquisition, it simply acts to determine the distribution of the number of parasites acquired by each individual host (Fig. 4.12).

The duration of exposure to infection $(\bar{t})$ and infective stage density (I_0) also influence the distribution of successful infections per host if the host population is heterogeneous with respect to susceptibility to infection. A constant degree of variability in susceptibility (caused by either behavioural, immunological or genetic differences) generates an increasing degree of aggregation of successful infections per host, as both the density of infective stages and the duration of exposure to infection increase.

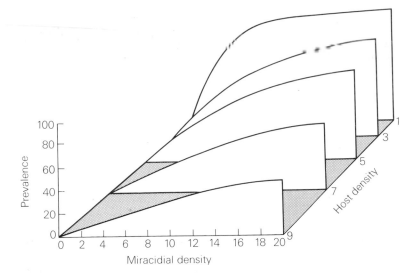

Fig. 4.11 Experimental determination of the relationships between the prevalence of snail infection (*Biomphalaria*) with *Schistosoma mansoni* and snail density plus miracidia density. (After Carter *et al.*, 1982.)

4.4.3 Transmission by ingestion

The rate of transmission of a parasite which gains entry to the host by ingestion is influenced by the feeding behaviour of the host. Ingestion may occur as a result of the host actively preying on infective stages (fish predating digenean cercaria), consuming food contaminated with infective agents (human consumption of vegetables contaminated with *Ascaris* eggs) or consuming an intermediate host which is infected with larval parasites (human consumption of fish infected with *Diphyllobothrium*) – a predator–prey association existing between final and intermediate hosts.

In such cases the relationship between the number of parasites acquired and infective stages density, or the density of infected intermediate hosts, may no longer be linear. The net rate of infection is determined by the feeding rate of the host which is non-linearly related to food or prey density. This is a consequence of either satiation effects or the amount of time taken to capture and consume a prey item. Host satiation results in a decrease in the rate of food consumption, and hence the rate of ingestion of parasites, irrespective of the density of infective stages or infective intermediate hosts. A predator will also only be able to consume a finite number of prey items during an interval of time due to the amount of time taken to hunt, capture and consume a prey item (commonly called the 'handling time').

The net effect of both influences is to limit the rate of parasite acquisition, two examples of which are shown in Fig. 4.13.

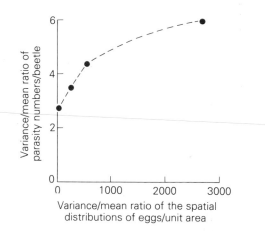

Fig. 4.12 The influence of the spatial distribution of *Hymenolepis* eggs on the distribution of infections within the beetle (*Tribolium*) intermediate host. (After Keymer & Anderson, 1979.)

4.4.4 Transmission by a biting arthropod

Many microparasites and macroparasites have indirect life cycles where transmission between

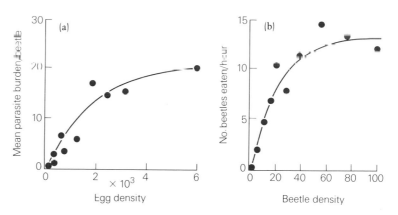

Fig. 4.13 Non-linear relationships between infective stage, (or infected host) density and the rate of parasite (or intermediate host) acquisition. (After Keymer & Anderson, 1979; Keymer, unpublished results.) (a) The acquisition of *Hymenolepis* by *Tribolium* beetles. (b) The rate of eating beetles by rats.

hosts is achieved by a biting arthropod, for example yellow fever, malaria, sleeping sickness and filariasis. The intermediate arthropod host, or vector, tends to make a fixed number of bites per unit of time, independent of the number of final hosts available to feed on. The transmission rate from infected arthropods to people (and from infected people back to susceptible arthropods) is therefore proportional to the 'man biting rate' ($\bar{\beta}$) multiplied by the probability that a given human is susceptible (or infected) and not simply proportional to the number of susceptibles (or infected) people (see equation 1). If the number of susceptible and infectious final hosts and vectors are represented by X, Y, and X' and Y' respectively, the net rate at which susceptible people acquire infection is $\bar{\beta}(X/N)Y'$. The quantity N represents the total number of people and the 'man biting rate' $\bar{\beta}$ consists of two components; the biting rate times the probability that an infectious bite leads to infection. Similarly, the net rate at which susceptible vectors acquire infection is $\bar{\beta}X'(Y/N)$.

If N uninfected final hosts are exposed to Y' infectious vectors, the number of hosts that acquire infection, Y, in a period of time, t, is approximately given by

$$Y = N[1 - \exp(-\bar{\beta}(Y'/N)t)]. \qquad (12)$$

The degree of transmission measured by Y is critically dependent on the biting rate $\bar{\beta}$, irrespective of the density of hosts, N, which are susceptible to infection (Fig. 4.14)

The prevalence of infection of many parasitic diseases within their vector populations is char-

acteristically low, even when the level of infection within the vertebrate host population is high. In regions of endemic malaria, for example, where more than 50% of the human population is infected, the prevalence of infection within the mosquito vector is typically 1–2%. In Bancroftian filariasis it is 5–8% and in blackfly infected with *Onchocerca volvulus* it is 1–5%. This is basically a consequence of the inverse relation between

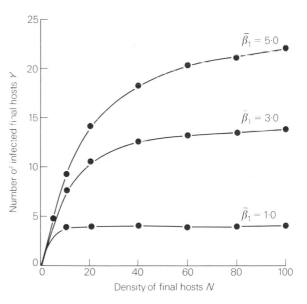

Fig. 4.14 The relationship between the number of final hosts that acquire infection (Y) when exposed to five infectious vectors for 10 units of time, and final host density (N) for different vector biting rates ($\bar{\beta}$) (see text).

standing crop (the number of infected hosts) and population turnover rates (the rates of loss of infected hosts) that arises in many biological systems.

There is often a substantial difference in the life expectancy of infection within the vertebrate host and vector, sometimes as a consequence of differences in host life expectancy. For a fixed transmission rate (as is the case for vector borne diseases where the biting rate determines transmission both from vertebrate to vector and *vice versa*), infections of long duration will give rise to higher prevalences than those of short duration. A single infection of malaria, for example, may last from a few months to many years (depending on the species of *Plasmodium* and the immunological status of the host), while the duration of infection in the vector is limited by the short life expectancy of the mosquito which is typically of the order of 2–15 days. This type of pattern is common amongst vector-borne protozoan and helminth diseases in humans. Viral infections, however, often have different characteristics, exemplified by yellow fever where the duration of infection in humans is roughly 10 days.

Transmission of a vector-borne disease is also influenced by the developmental period of the parasite in the vector; a period during which the host is infected but not infectious. This development delay is called the *latent* period and may often be significant in relation to the expected lifespan of the intermediate host. In malaria, for example, the latent period in the mosquito is 10–12 days, whereas the life expectancy is between 2 and 15 days, according to the species. The precise manner in which all these various factors (the biting rate $\bar{\beta}$, the expected lifespan of the vector $1/b$ and the latent period τ) influence the rate of transmission is not easy to determine.

However, in areas where the prevalence of infection in the human population (y) remains roughly constant through time (endemic regions) the prevalence of infectious vectors y' is given by

$$y' = \frac{[\bar{\beta}y\exp(-b\tau)]}{(b + \beta y)}. \tag{13}$$

The insertion of a range of parameter values into this expression (Fig. 4.15) illustrates how important vector life expectancy ($1/b$) and the

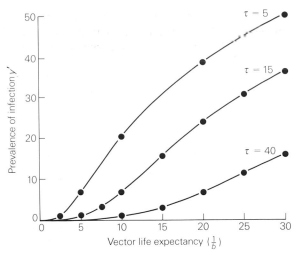

Fig. 4.15 The relationship between the prevalence of infectious vectors (y') and vector life expectancy $1/b$ for varying disease latent periods (τ) (see text).

duration of the latent period (τ) are as determinants of transmission success (measured by y').

Short vector life expectancy and long latent periods result in very low prevalences of infectious vectors even in areas of stable endemic disease where the disease is very prevalent within the human population.

4.5 REGULATION OF PARASITE ABUNDANCE WITHIN THE HOST POPULATION

The *perpetuation* of a parasite within its host population is distinct from the survival or *persistence* of an infection within an individual host. The mechanisms which control parasite population size within an individual host, however, are central to our understanding of perpetuation and stability within the host population as a whole. It is convenient to consider these issues under two headings.

4.5.1 The host as the basic unit of study

As noted earlier, the epidemiological study of many human infections, particularly viral, bacterial and protozoan diseases, is based on the division of the host population into a series of categories. Hosts are allocated to these categories on the basis of their infection status which, for example, may be either susceptible, infected or immune. The basic unit of study is therefore an individual host and the perpetuation of an infection is judged on whether or not infected hosts are present.

Vertebrate hosts mount responses to parasitic invasion which may be immunological in nature. If successful, these responses will either eliminate the parasite or constrain its population size within the host to a low level (see Chapter 8). Acquired immunity can cause second and later infections to be eliminated with no overt signs of disease (and without the host becoming infectious to others), so that hosts with acquired immunity in effect join an immune category that is protected from infection. In short, there are two main points relevant to our considerations of the regulation of the number of infected hosts: (1) hosts

may recover from infection; and (2) hosts that recover may possess a degree of protection, either transient or long term, to future infection.

Host immunity is clearly an important regulatory constraint on the spread and perpetuation of infection within the host community. The greater the proportion of immunes within the population (commonly referred to as the degree of herd immunity), the lower will be the potential for disease transmission. Host immunity essentially acts as a form of delayed regulation or negative feedback on the number of infected hosts. The prevalence of infection at one point in time is related to the proportion of hosts that will become immune at some future point in time, the length of the time interval being dependent on the average duration of infection within an individual host. A large number of infecteds, which eventually become immune, leads in due course to a reduction in the number of susceptibles. This reduction acts to decrease transmission success and thus limits the number of

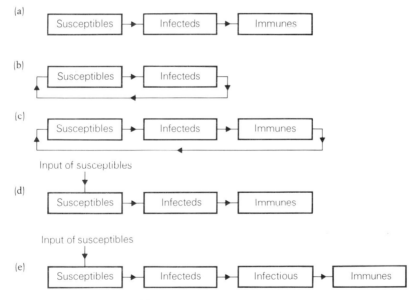

Fig. 4.16 Diagrammatic representations of the division of that host population into various compartments and the flow of hosts between these compartments (see text). (a) Host population of constant size consisting of susceptibles, infecteds (all assumed to be infectious) and immunes. (Immunity is lifelong.) (b) As (a) but no immune class, recovered infecteds passing back into the susceptible class. (c) As (a) but loss of immunity such that hosts pass back from the immune to the susceptible class. (d) As (a) but input of susceptibles due to new births or the arrival of immigrants. (e) As (d) but the infected class divided into two categories: those infected but not infectious and those infectious.

infecteds arising in the next generation. In other words, the number of infecteds at one point in time is inversely related to the number at some future point in time, host immunity acting as a regulatory constraint on the prevalence of infection (see Fig. 4.16).

The mechanisms of host immunity are dealt with in Chapter 8; our interest here centres on the population consequences and regulatory potential of acquired immunity.

Whether or not an infection spreads and perpetuates once it is introduced into a host population is dependent on a variety of factors. Initial spread is determined by: (1) the density of susceptibles; (2) the frequency of host contact; (3) the infectiousness of the disease (the probability that a contact between susceptible and infected results in transmission); and (4) the average time period during which an infected host is infectious. To illustrate these points consider the following

example. If a few infecteds are introduced into a population of hosts consisting of X, susceptibles, Y infecteds and Z immunes, the initial spread of infection is described by the following differential equations:

$$dX/dt = -\beta XY \tag{14}$$
$$dY/dt = \beta XY - \gamma Y \tag{15}$$
$$dZ/dt = \gamma Y. \tag{16}$$

For simplicity it is assumed that all infecteds are infectious. The parameter β represents the transmission rate (see equation 1), its magnitude being determined by the product of two components, namely the frequency of contact between hosts multiplied by the probability that a contact results in infection. The parameter γ denotes the rate of recovery from infection where $1/\gamma$ is the average duration of infectiousness.

The introduction of a few infecteds (numbering

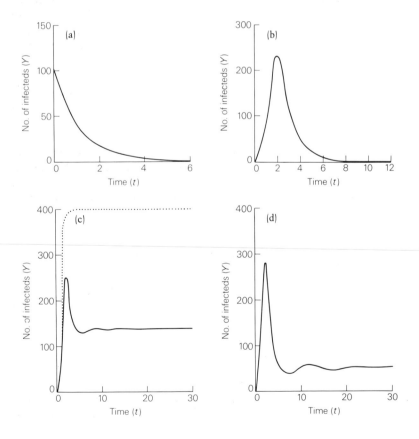

Fig. 4.17 (a) Host density below the threshold level for the occurrence of an epidemic ($X_0 = 500$, $Y_0 = 100$, $\beta = 0.0001$, $\gamma = 1.0$) (flow chart (a) in Fig. 4.16). (b) Host density above the threshold level ($X_0 = 500$, $Y_0 = 2$, $\beta = 0.01$, $\gamma = 1.0$) (flow chart (a) in Fig. 4.16). (c) The maintenance of endemic disease in a host population of fixed size where immunity is either absent (dotted line) (flow chart (b) in Fig. 4.16), or not lifelong (solid line) (flow chart (c) in Fig. 4.16). (d) Maintenance of endemic disease when immunity is lifelong by the input of new susceptibles (flow chart (d) in Fig. 4.16).

Y_0) into a population of N susceptibles will only result in an epidemic (representing the spread of the disease) provided that

$$N > \beta/\gamma \quad \text{(Fig. 4.16a \& b).} \quad (17)$$

This condition is a direct consequence of the fact that equation (1) must initially be positive if the number of infecteds is to increase above its introduction level Y_0. In other words, for an epidemic to occur, the density of susceptibles must be greater than a *critical threshold level*, N_T, which is determined by the product of the transmission rate (β) times the average duration of infectiousness of an individual host $1/\gamma$. A diagrammatic flow chart of the model described by equations (14) to (16) is displayed in Fig. 4.16a and the temporal changes of the number of infected hosts (Y) when equation (17) is, and is not, satisfied, are shown in Fig. 4.17a and b.

In our simple model we have assumed that once hosts recover from infection they are then immune for life, as is the case for certain human viral infections such as measles. In a closed community therefore, provided that equation (17) is satisfied, the disease will initially spread (leading to an epidemic) but will eventually die out once the supply of susceptibles is exhausted (Fig. 4.16a). The infection could perpetuate if some loss of immunity occurred, such that hosts passed back from the immune to the susceptible category (Fig. 4.16b & c; Fig. 4.17c).

More generally, however, disease perpetuation within the host population over a period of years will be dependent on the net inflow of suscep tibles into the community either as a consequence of new births or immigration (Fig. 4.17d). Thus in addition to the threshold density of susceptibles necessary for the occurrence of an epidemic, long-term endemic perpetuation will also depend on community size, larger communities tending to produce more offspring (susceptibles) than smaller ones. Epidemics of measles, for example, occur in school communities of a few hundred susceptible children, but the virus only becomes endemic on a continuous month–month and year–year basis in urban populations greater than 5 000 000 people. Even in very large communities the epidemiology of measles is characterized by cyclic, recurrent epidemic be

haviour (Fig. 4.7). Infections of this kind are probably diseases of modern societies; in primitive societies the net inflow of susceptibles into small communities was probably too low to maintain the diseases.

In brief, the perpetuation and stability of infectious diseases within their host populations revolves around a series of factors, some of which are interrelated. These factors are: (1) the ability of the parasite to persist within an individual host; (2) the length of the period during which an infected host is infectious; (3) the duration of acquired immunity to reinfection; (4) the infectiousness of the disease agents (transmissibility being dependent on both host and parasite characteristics); (5) the degree of herd immunity; and (6) the net rate of inflow of susceptibles into the host community.

Many viral and bacterial infections of humans are transient in nature and tend to induce lasting immunity to reinfection. These diseases only occur endemically, without periodic disappearance, in large communities with high inputs of susceptibles. Such infections characteristically exhibit wide fluctuations in prevalence through time, recurrent epidemic behaviour being a feature of their epidemiology.

Conversely, most protozoan diseases are able to persist within individual hosts for long periods of time and do not tend to induce lasting immunity to reinfection. These infections are able to survive endemically in small host communities with low inputs of susceptibles. The prevalence of protozoan infections is characteristically more stable through time than viral and bacterial diseases.

4.5.2 The parasite as the basic unit of study

The individual parasite forms the basic unit of study for helminth and arthropod infections (macroparasites), the pathology of these diseases being related to the number or burden of parasites harboured by the host.

In contrast to microparasites, the majority of helminths do not multiply directly within their vertebrate hosts (*direct reproduction*) but produce transmission stages (such as eggs or larvae) which pass out of the host as a developmental necessity (*transmission reproduction*). The number of parasites within an individual host (a *subpopulation* of parasites) is therefore controlled by the rate at which new infections arrive and the rate at which established parasites die. This is an immigration–death process in contrast to the birth–death one which governs the growth of microparasites within their hosts.

Macroparasites tend to produce lasting infections, with the host harbouring populations of parasites for long periods due to continual reinfections. Among many examples are the hookworm species of humans *Ancylostoma duodenale* and *Necator americanus*. For such parasites the rate of production of transmission stages and any resistance of the host to further infection typically depend on the number of parasites present in a given host. More importantly, helminths tend to be long-lived in their vertebrate hosts and do not usually induce lasting immunity to reinfection (Table 4.2).

Vertebrates are able to mount immunological responses to helminth invasion but these often act to restict rather than eliminate parasite establishment, reproduction and survival within the host. As such, an immune category of hosts,

Table 4.2 Life expectancies of various helminth parasites in humans and the time taken to reach reproductive maturity

Parasite	Life expectancy (years)	Development to maturity (days)
Ascaris lumbricoides	1–1	60–70
Ancylostoma duodenale	1–1	45–55
Necator americanus	3–5	28–49
Schistosoma mansoni	2–5	25–28

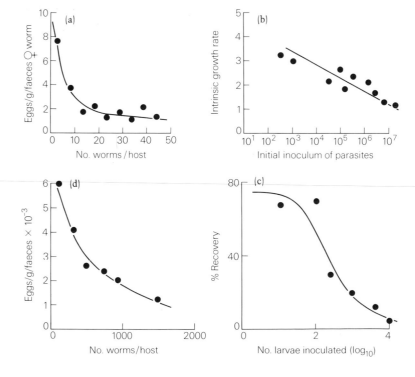

Fig. 4.18 Density-dependent survival and reproduction within parasite subpopulations. (a) Egg production by *Ascaris* in humans. (After Croll *et al.*, 1982.) (b) The intrinsic population growth rate of *Trypanosoma musculi* in mice. (Brett, unpublished data.) (c) Egg production by *Ancyclostoma* in humans. (After Hill, 1926.) (d) The survival of the hookworm *Ancyclostoma canium* in dogs. (After Krupp, 1961.)

totally protected from infection, rarely exists for helminth diseases. The important characteristic of host responses to helminth infections is the *density dependent* manner in which they act. The proportional reduction in establishment, survival and reproduction is greater in dense subpopulations of parasites than sparse ones (Fig. 4.18). These processes act as negative feedback mechanisms to constrain parasite population growth within individual hosts. Although immunological responses normally induce such effects, competition between parasites for finite resources, within or on the host, may also be important. This is particularly so within invertebrate intermediate hosts where competition between larval parasites for limited space and food resources is often severe (e.g. the production of cercariae within the molluscan host of digeneans).

Negative feedback or density-dependent mechanisms also occur in segments of helminth life cycles not involving the vertebrate host. For example, the rate of parasite-induced intermediate host mortality is often proportional to the burden of parasites harboured, hosts with high worm burdens dying more rapidly than those with light burdens (Fig. 4.19). Host death invariably results in the death of the parasites contained within and hence such responses cause density-dependent parasite mortality. Density-dependent constraints also act on the rate of acquisition of

parasites, particularly if the parasite gains entry to the host by ingestion, either as an infective stage, or within the body of an intermediate host. As we saw earlier, the rate of parasite acquisition in these cases is determined to a large extent by host feeding behaviour and is therefore much less dependent on the density of infective stages or infected intermediate hosts. Limitations on the rate at which hosts consume food items, due to satiation and handling time, act to regulate the intake of parasites at high densities of infective stages (and infected intermediate hosts) (see Fig. 4.13).

Any given parasite life cycle may contain more than one density-dependent process, particularly if the life cycle involves two or more hosts. The human schistosomes are good examples: adult worm survival and egg production are influenced in a density-dependent manner in the human host and cercarial production in the mollusc is virtually independent of the number of miracidia that penetrate the snail.

Despite many such examples, it is relevant to note that a single density-dependent process, in any segment of the parasite's life cycle, will act to restrict the population growth of all the various developmental stages throughout the entire cycle. A useful physical analogy is provided by water flow through a circular system made up of a series of interconnecting pipes of different

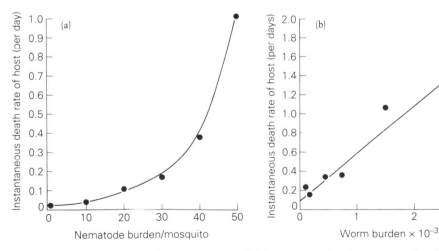

Fig. 4.19 The influence of parasite burden on host survival. (a) Mosquitoes (*Aedes trivittatus*) and the nematode *Dirofilaria immitis*. (After Christensen, 1978.) (b) Sheep and *Fasciola hepatica*. (After Boray, 1969.)

diameters. The rate of water flow through the entire system is determined by the segment of pipe with the narrowest diameter.

Regulation of parasite abundance by density-dependent processes within individual hosts acts to constrain the growth of the total parasite population within the community of hosts. The severity of such processes and the number that occur in the various segments of a parasite's life cycle determines the endemic level of helminth infection within the host population. This level is often measured by the average worm burden (average intensity) per host.

The average intensity, however, is a statistic of the frequency distribution of parasite numbers per host. The precise form of this distribution plays a major role in determining the net force of density-dependent regulation acting on the total parasite population. These distributions are invariably highly aggregated in form, where the majority of hosts harbour few parasites and a few hosts harbour most of the parasites. It is precisely in these few hosts that density-dependent effects will be most severe and they will thus influence the majority of the total parasite population. As the distribution of parasites becomes more aggregated, so the regulatory impact of density-dependent processes will be more pronounced.

In summary, macroparasite abundance is regulated by density-dependent processes acting on parasite subpopulations within individual hosts. The severity of these processes is a major determinant of the observed stability of many helminth infections (such as hookworms, roundworms and schistosomes) within human communities, even in the face of severe perturbations induced by climatic change or human intervention. In contrast to many microparasites, since a truly immune category of hosts rarely exists, threshold densities of susceptibles, the degree of herd immunity and the rate of input of susceptibles do not, in general, act to regulate the abundance of helminth and arthropod parasites.

4.6 POPULATION DYNAMICS

In the preceding section we considered the factors which exert a regulatory influence on the transmission of infection (or parasites) between hosts and the flow of parasites through their life cycles. These factors determine the ability of a parasite population to withstand perturbations and are particularly relevant to the interpretation of endemic patterns of infection. We now consider the broader question of how all the various rate-determining processes, which govern the population size of the many development stages involved in parasite life cycles, influence the overall population behaviour of an infection and its ability to perpetuate within the host population.

4.6.1 Transmission thresholds

The flow of parasites between hosts must exceed a certain overall rate if the infection is to perpetuate within the host population, in order to compensate for parasite losses throughout the life cycle. There exists a set of parasite transmission, reproduction and mortality rates, below which the infection will die out, and above which it will perpetuate. This level is defined as the *transmission threshold* and its determination is of great importance to the design and implementation of disease control policies.

The determination of this threshold and its biological interpretation are dependent on the basic unit of epidemiological study (i.e. either the infected host or the individual parasite).

4.6.2 Microparasites: direct transmission

In the case of microparasites, where the unit of study is the infected host, an infection will only perpetuate provided that one infected host, throughout its infectious lifespan, gives rise to at least one new infection when introduced into a population of susceptibles. The average number of new cases that arise from one infectious host, if introduced into a population of N susceptibles, is called the basic reproductive rate or number of an infection and is denoted by the symbol R. (Note that although called a rate, R is in fact a dimensionless quantity defined in terms of the generation time of an infection.) The transmission threshold is therefore given by the condition $R = 1$.

The determination of R in terms of the rate parameters which control disease transmission is straight forward. In the case of a directly transmitted viral infection such as measles (which induces lifelong immunity in recovered hosts) R is defined as

$$R = \beta N / (b + \gamma). \tag{18}$$

Here β is the disease transmission rate (see equation 1,), N is the population density of susceptibles, $1/b$ is the life expectancy of the host and $1/\gamma$ is the average period during which an infected host is infectious (see equations 14–16). R is simply the transmission or reproductive potential of the infection (βN), multiplied by the expected lifespan of the infectious host $[1/(b + \gamma)]$. The concept of a basic reproductive rate gives rise to the following important epidemiological principles.

1 The actual reproductive rate of an infection, $\hat{R}$, within a population of N hosts of which only a proportion q are susceptible, as opposed to its basic reproductive rate R (which measures its potential for reproduction or transmission in a population totally susceptible to infection), is

$$\hat{R} = Rq. \tag{19}$$

2 When the prevalence of infection y $(y = Y/N)$ remains fairly constant through time, the disease is at equilibrium within the host population and $\hat{R} = 1$.

3 The threshold host density, N_T, necessary for the successful introduction and initial spread of an infection (defined in equation 17) is related to the quantity R, where in a population of N hosts,

$$R = N/N_T. \tag{20}$$

4 The basic reproductive rate R is related to the average age at which hosts acquire infection, A, where

$$R = 1 + L/A, \tag{21}$$

and L is the life expectancy, $1/b$, of the host. This relationship often facilitates the estimation of R from epidemiological data. For example, in England and Wales during the period 1956–69,

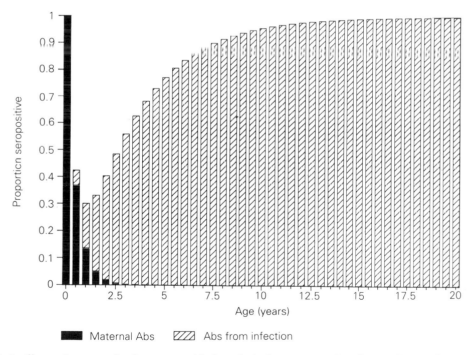

Fig. 4.20 An illustrative example of an age-stratified serological survey recording, by age, the population of a sample of hosts with antibodies to a particular viral infection. The example records the decay in maternally derived antibody (with a half-life of 6 months) and the rise in antibody induced by infection where the average age at infection is 5 years.

the average age at which children experienced an attack of measles varied between 4.5 and 6 years. The average age of infection, A, can be estimated from age-stratified serological data of the kind portrayed in Fig. 4.20. With a human life expectancy of roughly 70 years equation (21) yields R estimates of between 16.5 and 12.7.

5 If a vaccine is available for disease control, the proportion of the population, p, that must be protected at any one time for disease eradication is simply

$$p > 1 - 1/R. \tag{22}$$

In the case of measles, this would require the vaccination of roughly 92–94% of children given 100% efficacy for the vaccine. The magnitude of this figure provides an explanation of why measles and similar viral diseases are proving difficult to eradicate by mass vaccination in western Europe and the United States.

6 On a different scale, the proportion of susceptibles, s, left after an epidemic has passed through a small community, can be used to estimate R, since

$$R = [1/(1 - s)][\ln(1/s)]. \tag{23}$$

Thus if an epidemic of (say) chickenpox in a school leaves 30% of the children susceptible after its termination, R is 1.7. In order to have prevented this epidemic, 41% of the children would have had to have been vaccinated (equation 22).

7 Increased levels of vaccination will lead to a rise in the average age at which individuals experience their first attack of infection. Where the proportion of the population vaccinated is p, the mean age at first attack A is

$$A = L/[R(1 - p) - 1], \tag{24}$$

where L is host life expectancy.

8 Disease incubation periods, during which a host is infected but not infectious, will only significantly reduce the basic reproductive rate R if the incubation period is long with respect to host life expectancy. If the incubation period is of length $1/v$, then equation (18) becomes

$$R = \beta N f/(b + \gamma), \tag{25}$$

where f is the fraction of infected hosts that survive to become infectious, namely $v/(b + v)$.

9 The value of R determines both the relationship between the age of the host and the proportion of individuals who have experienced infection (the proportion serologically positive), and the equilibrium prevalence of infection y^* within the population (Fig. 4.21). The relationship between y^* and R is given by

$$y^* = [1 - (1/R)][b/(b + \gamma)]. \tag{26}$$

Highly infectious diseases (large βs) which persist in the host and give rise to long periods of infectiousness (small γs), have high R values. Conversely, diseases of low infectivity with short periods of infectiousness have small R values.

10 In the case of sexually transmitted infections

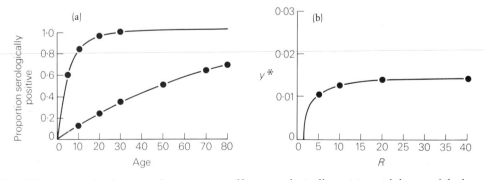

Fig. 4.21 (a) The relationship between the proportion of hosts serologically positive and the age of the host for various values of R (see text). (b) The relationship between the prevalence of infection and the basic reproductive rate R ($\gamma = 1$, $b = 0.014$). The level of the plateau is set by the duration of infectiousness ($1/\gamma$) and the life expectancy of the host ($1/b$) when the disease induces lifelong immunity.

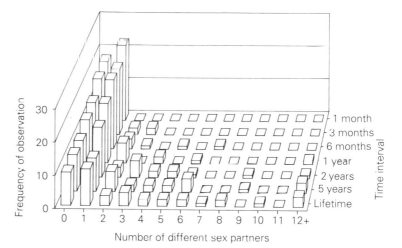

Fig. 4.22 Heterogeneity in sexual activity within a sample of male and female heterosexuals (age range 18–21 years) as recorded by the frequency distribution of the numbers of different sexual partners reported by individuals over various past time intervals (1 month to lifetime) (see Anderson, 1988). The distributions for all but the shortest time intervals are highly heterogeneous in character.

such as HIV, the aetiological agent of AIDS, the basic reproductive rate is defined as

$$R = \beta cD. \qquad (27)$$

Here β is the probability of transmission per sexual partner contact, D is the average duration of infectiousness of an infected host and, in the case of random choice of sexual partners, c is the mean rate of sexual partner acquisition. Survey data on human sexual behaviour, that records rates of sexual partner change, suggest that variability in sexual activity is high such that the variance, s_2 in the rate of partner acquisition is much greater in value than the mean. An illustration is presented in Fig. 4.22, which records frequency distributions of the rate of sexual partner acquisition over varying time intervals. Given heterogeneity in behaviour a better description of c in equation (27) is provided by the approximation $c = m(1 + cv^2)$ where cv is the coefficient of variation (s/M) of the rate of partner change. The definition of R now takes account of heterogeneity in the host behaviour that facilitates parasite transmission.

Such heterogeneity has a major influence on the magnitude of an epidemic of a sexually transmitted infection. As illustrated in Fig. 4.23, high heterogeneity (cv large) induces a smaller epidemic than low heterogeneity ($cv \rightarrow 0$). Conversely, high heterogeneity, for a fixed value of the average rate of partner change (m), increases the value of R by comparison with low hetero-

geneity (equation 27 with c defined in terms of m and cv). Hence if the variance in the host behaviour that facilitates transmission is high, an epidemic is more likely to occur and it will develop more rapidly, but it will be of smaller overall magnitude by comparison with an equivalent situation with low variability. The principle is applicable to all infectious agents where host behaviour is a major determinant of transmission success.

11 The simple definitions of the basic reproductive rate outlined in this section ignore much

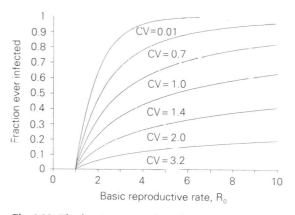

Fig. 4.23 The fraction ever infected in an epidemic of a sexually transmitted microparasite infection within a population of fixed size, as a function of the basic reproductive rate, R, and the coefficient of variation in sex partners per unit of time. (See May & Anderson, 1987.)

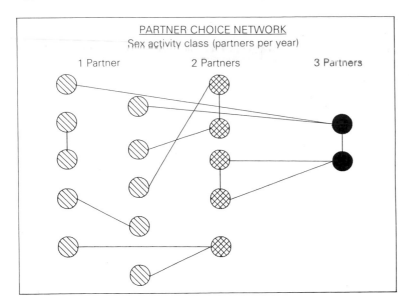

Fig. 4.24 An illustrative example of a network of sexual contacts, stratified according to sexual activity (defined as the number of partnerships formed per annum).

biological detail. A further aspect of host behaviour that is of major importance to transmission success concerns the manner in which different strata of the host population (e.g. age or sex classes, or groups defined on the basis of host behaviour such as sexual activity or water contact) mix. 'Who mixes with whom' is of great significance in the transmission dynamics of directly transmitted viral and bacterial infections (i.e. mixing by age class) and sexually transmitted infections such as HIV. In the latter case, group definition may be based on the level of sexual activity (rate of sexual partner acquisition as defined in Fig. 4.22) and the associated network of sexual contacts between groups determines the net transmission success of the infectious agent. A simple illustration of a sexual contact network is displayed in Fig. 4.24, where group definition is based on the number of different sexual partners per annum.

Whether mixing is assortative (like with like on the basis of sexual activity) or disassortative (like with unlike) has a major influence on the pattern of an epidemic of a sexually transmitted infection such as HIV. This principle is illustrated in Fig. 4.25 which plots simulated epidemics of HIV in a male homosexual community under the two assumptions of assortative or disassortative

mixing. Note that in the assortative case (those with high rates of sexual partner change choose their sexual partners mainly from the high activity groups) the epidemic rises more rapidly than the disassortative case, but is of smaller overall magnitude (Gupta *et al.*, 1989). More generally

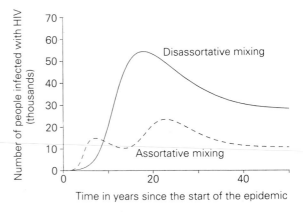

Time in years since the start of the epidemic

Fig. 4.25 Simulated epidemics of HIV in a male homosexual community under the assumptions of assortative (like with like on the basis of sexual activity) and disassortative (like with unlike) sexual mixing. Aside from the mixing pattern, all other parameters are the same in the two simulations. (See Gupta *et al.*, 1989.)

this example helps to highlight the importance of host behaviour as a determinant of the spread and persistence of an infectious agent.

4.6.3 Microparasites: vector-borne diseases

So far our attention has been restricted to directly transmitted microparasites. The concepts outlined above can be extended to encompass indirectly transmitted vector-borne diseases such as yellow fever and malaria. Without going into detail concerning derivations (see Anderson, 1980c, for fuller discussion), and ignoring for the moment disease latent periods, the basic reproductive rate of a vector-borne infection is

$$R = \frac{\beta^2 N_2/N_1}{(b_1 + \gamma_1)(b_2 + \gamma_2)}. \tag{28}$$

Here β represents the effective human-biting rate of the vector (effective in the sense of the proportion of bites that lead to infection). This parameter is squared since the action of biting is responsible for transmission from human to vector and vector to human. N_1 and N_2 represent the densities of humans and vectors respectively, $1/b_1$ and $1/b_2$ denote the respective life expectancies of humans and vectors, while $1/\gamma_1$ and $1/\gamma_2$ represent the duration of infectiousness in humans and vectors. R is again defined by the net transmission success both from human to vector and back again ($\beta^2 N_2/N_1$) multiplied by the product of the expected lifespans of infectious humans and vectors ($1/(b_1 + \gamma_1)$ and $1/(b_2 + \gamma_2)$). If we explicitly take into account latent periods in the vector and human, of length $1/v_2$ and $1/v_1$ respectively, then R becomes

$$R = \frac{\bar{\beta}^2 f_1 f_2 N_2/N_1}{(b_1 + \gamma_1)(b_2 + \gamma_2)}. \tag{29}$$

Here f_1 and f_2 represent the fraction of infected humans and vectors that survive to become infectious (namely, $v_1/(b_1 + v_1)$ and $v_2/(b_2 + v_2)$). As noted earlier, the fraction f_2 may be small due to the short life expectancy of certain vectors and the comparatively long latent periods of many viral and protozoan vector-borne diseases. In the human host, as in the case of directly transmitted infections such as measles, f_1 is normally close to unity in value.

Many of the epidemiological principles listed earlier for directly transmitted infections apply to vector-borne diseases. One important addition, however, concerns the threshold host density (or density of susceptibles) necessary for disease perpetuation. Note that this threshold condition now becomes

$$N_2/N_1 = (b_1 + \gamma_1)(b_2 + \gamma_2)/(\beta^2 f_1 f_2). \tag{30}$$

In other words, for the maintenance of a vector-borne parasite, the ratio of vectors to human hosts must exceed a critical level to maintain R greater than unity. This concept has been widely used in the design of control programmes for diseases such as malaria, where the aim of control has been to reduce vector density below this critical level (see Macdonald, 1957). The success or failure of such approaches, however depends on the level of accuracy achieved in estimating the many parameters which determine the value of R. Many problems surround the estimation of these parameters and, in addition, the biology of certain vector-borne diseases of humans is poorly understood at present.

The epidemiology of human malaria, for example, has received considerable attention over the past three decades, but many issues concerning both its biology and the approach to be adopted for disease control remain unresolved. The rapidity with which the prevalence of infection rises in areas of stable endemic malaria within the younger age classes of children, and its decline thereafter (Fig. 4.1b) suggest: (1) that the value of the basic reproductive rate R is very high in such areas; (2) that the duration of acquired immunity to reinfection is relatively short; and (3) immunity may be parasite strain specific such that with a highly genetically variable parasite population, herd immunity builds up slowly as hosts accumulate experience of the many strains circulating in a given location. It is not clear as yet how immunity is related to the number of infections experienced by an individual. The marked decline in prevalence in the adult age classes (Fig. 4.1b) suggests the existence of a degree of herd immunity. The maintenance of this, however, is dependent on the frequency with which humans are exposed to infectious bites. Vector control programmes reduce the frequency with which an

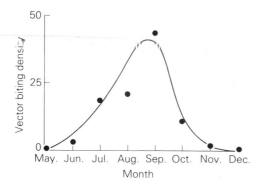

Fig. 4.26 Seasonal changes in the biting density of *Anopheles gambiae* in northern Nigeria. (After Garrett-Jones & Shidrawi, 1969.)

individual experiences infection, thus lowering the degree of herd immunity. As a direct consequence, the resurgence of malaria is often very severe when control ceases.

In certain areas, malaria prevalence appears to be much less stable as a consequence of low R values (caused by low vector densities; see Fig. 4.26) which fall below unity during certain periods of the year. In these areas the degree of herd immunity is comparatively low.

4.6.4 Macroparasites: direct transmission

The unit of epidemiological study for macroparasites is the individual parasite, and the definition of the basic reproductive rate outlined in the preceding section therefore requires modification. Macroparasites will only survive from generation to generation provided that a reproductively mature female (or in the case of a hermaphroditic species, a mature parasite), on average, gives rise to sufficient offspring to at least replace herself in the next generation. The basic reproductive rate R now becomes the number of female offspring produced by an adult female worm throughout her reproductive lifespan, which survive to reach sexual maturity in a population of N uninfected hosts. The transmission threshold is again defined by the condition $R = 1$.

Ignoring for the moment developmental delays, such as the time taken by a helminth to reach sexual maturity in the host, the basic reproductive rate of a directly transmitted hermaphroditic worm which is able to self-fertilize (e.g. the human tapeworm *Hymenolepis nana*) is

$$R = \hat{\beta} N \lambda / [(\mu_1 + b)(\mu_2 + \hat{\beta} N)]. \tag{31}$$

In this expression, $\hat{\beta}$ is the transmission rate (see equations 4 and 5), λ is the rate of egg production per worm, while μ_1, b and μ_2 represent the death rates of the adult parasite, host and infective stage respectively. R is simply the net rate of reproduction of the parasite, $\hat{\beta} N \lambda$ (where the transmission component, $\hat{\beta} N$, is essentially a form of reproduction since it places the parasite in a location where it is able to reproduce), times the product of the expected lifespans of the adult parasite within the host, $1/(\mu_1 + b)$ (responsible for the production of transmission stages) and the infective stage outside the host, $1/(\mu_2 + \hat{\beta} N)$ (responsible for transmission between hosts). Viewed in another light, R is simply the reproductive contribution of the adult parasite, $\lambda/(\mu_1 + b)$, multiplied the reproductive contribution of the infective stage, $\hat{\beta} N/(\mu_2 + \hat{\beta} N)$. Note the basic similarity of equation (31) with that derived for microparasitic infections (equation 18). The differences which arise are a direct result of the basic unit chosen for the epidemiological study of the parasite (either an infected host or an individual parasite).

Many of the epidemiological principles, outlined earlier for microparasites, apply equally to macroparasites. Some important differences occur, however, and these are as follows:

1 The *actual* reproductive rate of a parasite, $\hat{R}$, within a population which harbours a population of parasites, distributed between the individual hosts, will depend on the net force of density-dependent constraints on parasite reproduction and/or survival exerted by the established parasites or the host's immune system. As mentioned previously, this net force will depend on the frequency distribution of parasite numbers per host.

2 If the mean parasite burden per host and the frequency distribution of parasites remain approximately constant through time, the parasite population is at equilibrium and $\hat{R} = 1$.

3 When significant development delays are pre-

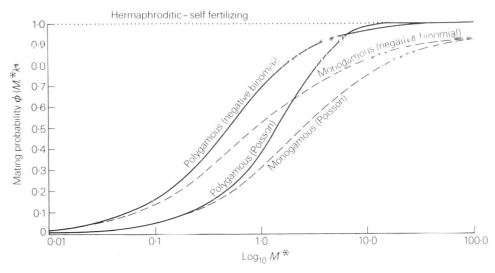

Fig. 4.27 The relationship between the probability that a female worm is mated (ϕ) and the mean parasite burden per host (M^*) for various sexual habits and distributed patterns of worm numbers per host (Poisson and negative binomial with parameter $k = 0.34$). (After May, 1977.)

sent in the parasite's life cycle, either between the arrival of an infective stage in the host and its attainment of sexual maturity (see Table 4.2), or between the production of a transmission stage and its development to the infective stage, R is reduced by a factor $f_1 f_2$, where f_1 is the proportion of worms that attain sexual maturity in the host and f_2 is the proportion of transmission stages that survive to become infective.

4 Many helminth parasites of humans, such as hookworms and schistosomes, have separate sexes. The production of transmission stages is therefore only achieved by female worms who have mated and R is reduced by a factor $r\phi$, where r is the proportion of female worms within the total parasite population (many helminths of humans appear to have sex ratios of approximately 1:1) and ϕ is the probability that a mature female worm is mated. The mating probability is dependent on a variety of factors, in particular whether the parasite is monogamous or polygamous (hookworms are thought to be polygamous while schistosomes appear to be monogamous) and the frequency distribution of parasite numbers per host (Fig. 4.27). The mating probability is of some significance to the dynamics of helminth

infections and will be discussed further in connection with breakpoints in disease transmission.

5 The mean worm burden per host (the intensity of infection) is linearly related to the value of R, while the prevalence of infection is determined by the mean and the frequency distribution of parasite numbers per host (Fig. 4.28). High mean burdens (resulting from high R values) may result in low prevalences if the distribution of parasites is highly aggregated within the host population (see Figs 4.27 & 4.28). The negative binomial probability distribution is a good empirical model of aggregated distributions of parasite numbers per host (Fig. 4.4), and a rough guide to the degree of worm clumping within the host population may be obtained from the following equation:

$$p = 1 - [k/(k + M)]^k. \tag{32}$$

This expression equates the mean parasite burden per host, M, and the prevalence of infection, p, with the parameter k of the negative binomial model which varies inversely with the degree of parasite aggregation (values of k in excess of ten imply that the worms are effectively randomly distributed within the host population; values close to zero imply a high degree of aggregation,

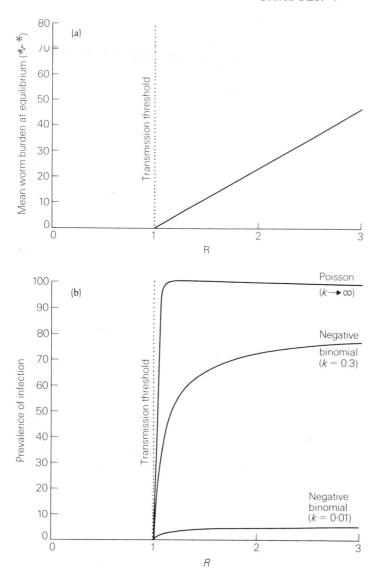

Fig. 4.28 (a) The relationship between R and the mean parasite burden per host at equilibrium (M^*). (b) The relationship between the equilibrium prevalence of infection and R for various distribution patterns of worm numbers per host (Poisson $k \rightarrow \infty$, negative binomial $k = 0.34$, negative binomial $k = 0.01$).

with the majority of parasites being haboured by a few hosts). Surprisingly, for some infections of humans, such as the intestinal nematode *Ascaris lumbricoides*, the degree of parasite aggregation (as measured by k) appears to be relatively independent of the human population sampled (e.g. its geographical location). An illustration of this point is presented in Fig. 4.29 where the relationship between mean worm burden of *A. lumbricoides* (determined by worm expulsion) and the prevalence of infection is recorded for various epidemiological studies carried out in different parts of the world. The solid line denotes the fit predicted by equation (32) with a k value of 0.543. Note how tightly the observed values cluster around the predictions of the negative binomial model. These data hint at a positive association between the value of k and mean worm burden, which may suggest the influence of density-dependent constraints on parasite population size at high densities (causing a decline in aggregation and a concomitant increase in the value of k).

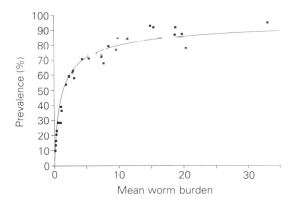

Fig. 4.29 The relationship between prevalence of infection (p) and the mean worm burden (M) for *Ascaris lumbricoides*. The squares are observed values and the solid line is the fit of equation (32) in the text with $k = 0.543$. (Guyatt *et al.*, 1990.)

The degree of clumping often varies in different age-classes of hosts, due to variability in their susceptibility to infection (usually caused by differences in host behaviour or acquired immunity between the various age-classes of the host population). Figure 4.30 records estimates of k for the hookworm *Necator americanus* and the roundworm *Ascaris lumbricoides* in different age-classes of two human populations.

6 The value of R determines the shape of age prevalence and age–intensity curves, high values resulting in a rapid increase in these two epi-

demiological measures as host age increases. The intensity of infection of many direct life cycle helminths appears to reach a plateau or equilibrium level in teenage or young adult age-classes before declining in older people (Fig. 4.38).

The mean worm burden at this plateau, M^*, is determined by the value of R and the force of density-dependent constraints on the survival and reproduction of parasite subpopulations within individual hosts. The relationship between these parameters for cases where density-dependence acts principally on parasite survival within the host is approximately given by

$$M^* = (R - 1)/(ld). \tag{33}$$

Here l denotes the expected lifespan of the adult parasites within the host (in the absence of density-dependent mortality), while d measures the severity of density dependent constraints of worm survival. The value of d is related to the degree of worm clumping within the host population.

7 Control measures (see Chapter 10), such as chemotherapy, increase the mortality rate of the adult parasites (the term μ_1 in equation 31), and therefore reduce the average worm burden by their action in reducing the value of R (see equation 33). If chemotherapy ceases, the value of R is inversely related to the time taken by the parasite population to return to its precontrol level. Figure 4.31 for example, portrays the rapid rise in the prevalence of *Ascaris* infections in

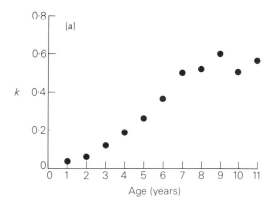

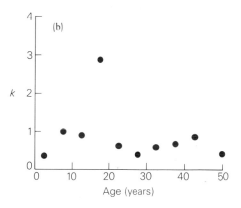

Fig. 4.30 The degree of parasite aggregation, measured inversely by the negative binomial parameter k, and host age. (a) *Necator* infections in an Indian community. (After Anderson, 1980a.) (b) *Ascaris* infections in Iran. (After Croll *et al.*, 1982.)

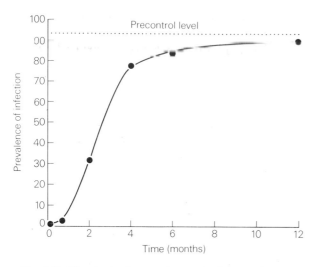

Fig. 4.31 The rise in the prevalence of infection of *Ascaris* after the cessation of chemotherapy in a human community in Iran. (After Arfaa & Ghadirian, 1977.)

Table 4.3 Population parameter value for the human hookworm *Necator americanus*. (From Anderson, 1980a.)

Parameter	Value
Human life expectancy $(1/b)$	50–70 years
Mature worm life expectancy $(1/\mu)$	3–4 years
Adult worm prepatent period	28–40 days
Proportion of L_3 larvae which gain entry to the host that survive to reach maturity	<0.1
Rate of egg production per female worm	15 000 per day
Average male to female sex ratio	1 : 1
Maturation time from release of egg to the development of the L_3 larvae	5+ days
Life expectancy of L_3 Larvae	5+ days
The degree of worm aggregation within the host population (measured inversely by the negative binomial parameter k)	0.01–0.6

a human community to its precontrol level, subsequent to the cessation of control. The time taken to return to the precontrol level is roughly 1 year, reflecting the high basic reproductive rate of infection in this community. Infections of this kind, which include human hookworms, will always be difficult to control by chemotherapy unless such measures are applied extensively within the population over many years. The precise number of years over which control must be applied should be in excess of the maximum lifespan of the longest lived stage in the parasite's life cycle (i.e. the adult worm in the case of the human hookworms and the egg in the case of *Ascaris*). The mean worm burden per host may be rapidly reduced by selectively treating the most heavily infected individuals (identified by means of faecal egg counts). High R values and tight regulatory constraints on worm populations growth, however, generate a high degree of population resilience to perturbation. The cessation of chemotherapy will, therefore, invariable result in the parasite population returning to its precontrol level.

8 The critical threshold density of uninfected hosts N_T, for the initial establishment and spread of directly transmitted helminths, may be derived from equation (31), where

$$N_T = (\mu_1 + b)\mu_2/[\hat{\beta}(\lambda - \mu_2 - b)]. \qquad (34)$$

For most directly transmitted helminths, the density of hosts is not a limiting factor in parasite transmission, due to the enormous reproductive capabilities of these parasites (λ is very large). The female of the human hookworm *Necator*, for example, produces approximately 15 000 eggs per worm per day, while that of *Ascaris* is capable of producing in excess of 200 000 eggs per worm per day. These two helminths occur endemically in low-density human communities.

9 Many rate parameters act to determine the value of R. The estimation of some of these is straightforward but certain parameters such as the rate of infection, $\hat{\beta}$, are more difficult to determine. Age–prevalence and age–intensity data, both from longitudinal and horizontal studies, often provide a basis for the determination of transmission or infection rates. Table 4.3, provides a rough guide to the parameters which determine the value of R for the human hookworm *Necator*.

4.6.5 Macroparasites: indirect transmission

Many helminths of importance to humans have indirect life cycles involving two or more hosts. The principles outlined above for directly transmitted helminths apply equally to those with indirect transmission. Additional complexities arise, however, and some of these are well-illustrated by reference to the epidemiology of human schistosomiasis caused by *Schistosoma mansoni*, *S. japonicum* and *S. haematobium* (see Chapter 2). These have two host life cycles, various species of snails acting as intermediate hosts. Transmission between human and snail, and snail and human, is achieved by means of two free-living aquatic larvae – the miracidium and cercaria respectively. Both larvae gain entry to their respective hosts by direct penetration. The parasites have a phase of sexual transmission reproduction in humans and one of asexual direct reproduction in the snail. In contrast to earlier examples the basic unit of epidemiological study differs in the two segments of the life cycle. The parasite acts as the unit in humans, while, because of the phase of direct asexual reproduction, the infected host is the unit of study for the segment involving the snail.

Ignoring for the moment the problem of worm pairing, the basic reproductive rate R of a parasite with this type of indirect cycle is

$$R = \frac{\lambda_1 \lambda_2 \hat{\beta}_1 \hat{\beta}_2 N_1 N_2}{(\mu_1 + b_1)(\mu_2 + \hat{\beta}_2 N_2)(b_2)(\mu_3 + \hat{\beta}_1 N_1)}, \quad (3.5)$$

Although involving a large number of parameters (determined by the number of distinct developmental stages in the parasite's life cycle), this rather formidable expression has a very simple interpretation which is directly comparable to those discussed earlier for other types of disease agent. The parameters λ_1 and λ_2 represent the rate of egg production per female adult worm and the rate of cercarial production per infected snail, respectively; N_1 and N_2 denote human and snail density; while $\hat{\beta}_1$ and $\hat{\beta}_2$ represent the transmission rates from cercariae to human and from miracidia to snail respectively (see Fig. 4.8c). The rates b_1, b_2, μ_1, μ_2 and μ_3 represent mortality rates of human hosts, infected snails, adult worms, miracidia and cercariae, respectively. R may therefore be interpreted as the product of the net rates of reproduction in humans $(\lambda_1 \hat{\beta}_1 N_1)$ and snails $(\lambda_2 \hat{\beta}_2 N_2)$, multiplied by the product of the expected lifespans of the parasite $(1/(b_1 + \mu_1))$.

Viewed in another light we may express this statement as

R = The total number of eggs produced per female worm $(\lambda_1/(\mu_1 + b_1))$ × the proportion of those eggs that produced miracidia which infect a susceptible snail $(\hat{\beta}_2 N_2/(\mu_2 + \hat{\beta} N_2))$ × the total number of cercariae produced by an infected snail (λ_2/b_2) × the proportion of those cercariae which establish as adult worms $(\hat{\beta}_1 N_1/(\mu_3 + \hat{\beta}_1 N_1))$.

The influence of R on the prevalence and intensity of infection, within both the snail and human populations, is identical to that discussed in connection with directly transmitted helminths. Additional points of epidemiological significance are as follows:

1 In the case of human schistosomes only a proportion, v_1 of the eggs produced will gain exit from the host and enter a habitat in which they are able to hatch to produce miracidia. R is therefore reduced by a fraction v. Similarly, developmental delays occur as a consequence of both the maturation of the parasites in the human host prior to egg production (roughly 25–28 days for *S. mansoni*) and the development of the parasite in the snail before cercarial release (roughly 28–30 days for *S. mansoni* in *Biomphalaria glabrata*). These developmental delays are termed *prepatent* or *latent* periods. The expression for R is therefore multiplied by a term $f_1 f_2$ where f_1 represents the fraction of parasites that penetrate the human host which survive to reach sexual maturity, and f_2 denotes the fraction of infected snails which survive the prepatent period to release cercariae. The fraction f_2 is often small since the snail host has a short lifespan (roughly of the order of 14–54 days) in relation to the prepatent period within the mollusc. Note the similarities of these patterns with arthropod transmitted viral and protozoan diseases such as malaria and yellow fever (see Section 4.4.4).

2 Schistosomes are dioecious and are thought to pair for life (monogamy). The uncertainties associated with worm pairing act to reduce the value of R in the manner described for directly transmitted helminths.

3 The threshold host density necessary for the spread of infection is now determined by the product $N_1 N_2$, of the human and snail densities (equation 34). The product term, in contrast to the ratio N_2/N_1 for vector-borne infections, arises as a consequence of the fact that transmission between both human and snail, and snail and human is achieved by means of free-living infective stages. Host densities are thought to be a limiting factor in the transmission of schistosome parasites; snail densities often dropping to very low levels during certain seasons of the year. The comparatively long lifespan of the parasites in

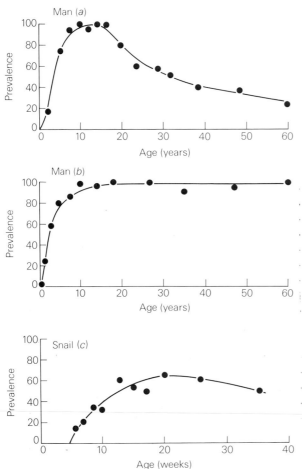

Fig. 4.32 The prevalence of schistosome infections in different age-classes of human (a) and (b), and snail (c) populations. (a) *Schistosoma haematobium* in Gambia. (After Wilkins, 1977.) (b) *S. mansoni* in Uganda. (After Ongom & Bradley, 1972.) (c) *S. haematobium* in *Bulinus nasutus productus* in West Africa. (After Sturrock & Webbe, 1971.)

humans, however, to some extent offsets these seasonal fluctuations. Molluscicides are widely used for the control of schistosomiasis, the aim being to reduce snail density such that the product N_1N_2 drops below the level required to maintain R above the transmission threshold ($R = 1$). A major problem in this approach, however, is the enormous reproductive capabilities of the snail which enables it to rapidly recolonize aquatic habitats once molluscicide treatment ceases (see Chapter 10).

4 The prevalence of schistosome infection in endemic areas, within both the human and snail populations, has a tendency to decline (although not always; Fig. 4.32b) in older age-classes of hosts (Fig. 4.32a). It is unclear at present whether the decline within the human population is due to an increasing degree of acquired immunity (the phenomenon of concomitant immunity) or to age-related changes in host behaviour (older people having less frequent contact with water contaminated with cercariae). The latter factor is probably the more important. The decline in prevalence in older age-classes of snails may be due to either a decrease in susceptibility to infection (Fig. 4.32c) or to age-related changes in the mortality rate of snails, perhaps induced by varying degrees of parasite pathogenicity within the different snail age-classes. Schistosome parasites are highly pathogenic to their molluscan hosts, decreasing snail reproduction and survival (Fig. 4.33b & c).

5 The numerical value of the basic reproductive rate R in any given area of endemic schistosomiasis, is clearly dependent on the many rate determining processes which govern the population sizes of the numerous developmental stages in the life cycles (equation 35). Certain of these rates are relatively easy to measure but others are more difficult, particularly the lifespan of the adult worms in man (μ_1) and the two transmission parameters β_1 and β_2. Transmission rates may, under certain circumstances, be estimated from age–prevalence data.

4.6.6 Breakpoints in parasite transmission

The problems associated with finding a partner to mate with create additional complexities in the

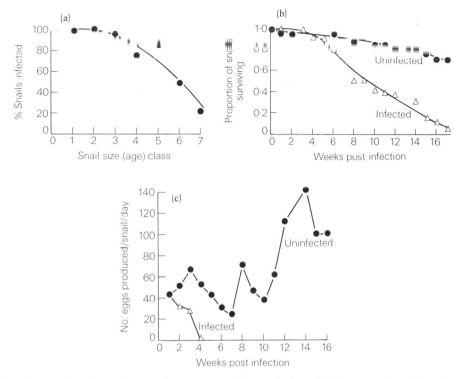

Fig. 4.33 The influence of snail size (age) on the proportion of *Biomphalaria* that become infected by *Schistosoma mansoni* when exposed to a constant number of miracidia (Anderson *et al.*, 1982) (size class 1 = diameter of 2–4 mm; size class 7 = diameter of 1–16 mm). (b) The influence of *S. mansoni* on the survival of *Biomphalaria*. (After Pan, 1965.) (c) The influence of *S. mansoni* on the rate of egg production by *Biomphalaria*. (After Pan, 1965.)

population dynamics of many helminth parasites. This is particularly true for dioecious species such as hookworms and schistosomes, but is also important for hermaphroditic species which are unable to self-fertilize.

When cross-fertilization is necessary, the fall of the mean parasite burden per host below a critical level results in mating becoming too infrequent to maintain sufficient production of parasite transmission stages for the perpetuation of the infection. This critical worm burden therefore defines a *breakpoint* in parasite transmission which is distinct from the *transmission threshold* $(R = 1)$ discussed earlier.

The precise level of this breakpoint is set by a variety of factors; in particular, the sexual habits of the parasite (whether the parasite is monogamous or polygamous, or whether an adult female worm requires mating more than once to main-

tain egg production throughout her lifespan) and the frequency distribution of parasite numbers per host.

The breakpoint and transmission threshold concepts are illustrated graphically in Fig. 4.34 by reference to the population biology of the human hookworm *Necator*. This species is polygamous and is invariably highly aggregated in its distribution within the host population (see Fig. 4.30). Below the point $R = 1$, the infection cannot maintain itself. Above this level, however, three equilibrium states exist: two are stable reflecting endemic disease (the solid line) and parasite extinction (the horizontal axis of the graph), while the third is the breakpoint which is unstable (the dotted line). If the worm burden falls below this point the parasite will become extinct, while above it the parasite population settles to the stable endemic disease level. This concept has

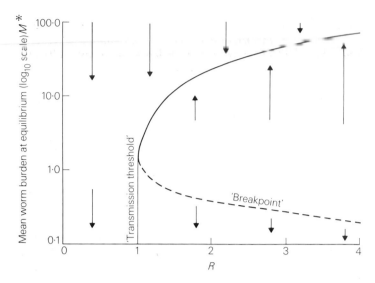

Fig. 4.34 The relationship between the equilibrium worm burden and R for a dioecious helminth. The graph illustrates the concepts of a transmission threshold ($R = 1$) and a breakpoint by reference to the population biology of the human hookworm *Necator*. (After Anderson, 1980a.)

obvious significance to parasite control and leads to the suggestion that the depression of helminth abundance below the breakpoint (say, by chemotherapy) would result in parasite eradication.

Unfortunately, however, recent studies indicate that the location of the breakpoint is critically dependent on the degree of worm clumping within the host population (Fig. 4.35). Highly aggregated parasite distributions, which are the rule for helminth infections (see Fig. 4.4), result in the breakpoint lying very close to zero worms per host (for example in Fig. 4.34 an R value of 3.0 gives a breakpoint of 0.3 worms per host). The intuitive explanation of this observation is that highly clumped worm distributions increase the frequency with which adult parasites encounter each other. The breakpoint concept is thus unfortunately of limited practical significance to disease control. This subject is discussed further by May (1977) and Anderson (1980a).

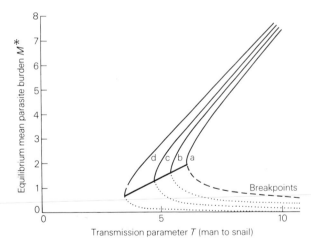

Fig. 4.35 The relationship between the equilibrium mean parasite burden and a transmission coefficient T which depicts the rate of transmission from human to snail [$\{\lambda_1\hat{\beta}_2N_2\}/\{(\mu_1 + b_1)(\mu_2 + \hat{\beta}_2N_2)\}$ in equation 34] for varying degrees of worm aggregation within the human population. The graph illustrates the breakpoint concept (dashed lines) by reference to schistosome infections in humans (May, 1977). a, Random; b–d, Negative binomial with k values: b, 1; c, 0.2; d, 0.05.

4.7 CLIMATIC FACTORS

Climatic changes have an important influence on the epidemiology of most infectious diseases of humans. Environmental factors such as temperature and rainfall vary seasonally in the majority of habitats, tending to induce regular cyclic fluctuations in the prevalence and intensity of parasitic infection. The action of climate on host and parasite, however, is independent of population abundance. Thus although it may be the cause of conspicuous changes in density, its action is non-regulatory in nature. Climatic parameters are therefore referred to as *density-independent* factors.

Climatic factors influence the population biology of human disease agents in the following principal ways.

● *Host behaviour.* Several changes in host behaviour, induced by the prevailing climatic conditions, often generate cyclic fluctuations in disease incidence. Such changes may be the result of differing work patterns associated with agricultural practices (the planting and harvesting of crops at different times of the year), or may result from social patterns influencing the behaviour of children (the vacation periods and school terms in Western societies). Agricultural practices are important to the transmission of helminth infections such as *Ascaris* and schistosomiasis, while the behaviour of children influences the epidemiology of many directly transmitted viral infections such as measles, mumps and chickenpox.

● *Intermediate host abundance.* Seasonal changes in the prevalence of many indirectly transmitted parasites are in part determined by the influence of climatic factors on the abundance of intermediate host populations. Seasonal fluctuations in the transmission of malaria and schistosomiasis, for example, are to a large extent the result of changes in the abundance of mosquitoes and snails respectively (see Fig. 4.26).

● *Infective stage longevity.* Climate has an important influence on the longevity of parasite transmission stages such as helminth eggs and larvae, the cysts of protozoa and free viral particles. Temperature, for example, is a major determinant of the survival of the miracidia and cercariae of schistosome flukes (Fig. 4.36a) and the L_3 infective larvae of hookworms. The longevity of transmission stages which live in

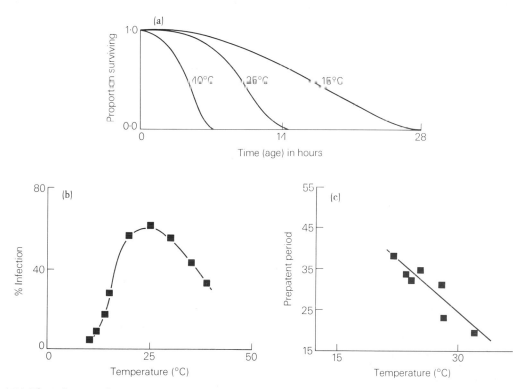

Fig. 4.36 The influence of water temperature on (a) the survival of *Schistosoma mansoni* miracidia (Anderson *et al.*, 1982); (b) the infectivity of *S. mansoni* miracidia to *Biomphalaria* (Chu *et al.*, 1966); and (c) the prepatent period prior to cercarial release of *S. mansoni* in *Biomphalaria*. (After Anderson & May, 1979b.)

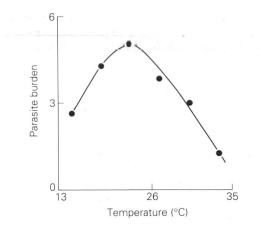

Fig. 4.37 The influence of air temperature on the rate of acquisition of larvae of the nematode *Dirofilaria immitis* by the mosquito *Aedes trivattatus* feeding on infected dogs. (After Christensen, 1978.)

terrestrial habitats, such as the eggs of *Ascaris* and larvae of hookworms are also markedly influenced by soil moisture content.

• *Infectivity.* In addition to their influence on infective-stage longevity, factors such as temperature and humidity have an impact on the infectivity of both transmission stages and infectious intermediate hosts. Temperature, for instance, controls the activity of schistosome miracidia and thus influences their ability to contact and penetrate the molluscan host (Fig. 4.36b). In addition, this factor also affects the rate at which infected snails produce cercariae.

Climate may play a role in determining the activity and infectiousness of arthropod vectors. For example, the optimum air temperature for the transmission of a filarial worm (*Dirofilaria immitis*) from dog to mosquito (*Aedes trivittatus*) is roughly 23°C; the biting efficiency of the vector decreases at lower or higher temperatures (Fig. 4.37).

• *Parasite development.* Temperature is an important determinant of the rate of parasite development either in the external habitat or within poikilothermic intermediate hosts such as snails or mosquitoes. The rate of development of human hookworms, from egg to infective larva is most rapid at around 25–30°C in moist conditions (roughly 5 days). If temperatures are below 17–

20°C, the ova and larvae of *Necator* cease development and death rapidly follows. *Ancylostoma* is able to develop at slightly lower temperatures than *Necator* and is thus found in certain temperate regions of the world.

Temperature also has a marked influence on the prepatent or incubation period of malaria in the mosquito vector and schistosome parasites in the molluscan host (Fig. 4.36c). The action of climatic factors on the rate of parasite development is a major determinant of the geographical distribution of many human diseases.

The influence of climate, whether acting on one or more of the parameters discussed above, is to alter a parasite's transmission success. If climatic change reduces the basic reproductive rate (R) it will result in a reduction in the prevalence of infection. Conversely, if it induces an increase in the value of R, disease prevalence will rise. Such changes often occur on a regular seasonal basis and sometimes result in the value of R falling below the transmission threshold ($R = 1$) during certain periods of the year (Fig. 4.38). In these instances the parasite will only perpetuate within the host population, on a year to year basis, provided that the period during which R is less than unity, is shorter than the maximum lifespan of the longest lived developmental stage in the disease agent's life cycle. The human hookworm, *Necator*, for example, has an

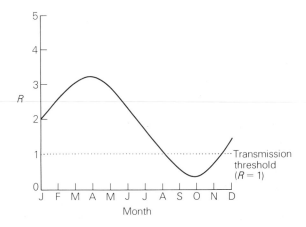

Fig. 4.38 A diagrammatic representation of seasonal changes in R, where the value of this parameter falls below unity at certain times of the year.

expected lifespan in humans of approximately 3.5 years and a maximum lifespan in excess of 5 years. For this parasite, therefore, seasonal changes in transmission success which result in R falling below unity during hot, dry seasons are of little consequence to the long-term stability of the infection. They do, however, induce observable seasonal fluctuations in the prevalence and intensity of infection.

In contrast, the longest-lived stage in the life cycle of the measles virus is in an infectious host which remains infective for approximately 5–7 days. Seasonal changes in host behaviour which result in R falling below unity for long periods of time do not allow this disease to persist endemically. Such patterns are frequently observed on islands which support small human communities.

The important conclusions to be drawn from the above examples are twofold. First, dramatic seasonal changes in climate will result in conspicuous fluctuations in disease prevalence. These will be most marked if R falls below unity during certain periods of the year and when the longest lived stage in the parasites life cycle has a lifespan of less than 1 year. Secondly, and most importantly, disease control measures will be most successful if applied intensively during the periods of the year when the value of R is at a minimum.

4.8 THE EPIDEMIOLOGICAL SIGNIFICANCE OF THE BASIC REPRODUCTIVE RATE R

The concept of a basic reproductive rate or number and its measurement are of central importance to the epidemiological study and control of infectious disease agents. The magnitude of R determines the ability of a parasite to perpetuate itself within the host population and, in conjunction with the density-dependent processes which act to regulate the spread of infection, determines the abundance and prevalence of infection. For any given parasitic species, R will vary from one geographical location to the next as a consequence of climatic factors and the prevailing sociological conditions within the human population.

The value of the basic reproductive rate directly reflects the degree of difficulty that will be en-

countered in attempts to control the spread of infection. Other things being equal, parasites with high R values will be much more difficult to control than those with low values. If disease control ceases before eradication is achieved, the time taken for the infection to return to its precontrol level is inversely related to the magnitude of R.

The measurement of R and the determination of the action of control measures on reproductive success are essential for the quantitative assessment of the impact of control policies on the abundance and prevalence of infection (see equation 23). Many rate parameters, such as parasite reproduction, survival and infection rates, determine the value of R. The development of methods to estimate these parameters from horizontal and longitudinal studies of disease prevalence and intensity is an important research priority in the epidemiological study of infectious diseases.

Developments in the research fields of immunology and molecular biology are beginning to provide new tools for epidemiological study. These include sensitive immunological techniques which enable quantitative measures to be made of host responses (both humoral and cell mediated) to specific parasite antigens (surface, excretory and secretory). These developments will facilitate the quantification of past and

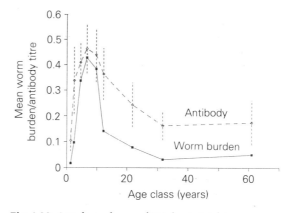

Fig. 4.39 Age-dependency of *Trichuris trichiura* infection intensity (—■—) and *T. trichiura* IgA antibody responses (—◇—) in arbitrary units. (From Bundy, 1988.) Antibody level (titre in serum) rises as intensity (mean worm burden) increases and declines as intensity decreases.

present parasite infection within individual hosts, via the collection of blood serum, lymphocytes or even saliva (which contains antibody markers of current or past infection) (Fig. 4.39).

Furthermore, techniques such as the polymerase chain reaction (PCR) offer great scope for the detection of both microparasites and macroparasites within hosts and it is likely that PCR methods will also facilitate quantitative measures in the near future. More generally, these new tools enable epidemiologists to focus more on the individual host and its parasites, rather than crude measures of the prevalence or average intensity of infection, such that observed distributions within populations of hosts, of parasite abundance and immunological markers of past or current infection, can be interpreted in the light of a knowledge of the genetic background of the host and the genetic constitution of the parasite population. A shift in emphasis from the population to the individual may help us to understand why distributions of infectious agents and immunological response to them, are invariably highly heterogeneous in character within human populations. This in turn will facilitate the development of better methods of control whether they be new vaccines or drugs, or the targeting of treatment or education to those at greatest risk to infection and from disease.

REFERENCES AND FURTHER READING

Anderson, R.M. (1980a) The dynamics and control of direct life cycle helminth parasites. *Lecture Notes in Biomathematics*, Vol. 39, pp. 278–322. Berlin, Springer-Verlag.

Anderson, R.M. (1980b) The control of infectious disease agents: strategic models. In G.R. Conway (ed.) *Pest and Pathogen Control: Strategy, Tactics and Policy Models*. London, John Wiley & Sons.

Anderson, R.M. (1980c) The population dynamics of indirectly transmitted diseases: the vector component. In R. Harwood & C. Koehler (eds) *Comparative Aspects of Animal and Plant Pathogen Vectors*, pp. 13–43. New York, Praeger.

Anderson, R.M. (1988) Epidemiology of HIV infection: variable incubation plus infectious periods and heterogeneity in sexual activity. *Journal of the Royal Statistical Society A*, **151**, 66–93.

Anderson, R.M. & May, R.M. (1978) Regulation and stability of host – parasite population interactions: I Regulatory processes. *Journal of Animal Ecology*, **47**, 219–247.

Anderson, R.M. & May, R.M. (1979a) Population biology of infectious diseases. Part I. *Nature*, **280**, 361–367.

Anderson, R.M. & May, R.M. (1979b) Prevalence of schistosome infections within molluscan populations: observed patterns and theoretical predications. *Parasitology*, **79**, 63–94.

Anderson, R.M. & May, R.M. (1982) The epidemiology of directly transmitted infectious diseases: control by vaccination. *Science*, **215**, 1053–1060.

Anderson, R.M. & May, R.M. (1991) *Infectious Diseases of Humans: Dynamic and Control*. Oxford, Oxford University Press.

Anderson, R.M., Mercer, J.G., Wilson, R.A. & Carter, N.P. (1982) Transmission of *Schistosoma mansoni* from man to snail: experimental studies of miracidial survival and infectivity in relation to larval age, water temperature, host size and host age. *Parasitology*, **85**, 339–360.

Anderson, R.M., Whitfield, P.J. & Dobson, A.P. (1978) Experimental studies on infection dynamics: infection of the definitive host by the cercariae of *Transversotrema patialense*. *Parasitology*, **77**, 189–200.

Arfaa, F. & Ghadirian, E. (1977) Epidemology and mass-treatment of ascariasis in six rural communities in Central Iran. *American Journal of Tropical Medicine and Hygiene*, **26**, 866–871.

Black, F.L. (1966) Measles endemicity in insular populations: critical community size and its evolutionary implications. *Journal of Theoretical Biology*, **11**, 207–211.

Boray, J.C. (1969) Experimental fascioliasis in Australia. *Advances in Parasitology*, **7**, 85–210.

Bray, R.S. & Harris, W.G. (1977) The epidemiology of infection with *Entamoeba histolytica* in Gambia, West Africa. *Transactions of the Royal Society of Tropical Medicine and Hygiene*, **71**, 401–407.

Bundy, D.A.P. (1988) Population ecology of intestinal helminth infections in human communities. *Philosophical Transactions of the Royal Society of London, B*. **321**, 405–420.

Buxton, P.A. (1940) Studies on populations of head-lice (*Pediculus humanus capitis*: anoplura). III. Material from South India. *Parasitology*, **32**, 296–302.

Carter, N., Anderson, R.M. & Wilson, R.A. (1982) Transmission of *Schistosoma mansoni* from man to snail; laboratory studies of the influence of snail and miracidial densities on transmission success. *Parasitology*, **85**, 361–372.

Christensen, B.M. (1978) *Dirofilaria immitis*: Effects on the longevity of *Aedes trivittatus*. *Experimental Parasitology*, **44**, 116–123.

Christensen, B.M. & Hollander, A.L. (1978) Effect of temperature on vector–parasite relationships of *Aedes trivittatus* and *Dirofilaria immitis*. *Proceedings of the Helminthological Society of Washington*, **45**, 115–119.

Chu, K.Y., Massoud, J. & Sabbaghian, H.C. (1966) Host–parasite relationship of *Bulinus truncatus* and *Schistosoma haematobium* in Iran. 3. Effect of water temperature on the ability of miracidia to infect snails. *Bulletin of the World Health Organization*, **34**, 131–133.

Crofton, H.D. (1971) A quantitative approach to parasitism. *Parasitology*, **62**, 179–193.

Croll, N.A., Anderson, R.M., Gyorkos, T.W. & Ghadirian, E. (1982) The population biology and control of *Ascaris lumbricoides* in a rural community in Iran. *Transactions of the Royal Society of Tropical Medicine and Hygiene*, **76**, 187–197.

Dietz, K. (1976) The incidence of infectious diseases under the influence of seasonal fluctuations. In J. Berger, W. Buhler, R. Repeges & P. Tautu, (eds) Mathematical Models in Medicine *Lecture Notes in Biomathematics*, **11**, pp. 1–5. Berlin, Springer-Verlag.

Fleming, A.F., Storey, J., Molineaux, L., Iroko, E.A. & Attai, E.D.E. (1979) Abnormal hemoglobins in the Sudan savanna of Nigeria. I. Prevalence of haemoglobins and relationships between sickle cell trait, malaria and survival. *Annals of Tropical Medicine and Parasitology*, **73**, 161–172.

Garrett-Jones, C. & Shidrawi, G.R. (1969) Malaria Vectorial capacity of a population of *Anopheles gambiae*. *Bulletin of the World Health Organization*, **40**, 531–545.

Griffiths, D.A. (1974) A catalytic model of infection for measles. *Applied Statistics*, **23**, 330–339.

Grove, D.I. & Warren, K.S. (1976) Relation of intensity of infection to disease in hamsters with acute schistosomiasis mansoni. *American Journal of Tropical Medicine and Hygiene*, **25**, 608–612.

Gupta, S., Anderson, R.M. & May, R.M. (1989) Networks of sexual contacts: implications for the pattern of spread of HIV. *AIDS*, **3**, 807–817.

Guyatt, H.L., Bundy, D.A.P., Medley, G.F. & Grenfell, B.T. (1990) The relationship between the frequency distribution of *Ascaris lumbricoides* and the prevalence and intensity of infection in human communities. *Parasitology*, **101**, 139–143.

Haswell-Elkins, M.R., Elkins, D.B. & Anderson, R.M. (1987) Evidence for redisposition in humans to infection with *Ascaris*, hookworm, *Enterobius* and *Trichuris* in a South Indian fishing community. *Parasitology*, **95**, 323–338.

Hill, R.B. (1926) The estimation of the number of hookworms harboured by the use of the dilution egg count method. *American Journal of Hygiene*, **6**, (supplement), S19–S41.

Keymer, A.E. & Anderson, R.M. (1979) The dynamics of

infection of *Tribolium confusum* by *Hymenolepis diminuta*: the influence of infective stage density and spatial distribution. *Parasitology*, **79**, 195–207.

Krupp, I.M. (1961) Effects of crowding and of superinfection on habitat selection and egg production in *Ancylostoma caninum*. *Journal of Parasitology*, **47**, 957–961.

Lie, K.J., Heyneman, D. & Kostanian, N. (1975) Failure of *Echinostoma linfoense* reinfect snails already harbouring that species. *International Journal of Parasitology*, **5**, 483–486.

Macdonald, G. (1957) *The Epidemiology and Control of Malaria*. Oxford University Press, Oxford.

Macdonald, G. (1965) *The epidemiology and Control of Malaria*. Oxford, Oxford University Press.

Macdonald, G. (1965) The dynamics of helminth infections with special reference to schistosomes. *Transactions of the Royal Society of Tropical Medicine and Hygiene*, **59**, 489–506.

May, R.M. (1977) Togetherness among schistosomes: its effects on the dynamics of infection. *Mathematical Bioscience*, **35**, 301–343.

May, R.M. & Anderson, R.M. (1979) Population biology of infectious diseases. Part II. *Nature*, **280**, 455–461.

May, R.M. & Anderson, R.M. (1987) The transmission dynamics of HIV infection. *Nature*, **326**, 137–142.

Muench, H. (1959) *Catalytic Models in Epidemiology*. Harvard, Harvard University Press.

Muller, R. (1977) *Worms and Disease: A Manual of Medical Helminthology*. London, Butler and Tanner.

Nawalinski, T., Schad, G.A. & Chowdbury, A.B. (1978) Population biology of hookworms in children in rural west Bengal. I. General parasitological observations. *American Journal of Tropical Medicine and Hygiene*, **27**, 1152–1161.

Ongom, V.L. & Bradley, D.J. (1972) The epidemiology and consequences of *Schistosoma mansoni* infection in West Nile, Uganda. I. Field studies of a community of Panyagoro. *Transactions of the Royal Society of Tropical Medicine and Hygiene*, **66**, 835–851.

Pan, C.-T. (1965) Studies on the host–parasite relationship between *Schistosoma mansoni* and the snail *Australorbis glabratus*. *American Journal of Tropical Medicine and Hygiene*, **14**, 931–976.

Pennycuik, L. (1971) Frequency distributions of parasites in a population of three-spined sticklebacks, *Gasterosteus aculeatus* L. with particular reference to the negative binomial distribution. *Parasitology*, **63**, 389–406.

Peters, W. (1978) Medical Aspects: Comments and discussion. II. In A.E.R. Taylor & R. Muller (eds) *The Relevance of Parasitology to Human Welfare Today*, pp. 25–40. Oxford, Blackwell Scientific Publications.

Riordan, K. (1977) Long-term variations in trypanosome infection rates in highly infected tsetse flies on a cattle route in south-western Nigeria. *Annals of*

Tropical Medicine and Parasitology, **71**, 11–20.

Rosenfield, P.L., Smith, R.A. & Wolman, M.G. (1977) Development and verification of a schistosomiasis transmission model. *American Journal of Tropical Medicine and Hygiene*, **26**, 505–516.

Schmid, W.D. & Robinson, E.J. (1972) The pattern of a host–parasite distribution. *Journal of Parasitology*, **58**, 907–910.

Stiven, A.E. (1964) Experimental studies on the epidemiology of the host–parasite system. *Hydra* and *Hydramaeba hydroxena* (Entz). II The components of a simple epidemic. *Ecological Monographs*, **34**, 119–142.

Sturrock, R.F. & Webbe, G. (1971) The application of catalytic models to schistosomiasis in snails. *Journal of Helminthology*, **45**, 189–200.

Tingley, G.A. & Anderson, R.M. (1986) Environmental sex determination and density-dependent population regulation in the entomogeneous nematode *Roman-omermis culicivorax. Parasitology*, **92**, 431–450.

Warren, K.S. (1973) Regulation of the prevalence and intensity of schistosomiasis in man: immunology or ecology? *Journal of Infectious Diseases*, **127**, 595–609.

Watkins, C.V. & Harvey, L.A. (1942) on the parasites of silver foxes on some farms in the South West. *Parasitology*, **34**, 155–179.

Wilkins, H.A. (1977) *Schistosoma haematobium* in a Gambian community. I. The intensity and prevalence of infection. *Annals of Tropical Medicine and Parasitology*, **71**, 53–58.

Williams, C.B. (1944) Some applications of the logarithmic series and the index of diversity to ecological problems. *Journal of Ecology*, **32**, 1–44.

Yorke, J.A. & London, W.P. (1973) Recurrent outbreaks of measles, chickenpox and mumps. II. Systematic differences in contact rates and stochastic effects. *American Journal of Epidemiology*, **98**, 469–482.

Chapter 5 / Biochemistry

C. BRYANT

5.1 INTRODUCTION

This chapter gives an account of the major biochemical processes that take place in parasitic protozoa and helminths. The reader may immediately point out that there are other sorts of parasites. Unfortunately, space is limited and even restricting discussion to those groups mentioned severely tests the author's capacity for condensation, for the parasitic protozoa and helminths are not homogeneous groups. The decade since the writing of the first edition of this chapter has seen the flowering of the study of parasite biochemistry, and with it has come the realization of the great *diversity* exhibited within and between taxons.

A single chapter is not much to devote to this diversity; of necessity it must be selective in its treatment. The following pages therefore concentrate on dynamic aspects of biochemistry and on concepts. It is assumed that the reader is familiar with basic mammalian biochemistry; any standard biochemistry text constitutes an adequate reference work. For many of the supporting data on parasites, the reader is particularly referred to the late Theodor Von Brand's heroic compendia *Biochemistry of Parasites* and *Biochemistry and Physiology of Endoparasites*; to Barrett's *Biochemistry of Parasitic Helminths*; to *Biochemical Adaptation in Parasites*, by Bryant & Behm; and to *Biochemical Protozoology*, edited by Coombs & North.

The following abbreviations are used in the text: ATP, ADP, AMP – adenosine tri-, di- and monophosphates; cAMP – cyclic AMP; CoA – coenzyme A; DNA – deoxyribonucleic acid; GTP, GDP – guanosine tri- and diphosphates; LDH – lactate dehydrogenase; MDH – malate dehydrogenase; NAD(P)(H) – nicotinamide adenine dinucleotide (phosphate) (reduced); PEP – phosphoenol pyruvate; PEPCK – phosphoenol pyruvate carboxykinase; PK – pyruvate kinase; RNA – ribonucleic acid; TCA – tricarboxylic acid.

5.2 ENERGY METABOLISM

All parasites need to convert organic molecules into their own substance; to do this, they need to maintain a supply of energy for macromolecular synthesis, growth, mechanical activity, differentiation and reproduction. Parasites characteristically exhibit rapid growth or multiplication, which make great demands on energy generating mechanisms. They also need protection from the immune response of the host, another major energetic cost. 'Energy metabolism', then, refers to those biochemical processes that result in the formation of ATP and other energy-conserving compounds which, in turn, are employed in many energy-dependent reactions. Energy metabolism is particularly important because the establishment of a parasite in a new host depends, in the short term, on its ability to sustain life in a harsh environment. Other crises, such as the evasion of the host's immune response, are subordinated to the immediate crisis of survival.

5.2.1 Environments and life cycles

The environments of parasites are legion. Parasitic protozoa and helminths may be found in all tissues of vertebrates and invertebrates. Each tissue has its own special characteristics. It may be rich in oxygen or carbon dioxide. It may provide the parasite with a well-regulated supply of metabolic precursors, as does the blood vascular system or, as in the intestine, availability of nutrients may depend on the host's diet. It is not surprising, therefore, that there is diversity in the pathways of energy metabolism of internal parasites, but there are common features about classes of environments that make a few generalizations possible. One is that, whether a parasite is aerobic or anaerobic, it requires a source of highly reduced organic compounds, an efficient mechanism for energy entrapment and the capacity to maintain its intracellular environments at the right level of oxidation.

Parasites often occupy more than one environment during their life cycles, adding further complications to the study of their biochemistry. Biochemical strategies that enable survival in one environment may not be appropriate to another. The parasite must have the genetic capacity to allow it to recognize and grow in more than one environment. There must be a complex programme for the expression of different genes at different times, leading to the elaboration of different biochemical pathways at each of the life stages. This is often achieved by sensitivity to 'trigger' stimuli. Many parasites embark on the adult stages of their life cycles in response to the high carbon dioxide concentrations, reducing conditions or the high temperatures encountered in their definitive hosts.

5.2.2 Energy stores

The most usual type of macromolecule retained as an energy source is some form of carbohydrate. In trichomonads and *Entamoeba*, glycogen accounts for 10–30% of dry weight, but there is little, if any, storage carbohydrate in trypanosomes or malarial parasites. This is probably because those forms that inhabit the vertebrate bloodstream, about which most is known, occupy an environment that contains ample glucose maintained at constant levels by the host's homeostatic mechanisms. In the insect host a very different form of metabolism occurs, which takes advantage of the availability of amino acids. Also of interest is a polyphosphate found in *Crithidia fasciculata*, a kinetoplastid flagellate. It may be an energy store, but may also be important in the regulation of metabolism.

Glycogen storage is a characteristic of helminth parasites, and is often found in large quantities, as much as 10% of dry weight in individuals of some tapeworms. Generally, glycogen is depleted during 'starvation' of helminth parasites in *in vitro* culture, which provides prima facie evidence that it is indeed an energy store. Some helminths also store trehalose, a soluble disaccharide, in their tissues. For example, in the acanthocephalan *Moniliformis dubius* and the larval nematode *Trichinella spiralis*, trehalose accounts for up to 2.5% of tissue solids.

Glucose, too, is universally present in parasites. It is generally not stored, but active transport mechanisms for its uptake from the environment are widespread in protozoa and helminths. In *Entamoeba*, glucose uptake is the rate limiting step in metabolism. There are active transport mechanisms for glucose uptake in cestodes and nematodes, while recent studies have shown that glucose uptake in trematodes may vary depending on the ecological niche occupied by the parasite. Thus, glucose uptake in *Fasciola hepatica* is passive, presumably because there is no shortage of glucose in its predilection site in the mammalian liver. On the other hand, in three different species of the fish fluke *Proterometra*, the possession of an active transport system for glucose uptake seems to be directly related to the availability of glucose in the immediate environment. The species that lives in the more external environment, where little glucose is available, has a well-developed active transport system. In all these parasites, glucose is an intracellular metabolic pool, maintained from the environment or by the breakdown of stored polysaccharide, and is either oxidized for immediate energy yield or converted to storage product.

In adult helminths and protozoa, lipids do not form an energy store. Almost all parasites except some larval helminths lack the enzymes necessary for their oxidation.

5.2.3 Regulation of energy metabolism

A useful aid in understanding metabolic regulation is the concept of the 'adenylate energy charge'. It is a measure of the energy resident in the adenylate system and is given by the following expression:

$$\frac{[ATP] + 1/2[ADP]}{[ATP] + [ADP] + [AMP]}$$

where the square brackets indicate the concentration of the adenylate in question. The value of the adenylate energy charge is about 0.8–0.9 for many mammalian tissues; in parasites it varies between 0.6 and 0.9, the lower values usually being obtained in investigations of anaerobic metabolism. It is less subject to fluctuation than the ATP/ADP ratio but in both cases, high values suggest that the organism is in good health and is probably engaged in synthetic processes. Many anabolic enzymes are activated, while catabolic enzymes are down-regulated, by the allosteric and substrate effects of high concentrations of ATP and low concentrations of ADP. At low adenylate energy charge levels the reverse is true.

Redox state, which can be practically determined as the ratio of the intracellular concentration of NAD(P) to that of NAD(P)H, is another useful indicator. A high concentration of NAD(P) favours catabolic processes, high NAD(P)H favours anabolic ones. Barrett (1991) observed NAD/NADH ratios in *Ascaris* muscle cytoplasm up to 2214 to 1, which indicates that the muscle is maintained at a high level of redox, achieved by coupling NADH oxidation to malic dehydrogenase.

Except in a few instances, metabolic pathways in parasites are regulated the same way as in other organisms. They respond to adenylate energy charge, to the redox state, their enzymes show allosteric activation and, generally, similar enzymes are involved in regulation. The metabolic pathways are subject to feed-forward and feed-back control and to product inhibition. Differences lie in the interrelationship of metabolic sequences, and the fact that, in different organisms, different properties of similar enzymes are enhanced or suppressed for the fine-tuning of metabolic pathways that adapt parasites to different environments.

5.3 ENERGY METABOLISM IN PARASITIC PROTOZOA

Protozoan parasites are not a homogeneous group and many of them (for example, the malaria parasites and the trypanosomes) are adapted to more than one host in a single life cycle. As hosts are as different as mammals and insects it is not surprising to discover that metabolism of a given parasite within each of its hosts is quite distinctive and that mechanisms exist for rapidly switching between modes when transmission to another host takes place. There are thus many different types of energy metabolism exhibited by parasitic protozoa, and there is space for little more in this chapter than a few generalizations.

Glucose is the major substrate for energy metabolism in many groups, such as malarial parasites, bloodstream salivarian trypanosomes, trichomonads, *Giardia* and *Entamoeba*. In these organisms, the emphasis is on glycolysis. Others (insect stages of *Trypanosoma cruzi*, *Leishmania* spp. and the salivarian trypanosomes) are capable of at least partial oxidation of fatty and amino acids by a specialized TCA cycle.

Not surprisingly, parasitic protozoa show many adaptations which are unique to this diverse assemblage. They include metabolic pathways with a high carbon flux and a high energy output, and specialized organelles for maintaining the appropriate enzymes in close proximity to one another. For example, in the bloodstream forms of the African trypanosomes the process of 'aerobic glycolysis' involves the interaction of two organelles and the cytosol. The enzymes of glycolysis, to the level of phosphoglycerate kinase, are contained in the glycosome, and are present in high concentrations; the cytosol contains phosphoglycerate mutase, and the mitochondrion, a simple tube without TCA cycle activity at this stage, contains the membrane-bound glycerophosphate oxidase complex. Glucose enters the glycosome, is rapidly metabolized to 1,3-diphosphoglycerate and glycerol-3-phosphate. The former leaves the glycosome and enters the cytosol, where it is converted to pyruvate (there is no LDH). Meanwhile, glycerol-3-phosphate leaves the glycosome for the mitochondrion where it is oxidized to dihydroxyacetone phosphate. This then re-enters the glycosome, while its electrons are transferred by the oxidase to oxygen. It is not clear whether this electron transfer results in the synthesis of ATP, but an important outcome is that it ensures the continual reoxidation of NADH. The net effect is to maintain

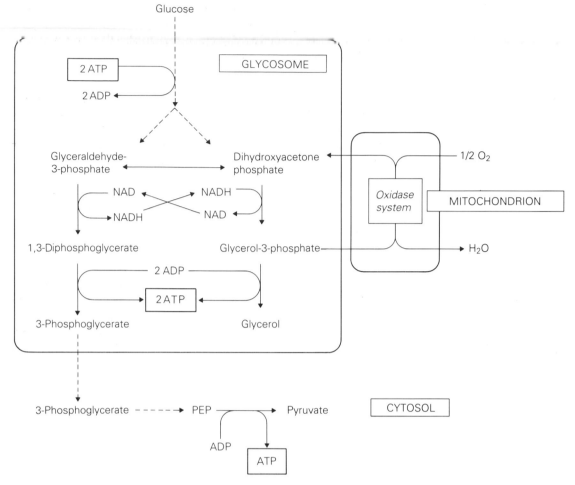

Fig. 5.1 Aerobic glycolysis in long-slender bloodstream forms of trypanosomes. Abbreviated metabolic scheme which shows the cooperation of enzymes from three cellular compartments within the cell. Dotted arrows indicate that several reactions have been omitted for simplification.

rapid glucose oxidation and ATP production in the glycosome and cytosol compartments (Fig. 5.1). The potential yield of ATP from this pathway is only 2 moles per mole glucose utilized, but, as glucose is plentiful, the lack of efficiency of energy conservation is amply compensated for by the high throughput of glucose carbon.

Glycosomes possess only a single membrane, and are unique to the kinetoplastids. Their origin is obscure but they are probably related to peroxisomes. They can sustain a very high rate of glycolysis because of their high concentrations of glycolytic enzymes and their impermeability.

The mitochondrion of trypanosomes is remarkable in that it changes its form and function at each stage of the life cycle. In the 'long-slender' bloodstream forms, there is little else but the glycerate phosphate oxidase complex. In later bloodstream forms – 'intermediate' and 'short-stumpy' – the mitochondrion may be branched and contain cristae, and possess the additional functions of acetate or succinate production from pyruvate. Fumarate reductase is present, and so are the enzymes of the TCA cycle although it does not function as such. The glycosome changes, too, and acquires PEPCK and MDH activities. The latter assumes the role of reoxidizing NADH as glycolytic activity diminishes.

These changes foreshadow the requirements of survival in the insect host. After ingestion, the parasite transforms to a procyclic trypomastigote and migrates to the salivary glands. It develops a highly branched mitochondrion, with large numbers of cristae, and a fully functional TCA cycle, and is able, at least partially, to oxidize proline and other amino acids from the insect's haemolymph.

Of particular interest among the metabolic adaptations of the parasitic protozoa are the hydrogenosomal pathways of the trichomonads. These organisms are classed as aerotolerant anaerobes and respire stored glycogen. Instead of mitochondria, they possess a unique subcellular inclusion, the hydrogenosome. The hydrogenosome has a double membrane, but lacks cytochromes and DNA, and its activity results in the production of molecular hydrogen. It is tempting to speculate that it owes its origin to an ancient symbiosis between a flagellate and a clostridial-like organism.

Although the details may vary between species of trichomonads, Fig. 5.2 illustrates the principles. Glycogen is metabolized by an augmented glycolytic pathway to PEP on the one hand and to glycerol on the other. PEP is further converted to pyruvate and malate, both of which may enter the hydrogenosome. Within the organelle, a series of redox reactions lead to the production of molecular hydrogen and the synthesis of ATP. If oxygen is present, it acts as the terminal electron acceptor and hydrogen is not produced. Oxygen reduction mostly occurs in the cytosol via a soluble NADH oxidase, and while there is no apparent energy gain it is thought that the process is important in oxygen scavenging, protecting the anaerobic enzyme battery within the hydrogenosome from damage by oxygen derived free radicals.

Like the trichomonads, neither *Entamoeba* nor *Giardia* possesses mitochondria, but neither do they possess any other organelle for energy metabolism. Many aspects of their metabolism appear to be primitive; *Entamoeba*, for example does not have a nucleolus, normal ribosomes,

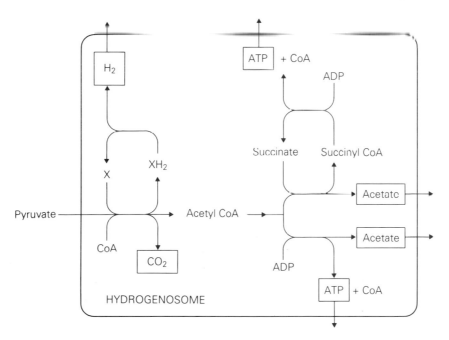

Fig. 5.2 The role of the hydrogenosome in trichomonads. Reactions that are written inside the box take place in the hydrogenosome. X is an unknown carrier that interacts with an unusual hydrogenase to produce molecular hydrogen.

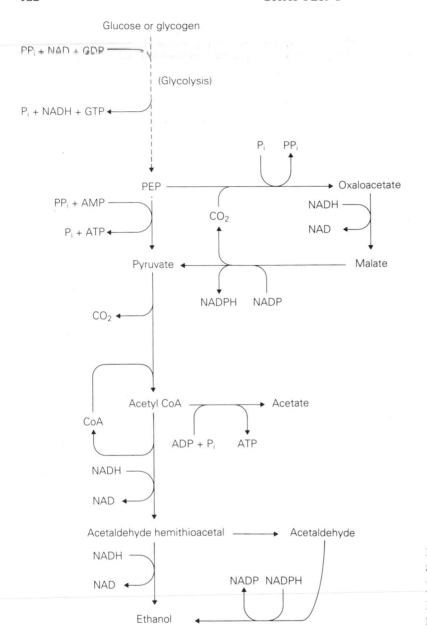

Fig. 5.3 Pathways of anaerobic glucose metabolism in *Entamoeba*. Those in *Giardia* are very similar but less completely known. P_i, PP_i – inorganic phosphate and pyrophosphate.

microtubules, true Golgi apparatus or an extensive endoplasmic reticulum. Both organisms have proved difficult to study because of the presence of ingested bacteria, whose enzymes may contribute to metabolism.

Entamoeba is an aerotolerant anaerobe, consuming oxygen either through NAD(P)H oxidases,

or a series of reactions centred on the pyruvate synthase complex. Glucose is the principal respiratory substrate and its transport into the cell appears to be the rate limiting step in metabolism. Glucose may be converted to glycogen and stored, or it may be catabolized to ethanol and carbon dioxide anaerobically, while aerobically, acetate

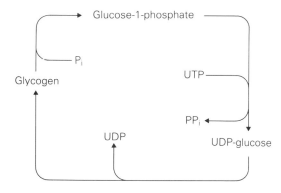

Fig. 5.4 The glycogen cycle generates pyrophosphate (PP$_i$) in *Entamoeba*, and may be important in metabolic regulation. UDP, UTP – uridine di- and triphosphates.

from that in other parasites, and includes low redox potential iron–sulphur compounds, with the properties of ferridoxins, active in both an aerobic and aerobic electron transport. A tentative scheme for aerobic electron transport in *Entamoeba* is shown in Fig. 5.5; it is unlikely that its activity results in energy conservation.

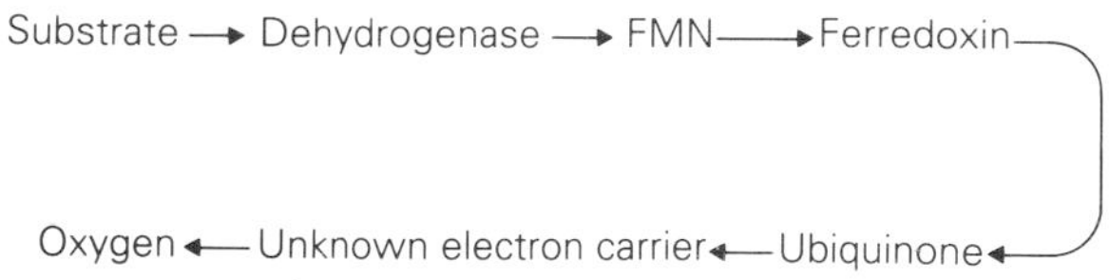

Substrate ⟶ Dehydrogenase ⟶ FMN ⟶ Ferredoxin

Oxygen ⟵ Unknown electron carrier ⟵ Ubiquinone

Fig. 5.5 Aerobic electron transport in *Entamoeba*. FMN – flavin mononucleotide.

is produced as well (Fig. 5.3). A glycogen cycle (Fig. 5.4) may contribute pyrophosphate which is important in glucose catabolism. It has been suggested that the pathway of glucose breakdown, with its reliance on pyrophosphate and thiol compounds for energy conservation, may be an ancient evolutionary relict, conserved because of the specialization of *Entamoeba* to parasitic mode of life.

Giardia, also an aerotolerant anaerobe, does not possess the unique enzymes found in *Entamoeba* although its respiratory end-products are similar, perhaps due to its occupation of similar habitats. There is no TCA cycle activity. It consumes oxygen, perhaps, like *Entamoeba*, in order to detoxify it, and grows best *in vitro* when oxygen concentrations are low. The pathway for glucose catabolism has not been fully elucidated, but is probably as shown in Fig. 5.3.

An important characteristic of both *Entamoeba* and *Giardia* is their ability to couple ATP formation directly to the cleavage of the thioester bond:

acetyl CoA + ADP + Pi → acetate + CoA + ATP.

5.3.1 Electron transport in parasitic protozoa

The 'anaerobic' protozoan groups, discussed above, do not possess mitochondria and do not appear to have haem proteins such as cytochromes or catalase. Electron transport is quite different

The other major groups of protozoans, the trypanosomes, leishmaniae and malarial parasites, all possess mitochondria containing cytochromes. They respire aerobically but there is much variation in electron transport between groups and between different stages in their life cycles. Many of them appear to have branched respiratory chains, like the helminths. For example, a branched electron transport chain with two terminal oxidases, has been proposed for *Plasmodium* species. Its role in respiration is unclear, since glucose metabolism in these organisms yields lactate as the only end product. It appears likely that oxygen utilization is associated instead with pyridine metabolism, via dihydroorotate dehydrogenase activity.

5.4 ENERGY METABOLISM IN PARASITIC HELMINTHS

All helminths utilize glucose as a respiratory substrate – indeed, tapeworms incubated *in vitro* are able to absorb almost all the glucose provided in incubation media. They have very active transport mechanisms which bind glucose at very low concentrations. Once glucose is absorbed, it is either converted into glycogen, to act as an energy store, or is metabolized directly via the glycolytic sequence of reactions as far as PEP. There are then a number of different options for its further metabolism. Each of the options may be found within a single taxonomic group of helminths –

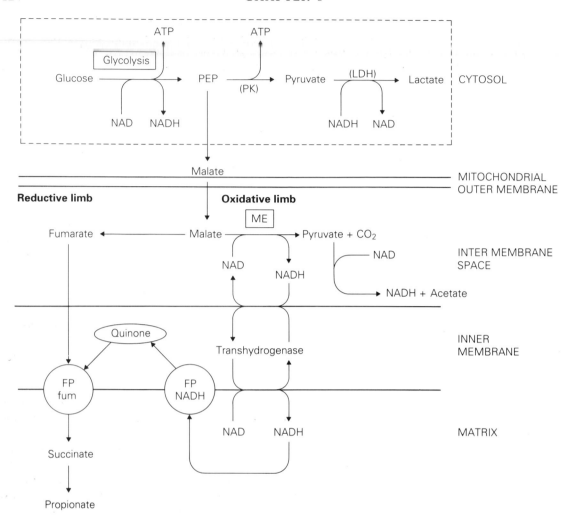

Fig. 5.6 Homolactate fermentation (in the dashed box) and the malate dismutation in parasitic helminths. In the malate dismutation, electrons are transferred from NADH (generated by malic enzyme activity and the pyruvate dehydrogenase system) across the membrane by the transhydrogenase. They are then passed to NADH dehydrogenase (FP NADH), quinone and fumarate reductase (FP fum). In this scheme there is no role for cytochromes. Another view suggests that cytochrome mediated electron transport intervenes at the level of the quinone. FP – flavoprotein; ME – malic enzyme.

there are no predictions that can be made about the type that will be encountered in a particular group.

5.4.1 Homolactate fermentation

True homolactate fermentation (Fig. 5.6) probably does not occur in helminths. In homolactate fermentation glucose is converted exclusively into lactic acid by glycolysis. The end-product, lactic acid, is excreted and energy generation is thus wholly independent of oxygen. ATP is synthesized at the phosphoglycerate kinase and pyruvate kinase steps, and NADH is reoxidized by lactate dehydrogenase, so that the pathway remains in redox balance. There is, however, evidence that other energy yielding processes also occur in the so-called helminth homolactate fermenters. For example, schistosomes display low levels of TCA cycle activity that contribute signi ficantly to energy metabolism. This is because the aerobic oxidation of glucose generates 18 times more ATP per mole of glucose than homolactate fermentation.

The same is true for the filarial worm, *Litomosoides carinii*. Overall, studies with ^{14}C-labelled substrates suggest that 2% of utilized carbohydrate may, in normal respiration *in vitro*, undergo complete oxidation to carbon dioxide and water. A simple calculation shows that 98 moles of glucose converted to lactate yields 196 moles of ATP, while 2 moles of glucose converted to carbon dioxide and water by the TCA cycle yields 72. From this, it is clear that at least 27% of energy generation in *L. carinii* could be aerobic.

5.4.2 Malate dismutation

The malate dismutation (Fig. 5.6) has been most extensively studied in *Ascaris* spp., *Hymenolepis diminuta* and *Fasciola hepatica*. Glucose is oxidized to the level of PEP. A carbon dioxide fixation step then occurs, catalysed by PEPCK, leading to the formation of oxaloacetate which is subsequently reduced to malate. This can be contrasted with what occurs in mammals. In mammals, PEP is converted to pyruvate, which in turn is converted to acetyl CoA by the pyruvate dehydrogenase system, and the subsequent metabolism of acetyl CoA is by the TCA cycle in mitochondria. The operation of this aerobic system permits the synthesis of large amounts of ATP and the ultimate reoxidation of reduced cofactors by oxygen. The role of PEPCK in mammals is the reverse of that encountered in helminths, the decarboxylation of oxaloacetate during gluconeogenesis. It serves as a bypass of the irreversible pyruvate dehydrogenase step.

Few, if any, parasites possess a TCA cycle as active as that of mammals. Generally, where the cycle is complete, it operates at very low activities. Important enzymes of the cycle are often missing or present in very small amounts. There are no rules as to which ones will be absent. Citrate synthase, aconitate hydratase and isocitrate dehydrogenase are often not detectable in trematodes and cestodes. But there are three enzymes of the TCA cycle which are almost invariably present in the mitochondria of parasitic helminths. They are MDH, fumarate hydratase and fumarate reductase. They catalyse the following reductive reactions, a reversed sequence of part of the TCA cycle:

$$\text{oxaloacetate} \xrightarrow[MDH]{} \text{malate} \xrightarrow[\substack{fumarate \\ hydratase}]{} \text{fumarate}$$

$$\xrightarrow[\substack{fumarate \\ reductase}]{} \text{succinate.}$$

In the mitochondria, there are also two important oxidative enzymes, malic enzyme (ME) and the pyruvate dehydrogenase complex, that catalyse the following reactions.

$$\text{PEP} + CO_2 + \text{GDP} \xrightarrow[PEPCK]{} \text{oxaloacetate} + \text{GTP},$$

$$\text{malate} + \text{NAD(P)} \xrightarrow[ME]{} \text{pyruvate} + CO_2 + \text{NAD(P)H.}$$

Malate then enters the mitochondrion, where it undergoes a dismutation. In a dismutation reaction, one molecule of a given compound is oxidized, while a second is reduced. The oxidation step is coupled to the reductive step and provides the reducing power to drive the reaction. In the malate dismutation, there is an additional need to maintain redox balance which demands the oxidation of one molecule of malate while two molecules are reduced. This is because in the oxidative arm of the pathway there are two reactions that each generate one reducing equivalent per molecule of malate oxidized, while the reductive arm has only one reaction that utilizes reducing equivalents. The products of the dismutation, usually acetate and succinate or propionate, are then excreted as the free acids (Fig. 5.6). The net result is the production of more ATP than is obtained fom homolactate fermentation on its own (Table 5.1).

The malate dismutation involves two subcellular compartments—the cytosol and the mitochondrion—which must both remain in redox balance. Lactate and malate, when produced in the cytosol, are equivalent in metabolic terms. The dehydrogenases that produce them have a similar role in regenerating NAD. Further, the

Table 5.1 ATP yield of different types of fermentation

End product	mol ATP formed per mol glucose
$CO_2 + H_2O$	36
Lactic acid	2
Ethanol	2
Alanine	2
Acetate	2
Acetate + succinate	3.7
Acetate + propionate	5.4
Acetate + propionate and other fatty acids	5

series of reactions that produce either lactate or malate are equivalent energetically, because PK and PEPCK each bring about the synthesis of ATP. PEPCK does it at one remove because its preferred cofactor is GDP not ADP, and it generates GTP. An additional reaction is thus necessary for conversion to ATP:

PK reaction:
    PEP + ADP → pyruvate + ATP.

PEPCK reaction:
    PEP + GDP + P_i + CO_2
                        → oxaloacetate + GTP.

PK and PEPCK stand at an important branchpoint of metabolism in helminths. The flow of carbon through PEPCK must be regulated so that the redox balance of each cellular compartment is maintained. The nature of the regulation is dynamic, as it is observed that under different environmental conditions, the proportions of the end-products change. PK and PEPCK are influenced allosterically by purine nucleotides and by metabolic intermediates. The concentrations of these change depending on the availability of terminal electron acceptors, bringing about adjustments to carbon flow through the different metabolic branches.

The malate dismutation is, of course, an over-simplification. Other substrates, such as pyruvate may also enter and be oxidized within the mitochondrion. There may be some TCA cycle activity and respiratory patterns may change during development. For example, the juvenile liver fluke, *Fasciola hepatica*, is an almost totally aerobic organism with tricarboxylic acid cycle activity. Three weeks later it has lost much of this and its energy metabolism depends on aerobic acetate formation. By about 2 months a proper malate dismutation has developed.

End-products of metabolism in parasitic helminths include ethanol, lactate, acetate, succinate and volatile fatty acids. No one helminth excretes all of these end-products, and it is difficult, given the present state of knowledge, to know why a particular set of end-products has been adopted by a particular organism. The best answer is that it is an accident of evolutionary history. The formation of such highly reduced end-products can be explained in terms either of 'energetic advantage' or of 'redox advantage'. The first implies that the specialized pathway to a particular end-product leads to energy conservation as, for example, ATP or GTP. The second means that reduced cofactors such as NAD(P)H, generated in other parts of the pathway, are re-oxidized during the formation of the end-products. Recycling of cofactors is thus made possible, and the various subcellular compartments are maintained at the appropriate redox level to permit the continued oxidation of glycogen, even under anaerobic conditions.

In *Ascaris* and a number of other nematode genera the malate dismutation is extended by a number of redox and condensation reactions. The main end-products of glucose catabolism are 2-methylvalerate and 2-methylbutyrate, with small amounts of propionate, acetate and other volatile fatty acids (Komuniecki *et al.*, 1981). The pathways for their production involve two reductive steps, in which NADH is consumed, an acyl CoA intermediate is formed, and the regeneration of CoA occurs at the end of the reaction sequences. They are thus remarkably similar to a reversed beta-oxidation pathway for fatty acid breakdown, absent from adult helminths. Helminths have only a very limited capacity for metabolizing fats, although it does occur in some larval nematodes.

Energy metabolism is variable; it varies between and within species and even within individuals. Variability in a system such as that of energy generation, that one feels intuitively ought to be conservative, is intriguing and deserves a little more discussion. Barrett (1981) remarks that end-products selected by helminths represent a compromise between substrate conservation and rate of working. In other words, a sufficiently high energetic yield can be obtained by incompletely oxidizing large amounts of respiratory substrate. Further, organisms that carry out mixed fermentations in various subcellular compartments have more opportunity for maintaining appropriate redox levels in those compartments, by altering carbon flow. This presumably provides flexibility in adapting to changing environmental circumstances.

The size of a parasite may be important in determining the type of energy metabolism and there is a good inverse correlation between the thickness of the body and the aerobic capacity of nematodes. As worms increase in size, poor diffusion of oxygen into the deeper tissues of large parasites may preclude complete dependence on oxygen as an electron acceptor. Fairbairn (1970) observed that, as the dissociation constants for lactic and succinic acids were high ($K_a = 13.87 \times 10^{-5}$ and 6.63×10^{-5}, respectively), they would be less likely to dissociate at physiological pH and would therefore pass more easily through a lipid membrane than the dissociated ionic species (protons and anions of acids). In other words, Fairbairn considered that there might be an energetic advantage in excreting certain acids. Yet another suggestion was made by Bowlus and Somero (1979). They pointed out that the properties of succinic acid were compatible with the enzyme systems that produced it and that succinate actually stabilizes protein structure. It

is difficult to know how to choose between these options. A sensible conclusion is that they all have a part of the truth.

5.4.3 Electron transport in helminths

The mechanism of electron transport in mammals involves the reoxidation of reduced cofactors by a system of enzymes and electron carriers. The latter include flavoprotein dehydrogenases, ubiquinone, cytochromes b, c and a and finally cytochrome oxidase which transfers the electrons to oxygen, with the formation of water. During electron transfer ATP formation occurs at the sites shown in Fig. 5.7.

All parasitic helminths show, *in vitro*, a measurable uptake of oxygen; while some of the oxygen is no doubt used for synthetic reactions, a part of it is used in respiration. It is widely accepted that while many adult helminths possess low activities of classical electron transport systems, their specialized electron transport systems are anaerobic. In addition, they may also be branched and therefore possess several terminal oxidases. In some helminths, for example, there may be three branches: one with cytochrome oxidase, the second with a b type cytochrome, and a third which generates hydrogen peroxide (Fig. 5.8).

Electrons are transported along the chain of electron carriers either to cytochrome oxidase, to the alternative oxidase or to the fumarate reductase system. The products of respiration are either water, the potentially dangerous hydrogen peroxide (removed either by catalase or

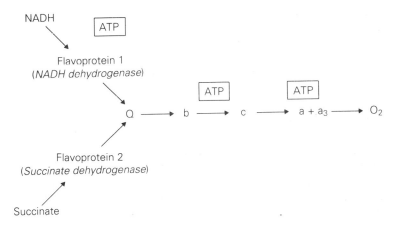

Fig. 5.7 A simplified diagram of the electron transport system in mammals, showing phosphorylation sites. Q is ubiquinone or coenzyme Q.

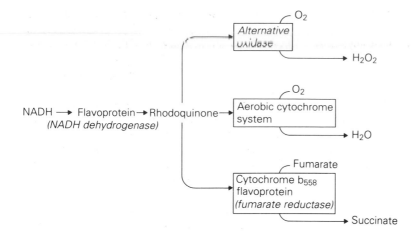

Fig. 5.8 A possible scheme for branched electron transport in parasitic helminths.

peroxidase, both of which are found in helminths at high activities) or succinate and its derivatives. By analogy with the mammalian system, it is presumed that there are three proton translocating sites for ATP synthesis.

Instead of ubiquinone, many helminths possess rhodoquinone, an extremely interesting anaerobic adaptation. The redox potential of rhodoquinone (E_0) is $-63\,mV$, which is considerably lower than ubiquinone ($+113\,mV$) and the fumarate/succinate couple ($+33\,mV$). Electron transport in the direction of fumarate is thus favoured. Hel-

minths also possess an enzyme complex called fumarate reductase, crucial to malate dismutation. Fumarate reductase is distinct from the enzyme that catalyses the reverse reaction in aerobic organisms, succinate dehydrogenase. The enzyme from *Ascaris* has been purified and found to consist of four major and two minor polypeptides, two of which have the same molecular weight as the two subunits of mammalian succinate dehydrogenase. It also contains large amounts of cytochrome b_{558}. A detailed account of helminth fumarate reductase and the evolution

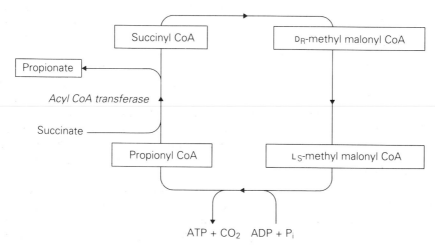

Fig. 5.9 Propionate production by the succinate decarboxylase system in helminth mitochondria. CoA is recycled by the transferase.

of electron transport systems is given by Behm (1991).

Finally, Köhler (1985) points out that acyl CoA transferase reactions and acyl CoA-dependent ATP conservation play an important role in energy metabolism in parasites. Propionate formation within mitochondria is a case in point and occurs in *Fasciola* and *Ascaris* as illustrated in Fig. 5.9. It is another example of the reversal of a cycle found in mammals, this time in ruminants that utilize propionate produced by the rumen microorganisms.

5.5 LIPIDS

Lipids can be divided into three fractions on the basis of their behaviour in experimental separation systems. The fractions are: (1) glycerides, i.e. esters of glycerol and fatty acids – and free fatty acids; (2) phospholipids; and (3) unsaponifiable lipids, which include waxes, sterols, terpenes and related substances. A major part of the structure of membranes comprises many different types of lipids.

While a great deal is known about the distribution of lipids and lipid-like compounds in parasites, their metabolism is a much neglected field of study. All parasites contain lipids in varying amounts and they include a wide variety of molecules, such as the acylglycerols, fatty acids, phosphoglycerides, waxes and sterols. There is no evidence that a particular site within a host requires a particular type of lipid profile, but there is some evidence that, in helminth parasites, lipid content reflects that of the host. The major fatty acids usually contain 16 or 18 carbon atoms, especially the unsaturated oleic and linoleic acids. In parasitic helminths, the range of fatty acids found often depends on host diet and can be changed by changing that diet. They may also reflect the constituent fatty acids of the host. For example, sharks contain C20 and C22 polyunsaturated fatty acids which help to keep them supple at sea-water temperatures, because unsaturated fatty acids have lower melting points than saturated ones. The same fatty acids are found in their tapeworms.

Phylogenetic relationships do not offer any indication of the types and amounts of lipids that a given parasite may possess. Thus the lipid content of five species of *Trypanosoma* ranges from 11 to 20% of dry weight, and for parasitic protozoa as a whole the average is about 14%. Helminth parasites show much greater variation: 1–34%.

The functions of the various lipid components in parasites are understood only by analogy with other organisms, but there is reason to suppose that there are a number of differences. Parasites like other organisms, contain cells with external and internal membranes. The former are especially important since they must protect the parasite from host attack and, at the same time, permit and perhaps control the uptake of nutrients across the membrane. Several unique pathways or enzymes have been identified in some parasitic protozoa. For example, acetate units for lipid synthesis in bloodstream and culture forms of *T. brucei* are preferentially derived from threonine by the action of threonine dehydrogenase and glycine acetyltransferase. This pathway does not occur in the host.

Parasite membranes differ in composition from those of their hosts. Membranes of plasmodial parasites have higher concentrations of unesterified fatty acids, triacylglycerols, 1,2-diacylglycerols, diacyl-phosphatidylethanolamine and phosphatidylinositol and lower concentrations of cholesterol, phosphatidylserine and sphingomyelin than their host's membranes. Parasite membranes thus have different properties from those of their hosts. Presumably, they must also possess quantitatively different synthetic machinery for supplying membrane constituents, or specific uptake processes for obtaining appropriate precursors. Specialized external membranes are present in some parasites. Adult *Schistosoma mansoni* possesses a lipid-rich, double, outer membrane which develops a few days after penetration of the host. It is constantly renewed, and it has a vital function as the interface between the parasite and the host.

In many animals, glycerides and free fatty acids form an energy store which may be drawn upon in times of starvation. While parasites have considerable quantities of glycerides and free fatty acids, they are, usually, not used in the synthesis of ATP. Parasites require an external source of

lipids for their maintenance in culture. For protozoa, it is often sufficient to add whole serum, the
essential components of which seem to be cholesterol and fatty acids. Trypanosomes are even
capable of absorbing fat droplets. Presumably
they are broken down to fatty acids, glycerol and
other alcohols. Parasitic helminths also take up
lipids and fatty acids, including acetate, from
incubation media.

No group of parasitic protozoa or helminths
seems capable of synthesizing long-chain fatty
acids *de novo*, i.e. from simple precursors. Although many incorporate acetate into long-chain
fatty acids, or carbon from glucose or glycerol
into the lipid fraction, the process usually involves the comparatively slight modification,
such as chain lengthening or the saturation of
certain unsaturated bonds in long-chain fatty
acids, of an existing molecule.

The apparent neglect of a resource widely exploited in the animal kingdom for energy metabolism needs explaining, but the proximate cause
is that parasites often lack all the enzymes necessary for the beta-oxidation of fatty acids and/or
an active TCA cycle. The latter is essential for
the oxidation of the acetyl CoA that is the end-
product of beta-oxidation. An exception to this
are the free-living larvae of some parasitic helminths, which are certainly capable of oxidizing
stored lipids in the production of energy, e.g. the
free-living larvae of *Strongyloides ratti* oxidize
palmitic acid by the beta-oxidation pathway.

Finally, there are many compounds in living organisms based on isoprene (2-methyl butadiene):

$$CH_2{=}\underset{\underset{CH_3}{|}}{C}{-}CH{=}CH_2.$$

They are called terpenes, and are found in oils like
camphor and geraniol, in rubber, and in carotenes
from plants. They are also found as side chains
in various vitamins and cofactors, including the
ubiquinones that are important in electron transport and folate metabolism. A ubiquinone has the
structure:

$$quinone{-}(CH_2{-}\underset{\underset{CH_3}{|}}{CH}{=}C{-}CH_2)_n H$$

where *n* refers to the number of isoprene units in
the side chain. There are many compounds with
isoprenoid side chains present in nematodes,
including ubiquinone itself, cholesterol, rhodoquinone, farnesol-like compounds and the poly-
propanol, solanesol. Adult filariae *in vitro* are
unable to carry out the synthesis of sterols.
Nematodes can synthesize ubiquinone, ecdysteroids, juvenile hormone and farnesol. Some of
these compounds play essential roles as hormones in the development of insects. Whether
they do so in helminths is not known.

5.6 METABOLISM OF NITROGEN COMPOUNDS

There are about 20 amino acids commonly found
in free-living organisms, and a small but variable
number of purines and pyrimidines. The same
range of molecules are found in parasites. Parasites usually depend on the host for a supply of
nitrogenous compounds as the ability to make
many of them has either been lost, or less likely,
never evolved in these groups.

5.6.1 Amino acid metabolism

All parasites synthesize amino acids. Trypanosomes, for example, are known to make alanine,
glycine, serine, aspartic and glutamic acids. The
malaria parasite synthesizes alanine, aspartic and
glutamic acids and the sulphur-containing amino
acids from a range of precursors. Some helminths
incorporate ammonia directly into pyruvate and
2-oxoglutarate to form alanine and glutamate,
respectively.

Amino acids are readily metabolized by transamination. The two most important reactions,
because they interact with pathways of energy
metabolism, catalysed by aminotransferases are:

2-oxoglutarate + amino acid → glutamate + oxoacid,
pyruvate + amino acid → alanine + oxoacid.

They are readily reversible and nearly universally
distributed in parasites. Degradation of amino
acids is catalysed by L-amino acid oxidases, thus:

L-amino acid + H_2O + O_2
→ 2-oxoacid + NH_3 + H_2O_2.

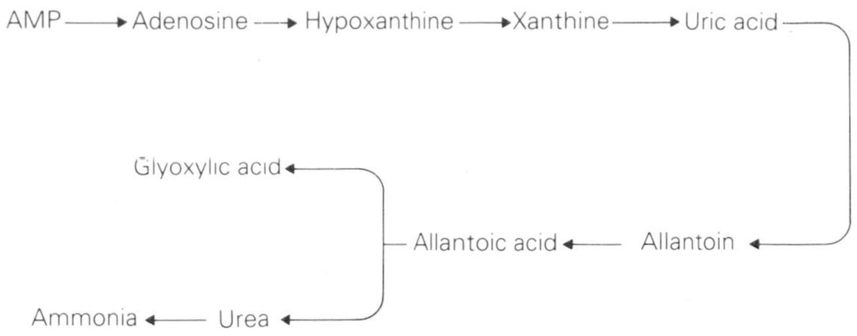

AMP ⟶ Adenosine ⟶ Hypoxanthine ⟶ Xanthine ⟶ Uric acid ⟶

Glyoxylic acid ◄—

Allantoic acid ◄— Allantoin ◄—

Ammonia ◄— Urea ◄—

Fig. 5.10 The probable pathway of purine degradation in parasitic nematodes, inferred from the detection of intermediates and end-products after the inclusion of purine compounds in incubation media.

Ammonia is excreted and the oxoacid may be oxidized during respiration.

Another end-product of amino acid and protein metabolism is urea, which in higher animals, is produced by the Krebs Henseleit (urea) cycle. There is no evidence that any parasite possesses a functional cycle, although they all have some of the enzymes associated with it. Any urea which is detected in parasitic helminths probably derives from purine degradation (Fig. 5.10).

Arginase, a common enzyme in parasites, splits arginine into ornithine and urea. Kurelec (1975) suggested a unique role for arginine metabolism in the liver fluke, *Fasciola hepatica* (Fig. 5.11).

It leads to the formation of proline, which is produced in large quantities by the fluke. The enzyme which catalyses the final step in the production of proline is several times more active than that of the host, and, unlike the mammalian enzyme, is not subject to end-product inhibition. This pathway may be important in the maintenance of redox balance by reoxidizing NADH, or proline may be important in nitrogen excretion. Another suggestion stems from the observation that proline, infused into the peritoneal cavities of rats, causes a bile-duct hyperplasia which is similar to that produced in the early stages of fascioliasis. It is therefore possible

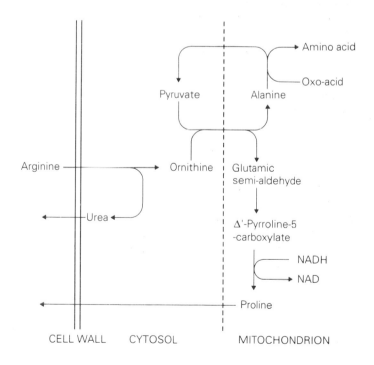

Fig. 5.11 A scheme for arginine catabolism and the production of proline and urea in *Fasciola hepatica*. Arginine is taken up from the environment, converted to ornithine and then, by interaction with a transamination cycle, to glutamic semialdehyde. This is eventually converted into proline in the mitochondrion. Proline and urea, products of the last and first steps of the pathway, are excreted.

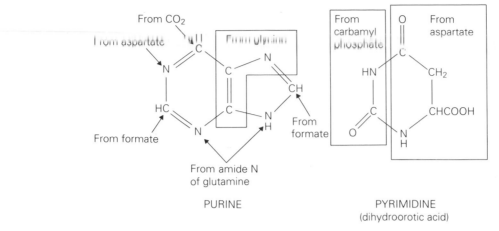

Fig. 5.12 The origins of the atoms that form the purine and pyrimidine rings during *de novo* synthesis.

that proline excretion is a metabolic strategy by which the parasite improves its environment.

5.6.2 Purines, pyrimidines and their salvage

Purines and pyrimidines are bases that possess the heterocyclic ring structures illustrated in Fig. 5.12. Adenine and guanine are purine derivatives; thymine, cytosine and uracil are pyrimidines. A nucleoside is a combination of a heterocyclic base with a sugar, often D-ribose. A nucleotide is the combination of a heterocyclic base with a sugar phosphate. ATP is a nucleotide. Heterocyclic bases are the backbone of the structure of DNA and RNA and provide the basis for the genetic code. Often, incorporated into coenzymes and vitamins, they participate in numerous metabolic reactions. Cyclic nucleotides act as specific signallers within the cell.

It is doubtful whether any protozoan or helminth parasite has the capacity to synthesize the purine ring *de novo*. A possible exception is the kinetoplastid flagellate, *Crithidia oncopelti*; but for the remainder detailed experiments lead to the conclusion that parasites rely on their hosts to provide a source of purines.

Parasites contain the same range of purines and their derivatives that are found in free-living organisms. In the absence of synthetic mechanisms, it is not surprising to find elaborate

systems for absorption and interconversion of existing purines. These are the so-called 'salvage' pathways so widely distributed in parasites. Cultivation experiments have demonstrated exactly which purines and purine-containing compounds are essential for growth, especially in the case of the blood dwelling malaria parasites and the schistosomes, as blood itself is a very well-defined medium. The salvage pathways are illustrated in Fig. 5.13. Purines are not, of course, invariably salvaged, as Fig. 5.10 shows.

There are some 'unique' enzymes of purine salvage in parasitic protozoa. For example, *Giardia duodenalis* has a guanine phosphoribosyl transferase that does not recognize hypoxanthine, xanthine or adenine as substrates, in contrast to the analogous enzyme in other organisms. *Leishmania donovani* promastigotes have a xanthine phosphoribosyl transferase, and *Eimeria tenella* possesses a hypoxanthine, guanine, xanthine–phosphoribosyl transferase as a single enzyme, not known in any other organism.

Most parasitic protozoa (except the trichomonads and *G. lamblia*) and all helminths seem to be capable of synthesizing pyrimidines *de novo* from carbamyl phosphate and aspartate (Fig. 5.12). The evidence for the presence of the pyrimidine synthetic pathway is largely circumstantial and, once again, depends heavily on culture experiments with defined incubation media. In parasitic protozoa and in *S. mansoni* the pathways

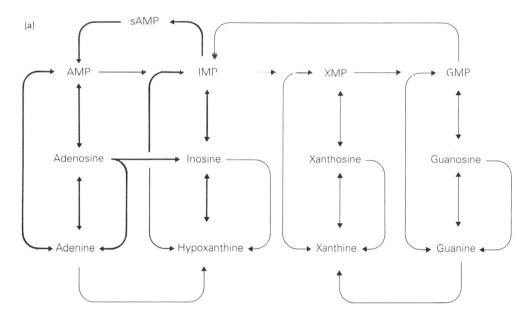

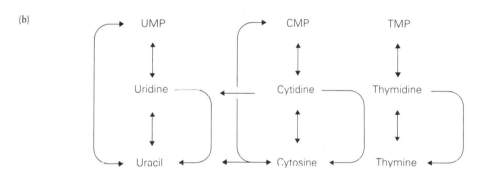

Fig. 5.13 Purine and pyrimidine salvage pathways in parasites. Reactions indicated by bold arrows have been detected in parasitic helminths. Different combinations of all reactions have been detected in parasitic Protozoa but not all are present in a single organism. AMP, GMP, IMP, XMP – adenosine, guanosine, inosine and xanthosine monophosphates; sAMP – succinyl AMP; UMP, CMP, TMP – uridine, cytidine and thymine monophosphates.

are qualitatively similar to those of their hosts.

Salvage pathways for pyrimidines also occur. They are of different importance in different parasites; malaria parasites make little use of them, whereas the trypanosomes rely equally on salvage and synthesis. Some of the enzymes involved in purine metabolism are peculiar to particular taxons. In kinetoplastids, for example, dihydroorotate oxidase (orotate reductase) is soluble, contains flavin, and appears to donate electrons directly to oxygen, forming hydrogen peroxide, whereas in most organisms it is bound to the mitochondrial membrane and joins the electron transport chain at the level of ubiquinone. Further, thymidylate synthetase in kinetoplastids and plasmodia is bound to dihydrofolate reductase to form a bifunctional enzyme complex. In mammals these two enzymes are separate.

Trypanosoma foetus, T. vaginalis and *G. lamblia* do not possess either enzyme; they are unique amongst parasitic protozoa in requiring exogenous thymidine. *Trypanosoma vaginalis* is different from all other protozoa, in that it cannot convert purine and pyrimidine ribonucleotides to the deoxyribonucleotides required for DNA synthesis, because it does not have a ribonucleotide diphosphate reductase. Deoxyribonucleotides must be supplied by a deoxyribonucleotide phosphotransferase acting on salvaged deoxyribonucleosides.

5.6.3 Nucleic acid metabolism

Nucleic acid synthesis is essential for rapidly growing cells and requires nucleosides and nucleotides. Nucleic acid synthesis (especially DNA synthesis) may be cyclic or stage-specific. In protozoa, which undergo rapid multiplication, DNA synthesis may be limited to specific phases of the life/cell cycle. In adult helminths, DNA and RNA synthesis is largely restricted to the reproductive tissues, except in the case of continuously-growing cestodes.

The DNA of parasites is unremarkable. There is no evidence for the presence of any unusual base, neither is there evidence to suggest that the organization of the genome is unusual. The argument that parasites should possess a diminished genome because they depend for many of their requirements on the genetic information of their hosts is not borne out by observation. In fact, parasitic nematodes and flatworms have more complex genomes than their free-living relatives.

With one exception, mitochondrial DNA from parasites is likewise unremarkable. The exception is DNA from the trypanosome kinetoplast. The kinetoplast is an analogue of the mitochondrion. Kinetoplast DNA occurs in relatively large amounts and is organized into maxicircles and minicircles. Maxicircles contain the equivalent of mitochondrial DNA, but the function of minicircle DNA, which is heterogeneous and rapidly evolving, is not understood. Maxicircles and minicircles are not associated with histones and occur in a concatenated mass that must be precisely divided between daughter kinetoplasts at cell division. It is therefore likely that the mechanisms of unravelling and replication are different from those in all other organisms.

5.7 OTHER PROCESSES

Dihydrofolate is essential for growth and reproduction in all organisms because it or its derivatives participate as cofactors in many methylation and other carbon transfer reactions. Intracellular sporozoa synthesize dihydrofolate *de novo* from the precursors GTP, *p*-aminobenzoate and glutamate (Fig. 5.14). Their mammalian hosts do not have this pathway; they recover folate from their diet and reduce it to dihydrofolate directly. Little is known about folate metabolism in helminths. Filarial worms do not appear to synthesize dihydrofolate *de novo*, but obtain 5-methyltetrahydrofolate as their major source of folate from the host.

The secondary messenger, cAMP, is essential in metabolic regulation. It causes the activation of some regulatory enzymes, and increases the phosphorylation state of some membrane proteins, which may be important in the regulation of neuromuscular activity. The intracellular concentration of cAMP is determined by the relative activities of adenylate cyclase and cAMP-phosphodiesterase.

Adenylate cyclase is present in *F. hepatica* in high activities. In this parasite, in *S. mansoni* and in *Ascaris suum*, it is activated by 5-hydroxytryptamine (but not epinephrine as in mammals) and is important in the regulation of carbohydrate metabolism and motility. Among the parasitic protozoa, cAMP inhibits cell division *in vitro* in trypanosomatids. In *Plasmodium falciparum* it may inhibit asexual multiplication

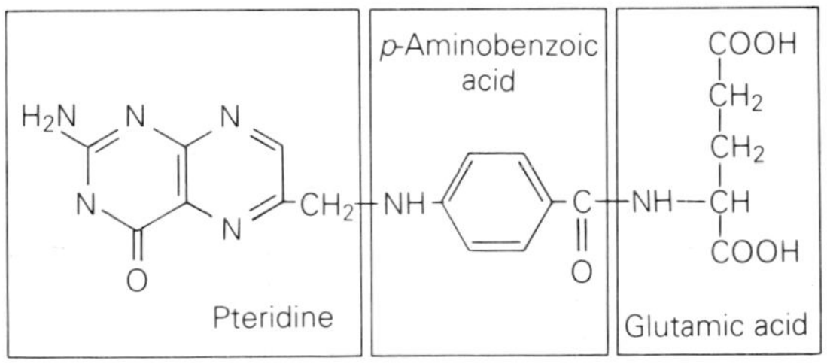

Fig. 5.14 The structure of folic acid.

and gametocyte formation. The major cellular receptors for cAMP in eukaryotes are the regulatory subunits of certain types of protein kinase. The trypanosomes have cAMP receptors, but their properties are different from those in other eukaryotes. Protein kinases, whether cAMP-dependent or independent, are important metabolic regulators in helminths. They have been found in filariae and other nematodes, and in schistosomes. They are probably ubiquitous and presumably have a similar regulatory function in helminth metabolism as in mammals.

Calmodulin is an intracellular receptor of calcium ions. It is widely distributed in the protozoan, animal and plant kingdoms. The calmodulin calcium ions complex activates many intracellular enzymes and processes. Calmodulin has been isolated from *Hymenolepis diminuta*. Its properties closely resemble those of mammalian calmodulin. Protozoan parasites also contain calmodulin which, in *T. brucei*, in vitro, probably controls the calcium induced changes in the activities of adenylate cyclase and endoribonuclease, and the release of variable surface glycoproteins which provide the organism with its elegant immunoevasion mechanism.

Polyamines include putrescine, spermidine and spermine. Their importance in the regulation of many cellular processes in protozoa is only now becoming clear. They bind ionically to nucleic acids and play a role in the cell cycle, in cell division and in differentiation. They may be especially important in parasitic protozoa because of their role of cell multiplication.

5.8 ENVOI

An important phenomenon exhibited by parasites is their capacity for metabolic variation. Both the protozoa and the helminths possess it in marked degree. Its enormous extent and significance has been highlighted in the last decade, as more and more reports of resistance to antiparasite drugs accumulate. Differences between strains (and the development of resistance) arise from the persistence of a particular selection pressure (such as antiparasite drugs) in the ecosystem. An excellent example is the resurgence of human malaria as resistance to chloroquine

spreads, due to widespread and indiscriminate use of that otherwise very useful antimalarial.

Metabolic variation is a problem compounded by the procedures of parasitological laboratories themselves. The establishment of a given parasite in culture, whether in *vitro* or by passage through living hosts, is a severe selection step, which permits the 'founder principle' full play. The parasite is removed, by cultivation, from access to the gene pool of its parent population, thus distorting the genetic profile of the cultivar. Further inadvertent selection by research workers, using procedures like cryopreservation that may cause high mortality, or by passage through a single strain of laboratory host, intensifies any differences. Bottlenecking the gene pool in this way leads to atypical frequencies of certain genes in the cultivated population and the emergence of a strain whose metabolic properties may be different from those of cultivars from other laboratories and from wild types. Generalizations based on work done with a single laboratory or field isolate may not be valid, and the alternative, that of working with several strains simultaneously, is expensive and often poses enormous logistical problems. It is a dilemma that has yet to be properly confronted.

None of the biochemical strategies described in this chapter can unequivocally be ascribed to adoption of the parasite habit. There is evidence that many of the same biochemical adaptations are also encountered in free-living lower organisms occupying habitats subject to low or fluctuating oxygen tensions and high carbon dioxide concentrations. If a particular metabolic capacity is absent, one may *suspect* that excision has occurred to effect economies in the energy expenditure of the parasite. For example, the loss of the ability to synthesize purines is a conundrum. Is this indeed an example of parasitic adaptation, the parasite relinquishing a costly suite of enzymes because of the universal availability of a source of purines from a host? Bearing in mind the importance of purines in constructing genetic material, this is giving up a real hostage to fortune. If purines, why not pyrimidines? Or any other major synthetic pathway? Perhaps it was the accidental loss of purine synthetic capacity early in evolutionary history that predisposed to parasitism? Fairbairn (1970) wrote of helminths 'No unequivocal loss of genetic capacity is known. Either . . . the genetic information is present but repressed, or an insufficient study of all stages of a life cycle has been made'. This last *caveat* should also be applied to protozoa. The only safe conclusion is that absence of evidence is not evidence of absence.

I acknowledge with gratitude research workers in the field of parasite biochemistry far too numerous to name here.

REFERENCES
AND FURTHER READING

Barrett, J. (1981) *Biochemistry of Parasitic Helminths*. London, Macmillan.

Barrett, J. (1991) Parasitic helminths. In C. Bryant (ed.) *Metazoan Life without Oxygen*, pp. 146–164. London, Chapman and Hall.

Barrett, J. (1991) Amino acid metabolism in helminths. *Advances in Parasitology*, **30**, 39–105.

Behm, C.A. (1991) Fumarate reductase and the evolution of electron transport systems. In C. Bryant (ed.) *Metazoan Life without Oxygen*, pp. 88–108. London, Chapman and Hall.

Bennet, E.-M., Behm, C.A. & Bryant, C. (eds) (1989) *Comparative Biochemistry of Parasitic Helminths*. London, Chapman and Hall.

Bowlus, R.D. & Somero, G.N. (1979) Solute compatibility with enzyme function and structure; rationales for the selection of osmotic agents and end products of anaerobic metabolism in marine invertebrates. *Journal of Experimental Zoology*, **208**, 137–151.

Bryant, C. & Behm, C.A. (1989) *Biochemical Adaptation in Parasites*. London, Chapman and Hall.

Bryant, C. & Flockhart, H.A. (1986) Biochemical strain variation in parasitic helminths. *Advances in Parasitology*, **25**, 276–319.

Coombs, G.H. (1986) Intermediary metabolism in parasitic protozoa. In M.J. Howell (ed.) *Parasitology – Quo Vadit?*, pp. 199–211. Proceedings of the VI International Congress of Parasitology. Canberra, Australian Academy of Science.

Cooms, G.H. & North, M.J. (eds) (1991) *Biochemical Protozoology*. London, Taylor and Francis.

Fairbairn, D. (1970) Biochemical adaptation and loss of genetic capacity in helminth parasites. *Biological Reviews*, **45**, 29–72.

Frayha, Q.J. & Smyth, J.D. (1983) Lipid metabolism in parasitic helminths. *Advances in Parasitology*, **22**, 309–387.

Fry, M. & Jenkins, D.C. (1984) Nematoda: aerobic respiratory pathways of adult parasitic species. *Experimental Parasitology*, **57**, 86–92.

Gutteridge, W.E. & Coombs, G.H. (1977) *Biochemistry of Parasitic Protozoa*. London, Macmillan.

Kita, K., Takamiya, S., Furushima, R., Ma, Y.-C. & Oya, H. (1988) Complex II is a major component of the respiratory chain in the muscle mitochondria of *Ascaris suum* with high fumarate reductase activity. *Comparative Biochemistry and Physiology*, **89B**, 31–34.

Köhler, P. (1985) The strategies of energy conservation in helminths. *Molecular and Biochemical Parasitology*, **17**, 1–18.

Komuniecki, R., Komuniecki, P.R. & Saz, H.J. (1981) Pathway of formation of branched-chain volatile fatty acids in *Ascaris* mitochondria. *Journal of Parasitology*, **67**, 841–846.

Kurelec, B. (1975) Catabolic path of arginine and NAD regeneration in the parasite *Fasciola hepatica*. *Comparative Biochemistry and Physiology*, **51B**, 151–156.

Lindmark, D.G. (1980) Energy metabolism of the anaerobic protozoon *Giardia lamblia*. *Molecular and Biochemical Parasitology*, **1**, 1–12.

Muller, M. (1980) The hydrogenosome. In G.W. Gooday, D. Lloyd & A.P.J. Trinci (eds) *The Eukaryotic Microbial Cell*. 30th Symposium of the Society of General Microbiology, pp. 127–142. Cambridge, Cambridge University Press.

Opperdoes, F.R., Misset, O. & Hart, D.T. (1984) Metabolic pathways associated with the glycosomes (microbodies) of the Trypanosomatidae. In J.T. August (ed.) *Molecular Parasitology*, pp. 63–75. New York, Academic Press.

Ramp, T., Bachmann, R. & Kohler, P. (1985) Respiration and energy conservation in the filarial worm, *Litomosoides carinii*. *Molecular and Biochemical Parasitology*, **15**, 11–20.

Rangel-Aldao, R. & Opperdoes, F.R. (1984) Subcellular distribution and partial characterisation of the cyclic-AMP binding proteins of *Trypanosoma cruzi*. *Molecular and Biochemical Parasitology*, **14**, 75–82.

Reeves, R.E. (1984) Metabolism of *Entamoeba histolytica* Schaudinn, 1903. *Advances in Parasitology*, **23**, 105–142.

Sherman, I.W. (1983) Metabolism and surface transport of parasitized erythrocytes in malaria. In E.J. Whelan (ed.) *Malaria and the Red Cell*, Ciba Foundation Symposium Vol. 94, pp. 206–216. London, Pitman.

Von Brand, T. (1973) *Biochemistry of Parasites*. London, Academic Press.

Von Brand, T. (1979) *Biochemistry and Physiology of Endoparasites*. Amsterdam, Elsevier/North-Holland, Biomedical Press.

Chapter 6 / Molecular Biology and Molecular Genetics

D. T. HART AND F. E. G. COX

6.1 INTRODUCTION

The term 'molecular biology' is as difficult to define as 'biochemistry' and in general the distinctions between molecular biology and biochemistry are becoming increasingly blurred. One has only to look at standard textbooks such as *Molecular Biology of the Cell*, by Alberts *et al.*, and *Molecular Cell Biology*, by Darnell *et al.*, and compare them with Stryer's *Biochemistry* to see the tremendous overlap.

In origin, molecular biology was an enabling technology in the field of biological chemistry and since its infancy has grown into an important contributor in its own right to modern biochemistry. In the field of molecular parasitology the enabling technologies of molecular biology serve a similarly important role and the existence of the journal *Molecular and Biochemical Parasitology* serves to illustrate the nature of the commonality. Molecular biology can therefore be dealt with separately from biochemistry and *Molecular Biology of the Gene*, by Watson *et al.*, emphasizes the central role of DNA and RNA, and the spectacular advances made as a result of the application of recombinant DNA technology to human biochemistry.

In this chapter, the central role of molecular biology, and in particular the advances made by the application of DNA and RNA technologies to elucidate parasite biochemistry, will be the unifying theme. This is also the approach taken by Hyde in his textbook *Molecular Parasitology* which is highly recommended for supplementary reading. Furthermore, in this chapter the application of DNA and RNA technologies to the manipulation and elucidation of parasite genomes, that is, the collection of a parasite's genes that constitute its genetic make-up, is appropriately termed molecular genetics.

The application of recombinant DNA technology has not only been used to study the genomic organization and regulation of gene expression in parasites but, equally important, it has been used in the study of gene products. Central to the molecular biology of the parasite is the 'DNA encodes RNA encodes Protein' concept and underlying this maxim is 'one gene – one gene product' which is a polypeptide.

In practical terms this means that once a parasite protein is recognized, the gene, and hence the DNA encoding it, can also be identified and most importantly *vice versa*! Since the race is on to sequence the entire human genome it is only a matter of time before host and parasite genetics are cross-fertilizing in protein–gene information. Recent advances in the technology mean that RNA can also be used in a similar way.

Parasite proteins, which have been considered in a different light in Chapter 5, are now the subjects of great molecular biological interest since manipulation of the genes offers an expedient route to antiparasitic vaccine and drug design. That is, parasite genes for antigens may be manipulated and expressed in bacterial systems to produce recombinant vaccines that allow parasite recognition by the host immune response. In a somewhat similar strategy, parasite enzymes can be expediently studied and compared to host counterparts, to be used as targets for the design of highly specific inhibitors that could serve as putative antiparasitic drugs.

The study of molecular biology is, therefore, of considerable practical importance and this chapter aims to introduce the reader to key aspects of parasite molecular biology and molecular genetics that have come to dominate parasitological research over the past decade. Central to an understanding of molecular parasitology is the realization that progress has only been possible because of enabling technologies drawn from many subject areas including immunology, enzymology, protein chemistry, molecular graphics and, of pre-eminence, recombinant DNA technology.

Molecular biology has pervaded all aspects of parasitology but the major advances can be summarized under the following broad headings: (1) an understanding of the basic biology and genetics of parasites; (2) the identification of key aspects of metabolism to serve as targets for chemotherapy; (3) the identification of surface antigens as targets for immune attack *via* vaccine development; (4) the identification of DNA sequences with potential use for diagnosis and in 'gene chemotherapy'; and (5) the use of RNA to resolve taxonomic relationships in both parasites and vectors.

As in the previous and subsequent chapters, it is not possible to do more than outline the general principles involved and to draw attention to a few specific examples. Many of these examples have been drawn from the protozoa, because most work has been done with these organisms, but the same general principles apply to helminths and it will only be a matter of time before these metazoans will be forced to reveal their molecular secrets.

It is assumed that the reader will be familiar with the basics of cell and molecular biology and anybody who is not is advised to refer to one of the aforementioned textbooks. In particular, Section 6.2 focuses on DNA and RNA technology and should be read very carefully since many of the examples that follow are dependent upon a clear understanding of these key technologies.

The 'take away' message from this chapter is that molecular biology has advanced our understanding of many aspects of molecular parasitology, such as protozoal molecular genetics, i.e. genomic organization, regulation of gene expression and metabolic adaptations to parasitism. This has allowed us to identify and begin to exploit differences between parasite and host which will facilitate the rational development of novel drugs and vaccines. It has also made possible the identification of parasites and even parasite-derived proteins that are of diagnostic importance and all this with a precision and expediency hitherto unattainable and even unimaginable.

6.2 DNA AND RNA TECHNOLOGY

6.2.1 Introduction to recombinant techniques

Our understanding of the molecular biology of parasites depends, to a very large extent, on the application of a number of techniques that have become so routine over the past few years, that most parasitologists are now familiar with the jargon of these 'recombinant' technologies. However, the application of recombinant techniques to parasitological studies has been far from simple and many protocols have had to be specifically modified for particular parasites. By way of a bonus, however, molecular and particularly recombinant studies on parasites have generated new insights into mechanisms of parasitism and have in turn brought about advancements in the respective technology. There is now a major bibliography of books concerned with the various techniques of molecular biology and most of the more important techniques are outlined in *Molecular Parasitology* by Hyde. What follows is a brief description of the main techniques in common use and what they are used for. The interested reader is recommended to go to standard textbooks, such as *Techniques in Molecular Biology* by Walker and Gaastra, for more information.

The universal starting point is the parasite itself. With the development of *in vitro* culture techniques, most if not all protozoa can be grown in considerable quantities and helminth worms are sufficiently large to use directly. DNA and RNA can be extracted from protozoa by direct cell lysis but helminths, particularly nematodes with their thick cuticles, may have to be specially treated, for example by freeze-thawing lysis, before the extraction protocol can even begin.

Isolation of the nucleic acids involves the inactivation of endogenous nucleases and the removal of proteins and carbohydrates using techniques that are only moderately modified for particular parasites. The partially purified material is then purified to homogeneity by centrifugation in caesium chloride which separates the nucleic acids by buoyant density. Due to their respective buoyant densities, proteins do not enter the gradient while DNA is 'buoyant' in the middle and RNA pellets at the bottom. Proteinase, DNAase or RNAase treatment of the RNA or DNA respectively can be done to further ensure purity.

The purity of the preparation can be verified by agarose gel electrophoresis and staining with ethidium bromide. DNA macromolecules are extremely large and run very slowly in agarose gels and can be detected near the origin while messenger RNA (mRNA) is much smaller, migrates more quickly and is seen as a smear behind the migration front. Ribosomal RNA (rRNA) is easily detected in two or three discrete bands near the middle of the gel while transfer RNA (tRNA) is hard to visualize. Messenger RNA can be separated from tRNAs and rRNAs by purification using affinity chromatography since only mRNAs have a poly-adenosine tail and this can be bound to poly-thymidine columns and subsequently eluted in a highly purified form.

Genomic DNA cannot be studied in its macromolecular forms and has to be digested into manipulable lengths using endonucleases which cut the DNA macromolecule at highly specific base sequences. The digestion fragments can be separated according to their size (i.e. molecular mass) by electrophoresis in agarose gels and visualized with ethidium bromide. Since an organism's genomic complement and organization is unique to that particular organism, genomic digests reveal banding patterns characteristic of each organism and these can be used as 'genomic fingerprints'. Particular DNAs can be identified by the use of probes, for example, oligonucleotides with a known sequence complementary (consensus sequence) to a specific gene, and screened against various DNAs using a technique known as southern blotting. In this simple technique, DNA is blotted onto a nitrocellulose or nylon membrane by the capillary flow of the transfer buffer and dried to irreversibly bind the DNAs to the durable membrane and give an exact and manipulable copy of the separated DNAs. Par-

asite DNAs complementary to the probe sequence will hybridize to the probe and, for example, when a radiolabelled probe is used, the corresponding parasite gene can be identified by detection of radioactive DNAs through exposing the blot to photographic film (autoradiography). Blots may even be re-probed, after extensive washing, with a variety of gene probes and all the homologous genes identified. Furthermore the ionic strength of the hybridization buffer can be used to check the degree of homology between the probe and parasite DNAs, and varying levels of stringency used to unequivocally identify specific DNAs. Therefore specific DNAs can then be identified, cut from the agarose gel and eluted.

RNA is more difficult to separate but it can be transferred to membranes by a technique known as northern blotting. In parasites, mRNA is often easy to obtain since it is present in large quantities. Indeed, parasites actively synthesizing a particularly abundant protein, for example, a surface antigen or key enzyme, may contain multiple copies of the corresponding mRNA but perhaps only one copy of the genomic DNA.

The use of mRNA may therefore have advantage over DNA and, by using the enzyme reverse transcriptase, the latter can be produced from the former and is called complementary DNA (cDNA). Complementary DNA, therefore, only represents the genes being expressed by the parasite at the time of the preparation of the mRNA and can, for example, be used to investigate differential gene expression during the parasite's life cycle. Indeed, parasite mRNA can be translated in cell-free systems (e.g. wheat germ or rabbit reticulocyte lysate) exogenously supplemented with radioactive amino acids to identify the translation products.

DNA and RNA can be sequenced by a variety of recombinant methods or by amplification in the polymerase chain reaction method and these aspects are discussed below.

Molecular techniques, therefore, provide everything that is necessary to study the ways a parasite controls gene expression and by manipulating the three components, DNA, RNA and polypeptide, it is possible to ask and to answer questions unthinkable a decade ago. The rest of this chapter is largely concerned with the ways in which mol-

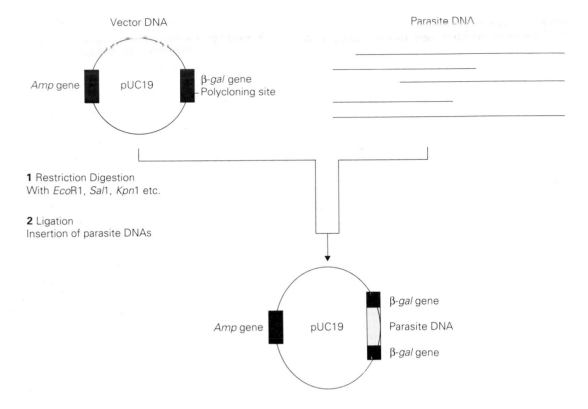

1 Restriction Digestion
With *Eco*R1, *Sal*1, *Kpn*1 etc.

2 Ligation
Insertion of parasite DNAs

3 Transformation of Host Bacteria and Establishment of Gene Library
Growth of transfectants on nutrient agar plates (containing Ampicillin and chromogenic substrate (X-*gal*))
   (a) Selection of transfectants (white colonies)
   (b) Selection of specific clones with gene probe

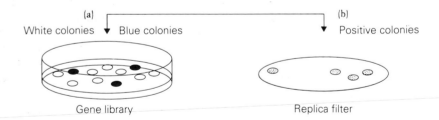

4 Manipulation of Recombinant DNA
Subcloning of specific clones
Sequencing of subclones
Mutagenesis of subclones etc.

Fig. 6.1 Molecular cloning strategies. *Amp*, ampicillin resistance gene; *β-gal*, *β*-galactosidase gene; X-*gal*, 5-bromo-4-chloro-3-indolyl-*β*-galactose. For an explanation see the text.

ecular techniques have been applied to some of these problems and how our understanding of the processes involved can be used to develop drugs, vaccines and diagnostic tools.

6.2.2 Gene cloning

Gene cloning involves four stages and can be performed directly upon parasite genomic DNA or DNA prepared from parasite RNA by reverse-transcription to yield cDNA. The various techniques used are discussed more fully in *A Laboratory Manual of Molecular Cloning* edited by Sambrook *et al.*

The four stages of molecular cloning are schematically shown in Fig. 6.1.

1 Digestion of the parasite DNA and cloning vector DNA. The foreign DNA is digested by specific restriction endonucleases into fragments that can be accepted by the cloning vector. The latter is also digested by the same restriction enzyme so that the ends of the parasite DNA and the cloning vector are compatible and can be ligated together.

2 Insertion of the parasite DNA into a cloning vector by ligation. The plasmids pBR322 and pUC19 are often used for parasite DNA fragments of fewer than 5000 base pairs. For larger double stranded DNA, *Escherichia coli* lambda phage can be used, and for single stranded DNA, *E. coli* phage M3 is used. Plasmids used as cloning vectors must possess markers that permit the selection of transformants. These genetic markers include ampicillin or tetracycline resistance, and histochemical markers such as enzymes that induce colour changes that can be visualized. Convenient plasmids, such as pUC19, are produced by genetic engineering, and there are now many 'custom designed' plasmids that provide almost unlimited possibilities for the insertion of parasite DNA. Plasmids with inserted promoters allow the controlled expression of mRNA thus facilitating studies on transcription and translation of parasite mRNA as well as the production of peptides by the host bacterium. Expression vectors, such as lambda gt11, are widely used to produce large quantities of fusion proteins, that

is, bacterial proteins associated with the foreign (i.e. parasite) protein.

3 Transformation of competent host bacterial cells with plasmid DNA. The recombinant plasmid with the parasite DNA inserted can be transfected into a suitable bacterial host in a number of ways. A simple method is to make the bacterium competent for the uptake of foreign DNA by calcium chloride hypotonic shock treatment after which up to 0.1% of the shocked cells take up the foreign DNA. The transfected or transformed bacteria are then grown on agar plates containing antibiotics to select transfectants and chromogenic dyes to select colonies that contain parasite DNA. A library of parasite genes can be created in this way and transformed cells with any desired parasite DNA can be selected by the use of specific DNA probes. Monoclonal antibodies (see Section 6.6) to the parasite protein can also be used to detect colonies that possess the required fusion protein and therefore the corresponding DNA.

4 Manipulation of the cloned DNA. Clones of specific parasite genes can be subcloned into another vector more suitable for DNA cloning so as to obtain the nucleotide sequence and, thus by reference to the genetic code, the amino acid sequence of the peptide of interest. This kind of molecular cloning is the most expedient way to obtain the primary structure of a protein since alternative protein (i.e. amino acid) sequencing techniques are laborious and can only be performed successfully in the absence of blocked or modified amino acids in the sequence. Nucleic acid and amino acid sequencing techniques are often used in a complementary fashion since partial amino acid sequences can be used to unequivocally confirm the identity of a putative gene identified by nucleotide sequence or, more expressly, to direct the synthesis of oligonucleotide primers in the search for the corresponding gene to a key parasite protein.

6.2.3 Site-specific mutagenesis

An invaluable technique for the investigation of the structure or function of a particular parasite protein is the use of site-specific mutagenesis.

This involves the introduction of specific mutations of the triplet code into the original sequence by means of specifically designed mutant oligonucleotide primers that can be annealed to the single stranded DNA in, for example, an M13 vector. This results in the creation of a heteroduplex consisting of both the homologous and mutant sequences which can be replicated in a suitable host bacterium. This provides a tool whereby the transcription and translation, as well as the function, of the mutated gene can be investigated and offers opportunities for the study of many aspects of parasite molecular and cellular biology such as the mechanisms controlling infectivity, drug action and drug resistance.

6.2.4 DNA sequencing

Once the parasite gene has been cloned or subcloned, or simply amplified by the polymerase chain reaction (see Section 6.2.5), it can be sequenced by chemical or enzymic methods. Enzyme methods provide the maximum amount of information for the minimum amount of manipulation. With the increase in commercially available vectors, capable of producing large

(a)

Template sequence with T7 primer hybridized
5' CTATAGTGAGTCGTATT**AAGCTAGCTAGCTAGCTAGG**..........3'
       |||||||||||||||||
5' GATATCACTCAGCATAA

DNA polymerase 1
+ adenine, thymidine and guanine
+ cytidine and dideoxycytidine*

Fragments produced
1 GATATCACTCAGCATAA**TTC***
2 GATATCACTCAGCATAA**TTCGATC***
3 GATATCACTCAGCATAA**TTCGATCGATC***
4 GATATCACTCAGCATAA**TTCGATCGATCGATC***
5 GATATCACTCAGCATAA**TTCGATCGATCGATCGATC***
6 GATATCACTCAGCATAA**TTCGATCGATCGATCGATCC***

(b)

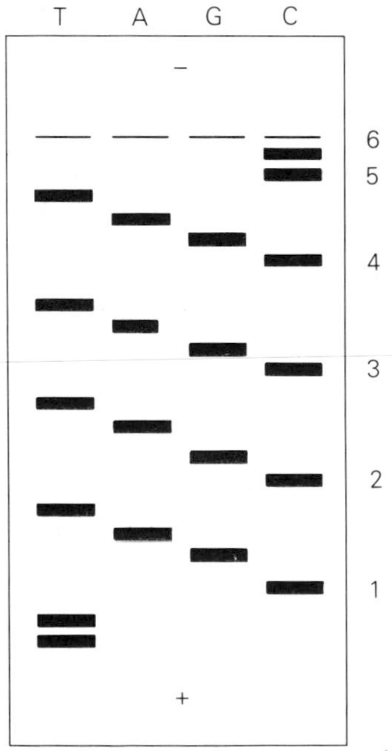

Fig. 6.2 Gene sequencing procedure. Read the sequence gel from the bottom up (since the smallest fragments migrate furthest) and compare it to the original sequence **TTCGATCGATCGATCGATCC**. The six dideoxyguanine fragments are numbered. For further explanation see the text.

amounts of single stranded DNA together with complementary primers, the enzyme method has become the most widely used sequencing route.

In outline, this method involves the replication of DNA, by DNA polymerase 1, from the start primer using the cloned parasite single stranded DNA as a template. The 'clever trick' in the method is the inclusion of radiolabelled dideoxy analogues of the four nucleotide bases which, because they lack the 3'OH group used in the phosphodiester linkage, stop the polymerase whenever they are incorporated into the sequence. This therefore produces a series of fragments varying in length but all terminating in a dideoxy nucleotide of guanine (G), cytosine (C), thymine (T) or adenine (A).

Thus from four separate reaction mixtures, each containing only one of the four complete sets of radioactive dideoxy analogues, it is possible to produce four complete sets of all the possible fragments generated by the template. These can be separated on a polyacrylamide gel and the terminating bases detected by autoradiography. The template sequence can be read from the direction of the smallest fragments (see Fig. 6.2a) and up to several hundred bases can be separated and read on polyacrylamide gels (Fig. 6.2b). Even large genes can be sequenced easily particularly if automatic sequencers, that use laser beams to read the four tracks simultaneously on continuously running gels, are used.

6.2.5 DNA amplification

A powerful cloning strategy which permits the selection and sequencing of specific genes is the polymerase chain reaction (PCR). This involves the *in vitro* amplification of DNA selected by specific primers. The primer is hybridized to, for example, the start and end of a small gene and the nucleotide sequence between the primers serves as a template for the enzyme DNA polymerase which doubles the number of copies of the selected DNA in each cycle of a chain reaction. As each copy can itself serve as a new template, a million fold amplification can be achieved in only 20 cycles! The PCR cycle involves (1) heat denaturation of the double stranded template, (2) primary annealing to the 3' and 5' regions of the

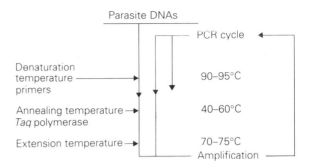

Fig. 6.3 Polymerase chain reaction (PCR) gene amplification. For an explanation see the text.

target DNA and (3) enzymic extension of the primer specified DNA (Fig. 6.3).

The use of the thermostable DNA polymerase (Taq polymerase) and automated thermal cycling devices has established PCR as a robust and routine technology with considerable potential and one that will be increasingly used not only in experimental work but also in rapid and accurate diagnosis which could be based upon minute quantities of parasite material.

6.3 GENE EXPRESSION AND REGULATION

Gene expression is the phenotypic manifestation of genome and is usually seen as the production of a particular protein, such as an enzyme. In eukaryotic cells, several genes operate together and are either activated or repressed in response to internal or external signals. Parasites are seemingly no different from other eukaryotic cells in this respect but some, particularly the trypanosomes, have evolved novel mechanisms that have actually changed the way in which molecular biologists and geneticists view the process of gene expression. In this section some of the fascinating examples of trypanosomal molecular biology and molecular genetics are elaborated.

6.3.1 Antigenic variation in trypanosomes

The African trypanosomes have complex life cycles involving phases in the tsetse fly vector and in the mammalian host, and the expression

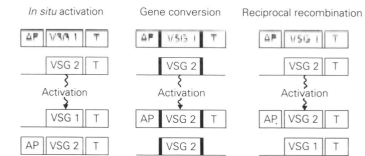

Fig. 6.4 Mechanisms of gene activation involved in trypanosome antigenic variation. VSG gene; AP, active promoter; T, telomere; ■, homology regions. For an explanation see the text.

and regulation of stages with specific genes such as those for cell surface coat procyclin in the insect vector and the variant surface glycoprotein (VSG) in the blood. Trypanosomes are able to survive in their mammalian hosts by producing a sequence of variant surface antigens (see Chapter 8, Section 8.5.4) and the regulation and expression of these glycoproteins has been intensively studied. The African trypanosomes possess some 1000 VSG genes and the highly ordered expression of these genes is regulated at several levels.

6.3.2 Telomeric gene expression

The ends of eukaryotic chromosomes are known as telomeres and these serve as the principal expression sites for genes and in trypanosomes the telomeric location of a particular VSG gene correlates directly with its expression. The mechanism whereby a VSG gene is placed in a telomeric site therefore represents the first level of regulation. Trypanosomes possess more than 200 telomeric expression sites although fewer than 20 of these are used to express surface antigens and only one site is active at any one time. The mechanism of telomeric activation is thought to involve one of three mechanisms, in situ activation, gene conversion or reciprocal recombination (Fig. 6.4).

6.3.3 Discontinuous transcription of mRNA

Another level of regulation of gene expression in trypanosomes occurs once the gene has been placed in the telomeric expression site and involves the transcription of the gene by RNA polymerase to produce mRNA. This transcriptional process is discontinuous as found in all the kinetoplastid flagellates examined. The mechanism of discontinuous transcription involves the co-transcription of two precursor RNAs, one derived from a common mini-exon or spliced leader exon and the other from the main exon, followed by a process of trans-splicing. Trans-splicing involves the addition of a small conserved RNA sequence to the 5′ end of transcribed RNA in such a way that there is a unique segment of the mRNA that is not found in the corresponding gene (Fig. 6.5a). It was originally thought that trans-splicing only occurred in kinetoplastid flagellates but it is now known to occur also in a number of nematodes including *Onchocerca volvulus*, *Ascaris lumbricoides* and *Brugia malayi* and mini-exon type sequences have been found in several protozoa and nematodes (Fig. 6.5b).

The biological significance of this process is unknown but its occurrence in these diverse groups of parasites makes it a possible target for the development of a novel type of gene chemotherapy which could exploit the essential transcription of mini-exon genes in effective drug design. The recent use of antisense oligonucleotides as antitrypanosomal agents is a rather remarkable example of how this could be achieved.

Antisense DNAs and RNAs, that is, 3′–5′ sequences rather that the normal 5′–3′ sequences, have considerable potential as specific inhibitors of gene expression. In nature bacteria have been shown to use antisense RNAs and DNAs to regulate gene expression, and *in vitro*, constructed plasmids coding for antisense RNAs have been used to selectively inhibit gene expression in *E. coli*, *Xenopus* oocytes, *Drosophila* embryos and transfected mammalian cells.

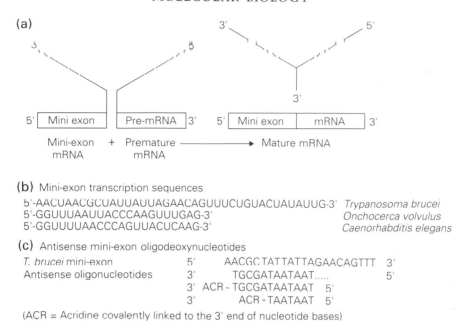

(a)

| 5' | Mini exon | Pre-mRNA | 3' |

5' Mini exon mRNA 3'

Mini-exon mRNA + Premature mRNA ⟶ Mature mRNA

(b) Mini-exon transcription sequences

5'-AACUAACGCUAUUAUUAGAACAGUUUCUGUACUAUAUUG-3' *Trypanosoma brucei*
5'-GGUUUAAUUACCCAAGUUUGAG-3' *Onchocerca volvulus*
5'-GGUUUUAACCCAGUUACUCAAG-3' *Caenorhabditis elegans*

(c) Antisense mini-exon oligodeoxynucleotides

T. brucei mini-exon	5'	AACGC TAT TAT TAGAACAGTTT	3'
Antisense oligonucleotides	3'	TGCGATAATAAT.....	5'
	3' ACR - TGCGATAATAAT	5'
	3' ACR - TAATAAT	5'

(ACR = Acridine covalently linked to the 3' end of nucleotide bases)

Fig. 6.5 A model for discontinuous transcription of trypanosomatid mRNA. For an explanation see the text.

The 35 nucleotide mini-exon at the 5' end of all trypanosomal mRNAs is a prime target for oligonucleotide directed therapeutic intervention since the transcription of mini-exons is essential for gene expression. Similarly the occurrence of an analogous mini-exon sequence in nematodes suggests that antisense agents may have wide antiparasitic potential.

Recent work has indeed demonstrated the effective inhibition of both *in vitro* translation of trypanosomal mRNA and *in vivo* gene expression in cultured trypanosomes using antisense mini-exon oligodeoxynucleotides (Fig. 6.5c). Antisense oligonucleotide therapy is not without major limitations and the uptake of oligonucleotides to date is low and degradation by parasite exonucleases high. Modified oligonucleotides, such as the acridine derivatives of the antimini-exon nucleotides, have proved to be most active.

6.3.4 RNA editing in kinetoplastid flagellates

The kinetoplast, which is characteristic of members of the order Kinetoplastida (see Chapter 1, Section 1.4) is a mitochondrion-associated organelle composed of about 50 maxi-circles and 10 000 mini-circles of mitochondrial DNA. The maxi-circles encode several mitochondrial proteins but the function of the mini-circles is less clear. However both maxi-circle and mini-circle DNA encode a unique molecule, 'guide RNA' (gRNA), which is involved in the regulation of gene expression.

In kinetoplastid flagellates, the mitochondrial genome is very condensed and specific genes are much shorter than those of other eukaryotes. These condensed genes are called 'cryptogenes' and when they are transcribed they produce cryptic premature mRNAs that require editing into mature mRNA before translation. This RNA editing involves the addition, or less frequently deletion, of non-coded ribonucleotide residues such as uridylyl to the cryptic RNA. This is thought to be carried out by a hypothetical 'editsome' particle, which possesses the enzymic activity for cleavage, addition, deletion and ligation, directed by the gRNA. The extent to which kinetoplastid mRNA undergoes RNA editing seems to correlate with the complexity of the parasite's life cycle and the changes that occur in mitochondrial respiration, for example subunit III of the respiratory cytochrome oxidase is extensively edited in African trypanosomes compared with their crithidial and leishmanial counterparts.

The functional changes that result from RNA editing are not known but some clues can be derived from other organisms. For example, in plant chloroplasts, and in other eukaryotic systems, insertions, deletions and modifications of bases are used to: (1) generate new start and stop codons; (2) change the DNA sequence so that the codon for one amino acid is switched to the codon for another; and (3) change the sequence to another reading frame.

Once again the discovery of such unique molecular biology editing processes presents possibility of exploiting the editing mechanism as a target for chemotherapy.

6.3.5 Transfection systems in kinetoplastid flagellates

Transfection in eukaryotes involves the uptake of foreign DNA and its incorporation in the genome of the recipient cell. Transfection systems developed for yeast and mammalian cells have added considerably to our understanding of eukaryotic molecular genetics. Transfection is simple and involves the electroporation of foreign DNA, a technique that exploits the transient induction of permeability in membranes when cells are pulsed with a high voltage, for example 3.0 kV per cm, for short periods of about 10 ms.

The development of transfection systems for parasites has been hindered by the lack of a suitable vector which could carry foreign DNA and our poor understanding of parasite promoters and terminators. Several possible vectors have been suggested including viruses that occur naturally in some parasites such as *Leishmania* or the extrachromosomal plasmid-like DNA amplified in *Leishmania* under drug pressure (see Section 6.7.3).

An alternative approach is to construct recombinant DNA custom designed plasmids containing the correct promoter, termination, mini-exon and polyadenylation sequences and this has been done for several *Leishmania* parasites. Stable transfectants of *L. enriettii*, a natural parasite of guinea pigs, have been obtained when the gene that confers resistance to the antibiotic neomycin was introduced. Leishmanial cell lines have been derived which are resistant to neomycin and possess stable extrachromosomal plasmid-like DNA which is transcribed to produce neomycin resistant mRNA correctly polyadenylated, trans-spliced and translated. The 'methotrexate amplification' plasmid-like DNA from *L. major* has also been used and, using this system, and the gene for neomycin resistance, stable transfection efficiencies approaching those obtained with mammalian systems have been achieved.

The major disadvantage of extrachromosomal transfectants is that the introduced gene is not incorporated into the nuclear DNA, nevertheless they do offer the opportunity to study many aspects of parasite biology such as the mechanisms of drug sensitivity and resistance.

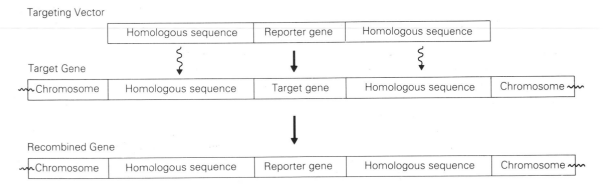

Fig. 6.6 Homologous gene recombination. For an explanation see the text.

6.3.6 Homologous gene recombination

Homologous gene recombination involves the replacement of a particular gene with an alternative DNA which has upstream and downstream homology to the original. For example, a reporter or mutated gene could be incorporated into the genome (Fig. 6.6). This system is widely used in yeast and mammalian molecular genetics and has yielded vast amounts of information particularly about gene recombination events (e.g. transfer of drug resistance genes). There is now evidence that homology gene replacement can be used to successfully transfect parasitic protozoa. Neomycin-resistant genes have been introduced into the genome of trypanosomes by transfection. The recombinant construct of the neomycin gene had homologous flanking sequences to the trypanosomal gene for tubulin.

Transfection of parasite genomes will allow the dissection of genes and the establishment of the role of any particular gene, for example, the 'molecular genetics' basis of transmission or infectivity. Such molecular dissection techniques should also aid in the identification and exploitation of key genes that could serve as targets for drug or vaccine design.

6.3.7 Genetic exchange in malaria parasites and trypanosomes

Genetic exchange by recombination of genes is characteristic of sexually reproducing organisms and from a practical viewpoint is important in the spread of drug resistance. This has been most clearly shown in malarial parasites where the genes for chloroquine and pyrimethamine resistance are independently inherited and are passed on to hybrids of resistant and susceptible strains thus facilitating the spread of resistance. Genetic exchange in asexually reproducing organisms is more difficult to understand but there is now evidence that this does occur in trypanosomes and possibly other asexually reproducing protozoan parasites.

The evidence for genetic exchange in trypanosomes comes from studies on isoenzymes (see Section 6.9.1) and genomic DNA in which the patterns seen are compatible with the exchange

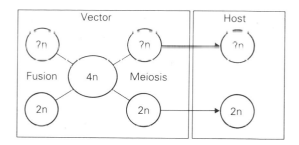

Fig. 6.7 A simple model of genetic exchange in trypanosomes. For an explanation see the text.

of genetic information. Indeed, studies on cloned lines of *Trypanosoma brucei*, which produce hybrid progeny, demonstrate unequivocal evidence of recombination progeny when passaged through tsetse flies (Fig. 6.7).

The existence of genetic exchange in trypanosomes is of considerable importance in disease control since it offers the parasite endless possibilities for the transfer of genes for novel antigenic variants or for new mechanisms of drug resistance.

6.4 CHROMOSOMES AND GENE MAPPING

Chromosomes are the condensed structural arrangement of an organism's genes and constitute a macromolecular mechanism to regulate differential gene expression since gene expression occurs preferentially at the ends of chromosomes known as telomers. Genes recombined in these sites will be highly expressed and much of what is known about the molecular genetics of higher eukaryotic organisms has been based upon a detailed understanding of structural arrangement of genes in the chromosome (e.g. the mapping of genes on particular chromosomes) and the molecular mechanism to shuffle genes into and out of the telomeric sites (e.g. molecular genetics of recombination).

The best studied examples in humans (and mice) are the histocompatibility genes, however, these will soon be superseded by the Human Genome Project in which, in a concerted effort supported by most of the major research funding organizations in the world, the aim is to map the entire human genome by the year 2000. Interestingly, in a detailed on-going study of the free-living nematode *Caenorhabditis elegans* the genes of helminths are being mapped. Chromosomal mapping in parasitic helminths should prove to be relatively easy since they possess between 3 and 12 condensed chromatid structures but so far there has been little interest in the construction of such gene maps.

In free-living and parasitic protozoa the condensation of chromosomes is a seemingly rare molecular event and the absence of condensed chromatid structures has until recently precluded chromosomal mapping. The advent of chromosome separation techniques which use pulsed field gradient gel electrophoresis (PFGE) has facilitated the study of protozoan chromosomes. This technique basically involves the electrophoresis of DNA in an electric field that is constantly changed in direction causing the chromosome-sized networks of DNA to migrate according to their size, shape and flexibility. With a number of modifications, PFGE has been applied to parasite DNA and is capable of separating chromosomes between 10 and 10 000 kilobases (kb). It has therefore been particularly useful for parasitic protozoa which have chromosomes in the region of 50–3000 kb.

PFGE has suggested that the malaria parasite, *Plasmodium falciparum*, has at least 14 chromosomes and detailed gene maps of most of these chromosomes have now been prepared. Of particular importance is the identification of a single locus for chloroquine resistance gene(s) on chromosome 7, a *P. falciparum* version of the multi-drug resistance gene (pfmdr) on chromosome 14 and genes for the expression of several key surface antigens on chromosome 5. Such genomic studies on *Plasmodium* should therefore aid in the elucidation of essential parasite adaptation such as host cell recognition and invasion (e.g. adherence to endothelial cells), and drug resistance, as well as antigen presentation and variation.

Studies on the chromosomes of *Leishmania* have been less rewarding and have raised more questions than they have solved. It is thought that there are about 16 leishmanial chromosomes but this is not easy to confirm because of the considerable degree of polymorphism that exists in this genus. It also appears that *Leishmania* may not obey the normal rules of eukaryotic genetics once again emphasizing the molecular genetics interest in the kinetoplastid flagellates.

In the African trypanosomes, the exact number of chromosomes is even less clear, however it is known that there are a number of groups of chromosomes, large, intermediate and small. Within the group of small chromosomes there are about 100 mini-chromosomes which carry genes for the surface coat and are thought to provide a virtually unlimited source of telomeric gene expression sites that are essential for antigenic variation (see Section 6.3.1).

Molecular genetics studies are therefore likely to become of unquestionable importance as it becomes possible to pinpoint genes which govern such factors as virulence, susceptibility and resistance to drugs and, most importantly, to determine the geographic distribution and spread of particular genetic strains providing vital information for epidemiologists.

6.5 POLYPEPTIDE ISOLATION AND IDENTIFICATION

No discussion of molecular parasitology can be complete without some consideration of the protein gene products. Since DNA encodes RNA encodes protein (polypeptide) it is a conventional dictum that for every gene there is a gene product and that, since only a fraction of the genes are expressed at any one time, the identification of polypeptides is important in the understanding of molecular parasitism. It follows that if the sequence of a gene is known the amino sequence can be deduced and, *vice versa*, if a polypeptide is identified the gene can be studied. The study of both polypeptides and genes has been facilitated by *in vitro* techniques to synthesize oligopeptides and oligonucleotides cheaply and in large quantities.

Gene products are particularly important since in parasitology they constitute the enzymes that are putative targets for drug attack and also represent the main antigens recognized by the immune system as well as the cytokines that act as signals between the cells of the immune system. Parasite enzymes and surface antigens that are very different from those of the host are particularly important as these are the most likely targets for specific drug action and vaccine design.

The isolation and identification of parasite enzymes is relatively easy as they can be released from lysed protozoal cells or organelles and characterized in a number of well-defined ways. Structural proteins, isolated from whole parasites or, more usually, membranes, can be cleaved at specific amino acids to produce relatively small peptides which can be separated by electrophoresis. The various peptides then separate out according to their molecular mass and can be visualized by staining with Coomassie blue, or a similar dye. As many peptides have the same molecular mass but vastly different charges, and therefore structures, it is often better to use a two-dimensional gel in which the current is first run in one direction and then at right angles to it. This gives much better separation of complex proteins. Peptides of interest can be cut from the gels or transferred to a membrane for further study in a process known as western blotting. Individual peptides can be identified by labelling with radiolabelled probes that recognize particular amino acid sequences or with labelled antibodies specific to that oligo- or polypeptide. Western blotting techniques form the backbone of much of modern parasite immunology and this will be discussed further in Section 6.8.

6.6 MONOCLONAL ANTIBODIES

Monoclonal antibodies must be considered here because of the important role they have played in molecular parasitology studies. Antibodies bind specifically to appropriate protein antigens and can therefore, if labelled with a fluorescent dye, enzyme or radioactive marker, be used to identify particular polypeptides. Alternatively, antibodies can be used in affinity columns to bind polypeptides that can subsequently be eluted. Antibodies have, therefore, become essential tools for many biochemical and molecular parasitology studies.

Antibodies are usually raised by immunizing a suitable animal with the protein under investigation and, after a period, using the immune serum. The limitation is that antibodies obtained in this way recognize the whole protein and not particular components that might be of interest. Monoclonal antibodies recognize only a single peptide and are, therefore, much more specific. Monoclonal antibodies are raised by immunizing an animal with an antigen and then removing and separating individual lymphocytes, each of which is capable of producing antibody with a unique specificity. Lymphocytes do not divide so have to be fused with a dividing cell line which, when carefully selected, gives an endless supply of the antibody required. Because of their absolute specificity and ease and reproducibility of preparation, monoclonal antibodies are widely used to identify and purify particular polypeptides. They can, for example, be used to screen the products of recombinant DNA technologies, to identify antigens in or on the surface of the parasite or to concentrate polypeptides from a diluted source.

6.7 RATIONAL DRUG DESIGN

The principles of drug development are discussed in Chapter 9 (Section 9.5) and in this section the emphasis will be on the applications of molecular techniques to antiparasitic drug design and an integrated scheme is presented (Fig. 6.8).

The rational approach to drug design (see Chapter 9, Section 9.5.2) aims to exploit differences in the molecular structure and function between parasites and their hosts and, to this end, exploits the techniques of molecular biology and molecular graphics. There are many parallels between the design of drugs and the development of vaccines (see Section 6.8).

The first step is to identify the target molecule, usually a protein, and to determine its role in the life cycle of the parasite. Having done this, the second step is to characterize the molecule which, if it is a protein, can be done using a combination of biochemical and recombinant DNA technologies (see Section 6.5). If possible, the purified protein should be crystallized and the molecular coordinates determined by X-ray crystallography and molecular models, constructed using computer simulation and molecular graphics programmes.

Molecular models not only provide information about the structure of target molecules but can also be used to redesign existing drugs, as

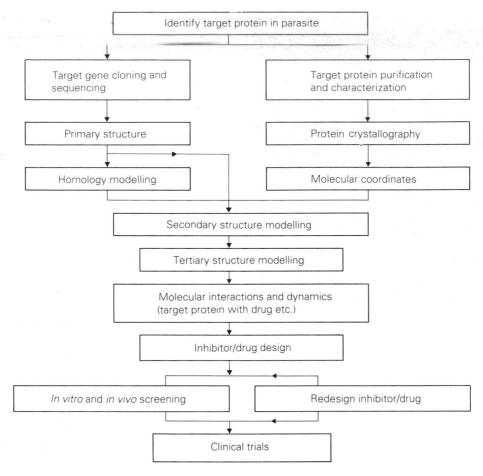

Fig. 6.8 Scheme for rational drug design.

in the case of nifurtimox (see Section 6.7.2). The use of these techniques is discussed in Sections 6.7.1 and 6.7.2 below and a more biochemical approach to our understanding of the action of the antimalarial drug, chloroquine, is discussed in Chapter 9, Section 9.4.1.

6.7.1 Antiglycosomal drug design

Glycosomes are organelles involved in glycolysis that are unique to the kinetoplastid flagellates and, therefore, obvious targets for drug design. The genes for the glycosomal enzymes have been cloned, their sequences determined and molecular models constructed from molecular coordinates obtained from both crystallographic studies and homology with known crystal structures. These molecular modelling advances have supported suggested mechanisms of action of the antitrypansomal drug, suramin. It has been suggested for some time that suramin may inhibit glycolysis and that the target may therefore be a glycosomal enzyme.

Molecular modelling of the glycosomal enzyme phosphoglycerate kinase (gPGK) suggests that this hinge-bending enzyme may bind the negatively charged drug suramin to positively charged surface motifs which straddle the active site and possibly block the hinge-bending mechanism. Indeed four positively charged residues (amino acids) on the surface of gPGK have molecular coordinates that could facilitate the binding of

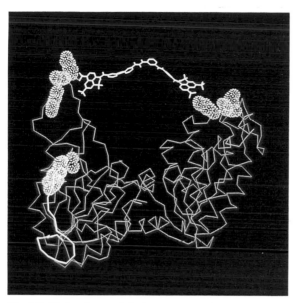

Fig. 6.9 Molecular model of trypanosomal phosphoglycerate kinase (PGK). The alpha-carbon model of PGK is shown at the bottom and the two-domain structure of the enzyme has the active site residing in the cleft formed by the carboxy-terminus (left) and the amino-terminus (right) domains. The positively charged amino acids that constitute the unique structural motif are shown as space fill residues and the antitrypanosomal drug, suramin, is shown forming a bridge between the two domains. The complementarity in structure between the positively charged sulphonic residues of suramin could result in the inhibition of the hinge-bending mechanism of action of PGK.

the negatively charged suramin molecule to the enzyme since both the surface motif and drug charges are some 40 Å apart (Fig. 6.9).

These positive motifs occur on several of the glycosomal enzymes in the *T. brucei* group of trypanosomes but are absent from other trypanosomatids. This may explain why suramin is effective against these African trypanosomes but not against other kinetoplastid flagellates.

6.7.2 Antitrypanothione metabolism drug design

Another example of a unique target is the mechanism for detoxification of oxygen metabolites and the maintenance of redox potential in try-

panosomes. Kinetoplastid flagellates lack the enzymes catalase and glutathione peroxidase which act as scavengers of toxic peroxides (see Chapter 8, Section 8.2.2) but do possess a unique counterpart, trypanothione reductase.

The gene for trypanothione reductase has been cloned and sequenced from several different trypanosomes and site-directed mutants have been constructed to identify the key amino acid residues that give the molecule its unique substrate specificity. Such studies have helped to elucidate, at least in part, the mode of action of a number of antitrypanosomal drugs (Fig. 6.10). For example, trypanothione is a dimer of glutathione linked by the polyamine spermidine which may therefore explain the trypanocidal effects of Eflornithine (difluoromethylornithine, DMFO) which inhibits ornithine decarboxylase, an enzyme involved in polyamine synthesis (see Chapter 9, Section 9.4.1).

Furthermore, pentamidine and berenil inhibit adenosine decarboxylase which produces decarboxylated s-adenosylmethionine, a key intermediate in spermidine synthesis, and the arsenical drug melarsoprol seems to actively sequester trypanothione and thus inhibits trypanothione reductase. Therefore perturbation of trypanothione metabolism seems to be the basis of the mechanism of action of a wide range of antitrypanosomal drugs.

This mechanistic knowledge has been exploited and used in the rational redesign of nifurtimox which is used in the treatment of the American form of trypanosomiasis, Chagas disease. Nifurtimox inhibits mammalian glutathione reductase and, as a result of the discovery of trypanothione reductase, has been redesigned by the addition of several basic side chains, to perturb more efficiently the trypanosmal counterpart.

6.7.3 Mechanisms of drug resistance

In the same way that molecular techniques can be used to provide information about the ways in which drugs work, they can also be used to help to elucidate the mechanisms of drug resistance with the overall aim of being able to prevent or reverse this ever growing problem. Drug resistance is discussed in Chapter 9, Section 9.4.3 and

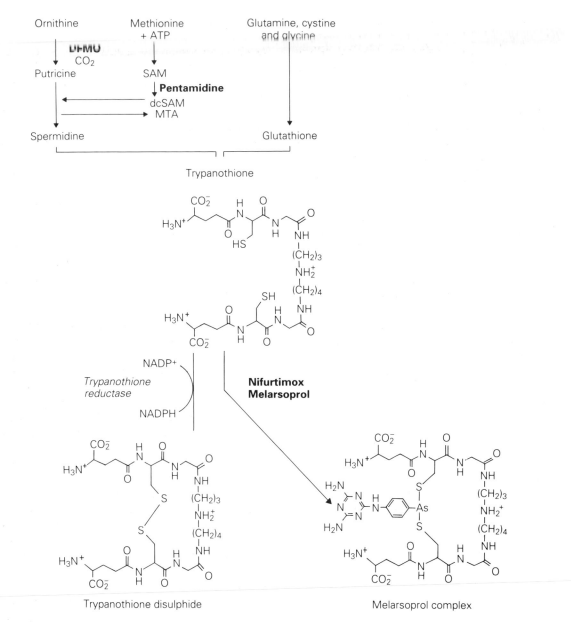

Fig. 6.10 Trypanothione metabolism. For an explanation see the text.

this section is more concerned with the application of molecular techniques.

Drug resistance can occur at several levels of molecular and cellular organization, e.g. a parasite may become resistant by reducing the effective accumulation of the drug either by selective exclusion or by active extrusion. If the drug is taken up by an energy dependent translocator or receptor mediated process, mutants defective in translocation activity would be resistant to the drug, as would mutants that were able to increase the extrusion of the drug by the over expression of genes controlling the efflux translocator.

One of the mechanisms involved in chloroquine resistance in malaria parasites is an increase in the efflux of the drug by an ATP dependent transporter. In order to be effective, chloroquine has to accumulate within the parasite but malaria parasites have been shown to efflux the drug by amplifying the gene for an active efflux transporter protein similar to those described as multi-drug-resistant proteins, known as P-glyco-proteins, in certain drug-resistant cancer cells. The existence of such a mechanism would explain cross-resistance to a range of diverse drugs and also the ability of certain calcium channel blockers to reverse resistance (since the co-transport of calcium is suggested to be involved in the efflux mechanism).

Chloroquine resistance, however, probably involves more than one mechanism (see Chapter 9, Section 9.4.3) and mutations of the novel malarial haemoglobin polymerase, which seems to be the target enzyme of the drug, is an obvious possibility. Nevertheless, several protozoa have amplified extrachromosomal DNAs that code for possible extrusion proteins suggesting that there may be a common extrusion mechanism of drug resistance operative within these parasites.

Another mechanism of drug resistance is the increased gene expression of the target protein in response to drug pressure. Folate is an essential cofactor in microbial metabolism and, the bifunctional enzyme, dihydrofolate reductase–thymidine synthetase, plays a key role in folate metabolism. Several effective antimicrobial drugs, including methotrexate, are effective inhibitors of this enzyme system. Resistance to methotrexate in *Leishmania* is due to the overproduction of this enzyme system and highly amplified, extrachromosomal circular DNAs (plasmid-like) containing many copies of the gene have been identified.

Interestingly, *Leishmania* seems to possess at least six multicopy circular DNAs that have arisen from gene amplifications and, as at least some of these are directly involved in drug resistance, this suggests that gene amplification may be a common mechanism of drug resistance in the leishmanias.

6.8 VACCINE DEVELOPMENT

Vaccines are vital components of any scheme to prevent or control disease and spectacular advances have been made in eliminating or containing many viral and bacterial infections. Nearly all the vaccines in current use are derived from killed or inactivated pathogens and one of the main challenges for molecular biologists is to develop effective, safe and cheap vaccines, produced either synthetically or by using recombinant technology. There is no doubt that this can be done and a recombinant vaccine against human hepatitis B is already available.

Theoretically, what is required is the identification of the protective antigen(s), determination of its structure and its production in a recombinant form. Most parasite antigens that are recognized by the immune system are surface molecules and, particularly for helminths, the first step is to isolate these using standard protein purification techniques. Not all surface molecules are, however, involved in protection and those that are can be recognized by exposing the parasite, or isolated surface molecules, to immune sera and identifying the surface proteins detected by the antibodies. This can be done by western blotting and subsequent exposure to labelled immune sera.

Once a potentially antigenic macromolecule has been identified, it can be eluted from the gel and characterized, used to raise monoclonal antibodies that will recognize the same antigen, and be used for affinity purification. Internal antigens can be identified in material sectioned for electron microscopy using antibodies that have been labelled with electron dense materials such as colloidal gold. These techniques have been used to provide vast amounts of antigen for experimental purposes but success in developing vaccines against parasites has been very limited.

Ideally, the gene for the antigen(s) should be inserted into a vector which would multiply in the host without itself causing any detrimental effects and suggested vectors include the vaccinia virus or attenuated strains of *Salmonella* both of which have been used experimentally with some success. Unfortunately, the parasite genome is extensive in surface antigen genes, parasites are antigenically very complex and the macromolecules that bring about protection are sometimes also involved in pathology and the evasion of the immune response. These aspects are discussed in Chapter 8 (Section 8.9) and here only the more relevant molecular aspects will be described.

The malaria parasite, *Plasmodium falciparum*, has received the most attention and all the theoretical requirements have been fulfilled. The candidate antigen, the sporozoite circumsporozoite antigen (see Chapter 8, Section 8.5.7) has been identified, the gene for it cloned and the recombinant product correctly expressed, but despite promising preliminary results the vaccine has been only partially protective. Other antigens from the malarial parasites have also produced equivocal results. However, it is encouraging that so much progress has been made in a very short period of time and the possibilities of a recombinant vaccine, even though it might not be completely protective, look good.

The most promising recombinant vaccine is one against the larval stages of the sheep tapeworm, *Taenia ovis* (see Chapter 8, Section 8.6.4). The antigen involved is derived from the onchosphere which is used as a source of the mRNA used to prepare a cDNA library. Conventional cloning methods failed to produce a protective antigen but a plasmid vector containing a schistosome glutathione-s-transferase was successful. This illustrates some of the difficulties involved in developing antiparasite vaccines and the continual need for novel approaches.

Recombinant vaccines against schistosomes are now undergoing trials with promising results. There is no doubt that recombinant vaccines will be the vaccines of the future but it will be some time before one is available for use against any human disease. In the meantime, a synthetic antimalaria vaccine, based on findings employing molecular techniques, is now undergoing human trials (see Chapter 8, Section 8.9).

6.9 MOLECULAR DIAGNOSIS AND TAXONOMY

Precise identification of the cause of an infection is an essential for treatment and plays a major role in our understanding of the epidemiology of

disease and the implementation of control measures (see Chapter 10, Section 10.2.4). Techniques for the identification of a particular parasite and the ability to distinguish it from another similar one bridge the gap between diagnosis and taxonomy and a variety of techniques have been used both for diagnosis and to establish evolutionary relationships between organisms. However, none of the techniques presently available is entirely satisfactory and molecular techniques are gradually being developed as alternatives.

6.9.1 Diagnosis

Diagnosis of parasitic infections has traditionally been based on finding the causative agent and this is still the most satisfactory method particularly when dealing with infections in individuals. However, this technique is not satisfactory if the parasites causing different diseases, or differing in pathogenicity, are morphologically indistinguishable as in the case of the African trypanosomes, the leishmanias and various kinds of amoebae. Serological methods, which rely on the identification of specific antibodies, are suitable both for individuals and populations but suffer from the major drawback that they often cannot distinguish between patent and past infections and also cannot be used routinely to distinguish between morphologically similar organisms.

One of the earliest biochemical techniques involved the use of isoenzymes. Isoenzymes are enzymes that perform the same function but have different mobilities when run in an electrophoretic gel. As such behaviour is genetically controlled, parasites that possess different patterns of isoenzymes must be genetically distinct and populations of parasites with the identical patterns are known as zymodemes. The technique used is very simple and using a battery of zymodeme typing enzymes, large numbers of samples from different isolates of parasites can be typed in parallel.

This method was first used extensively in studies on the different species and subspecies of rodent malaria parasites in which the characteristic isoenzyme patterns correlated well with morphological, biological and geographical characteristics thus providing the confidence required for the elucidation of more difficult problems. Most work has been done on separating the various species and subspecies of *Leishmania* and the findings correlate well with the less easily measured clinical manifestations.

Isoenzymes have been used to distinguish between the different forms of African trypanosomes and have helped to establish that *T. brucei gambiense* has a reservoir host, the pig. In the case of *T. cruzi*, it has been possible to distinguish at least three distinct zymodemes also differing in a number of biological characteristics such as the extent to which they infect humans. Isoenzymes are the only way to distinguish between potentially pathogenic and non-pathogenic strains of *Entamoeba histolytica*.

Diagnosis based on gene products, such as enzymes, is limited by the fact that these can vary from stage to stage of the infection. On the other hand, genomic DNA is dependably constant and specific diagnosis can be achieved by working directly with the parasite DNA. In its simplest form, the whole DNA is cut up, using particular restriction enzymes, and the fragments electrophoresed and the bands stained with ethidium bromide. This produces reproducible patterns of banding which have been used to characterize a number of protozoa, particularly the *Leishmania* complex where the DNA used is the kinetoplast mini-circle DNA (kDNA) which differs considerably between species. Such restriction endonuclease cleavage patterns are now used, together with the isoenzyme patterns, for the definitive identification of *Leishmania* species and subspecies and are required for the description of new taxa. Such techniques have also been used to distinguish between pathogenic and non-pathogenic species of the facultative amoebae of the genus *Naegleria*.

DNA probes (see Section 6.2) are being increasingly used in parasitological studies. One of the problems with these techniques has been the necessity to use radioactive labels, an insurmountable problem in developing countries and in field studies, but these can be replaced by non-radioactive techniques, for example enzymes that change the colour of a chromogenic substrate. The starting point is the selection of a characteristic DNA sequence to which to target the probe and this must be sufficiently specific to permit the level of discrimination required but not so specific as to produce too many fine distinctions. For example, in some circumstances it might be necessary simply to identify the presence of a member of the genus *Plasmodium*, in others, a diagnosis to the level of species might be required and in yet others it might be necessary to distinguish between strains of the same species.

DNA probes have been extensively used for the diagnosis of leishmaniasis. Initial studies which distinguished between the New World species, *L. braziliensis* and *L. mexicana*, have since been extended to all species. The technique is very simple and a small biopsy sample is applied to nitrocellulose paper, denatured and hybridized to a specific *Leishmania* probe and the degree of hybridization detected using one of a variety of labelling techniques. In general, sequences can be detected in lysed parasite material collected onto any suitable substrate.

DNA probes can also be used for the diagnosis of malaria and are particularly useful where only small numbers of parasites are present in the blood. A specific application is the identification of malaria sporozoites in squashes of mosquitoes and this has provided invaluable epidemiological data which would have been impossible to obtain in any other way. Commercially prepared probes are now becoming available for the identification of many parasites.

DNA techniques can be made even more sensitive if the amount available is expanded using the polymerase chain reaction and there is no reason why an absolute diagnosis cannot be made on a single parasite!

6.9.2 Taxonomic and systematic relationships

The determination of systematic relationships is of more than mere academic interest. The various taxonomic groupings described in Chapters 1, 2 and 3 provide a universally accepted framework which scientists use to communicate with one another. However, systematic relationships also imply genetic similarities and genetic similarities imply similar biochemical and immunological characteristics. For experimental work, it is, therefore, essential to know how closely one organism is related to another before any broad conclusions can be drawn and in formulating control measures it is important to know the extent to which data collected for one species or subspecies is relevant to another.

For many of the helminth worms and most vectors, morphological data can be used to establish relationships. However, in the case of species complexes such data is inadequate (see Chapter 3, Section 3.7) and it is necessary to resort to cytogenetics. For the protozoa, and other parasites with few morphological characteristics, data obtained at the genetic level is essential for any understanding of the relationships between genera, species or subspecies. The most useful taxonomic macromolecules are rRNA. Ribosomal RNA consists of large and small units, the larger consisting of two subunits, 28S and 5S, and the smaller of one 16S unit. The 5S RNA is the one most extensively used.

Ribosomal RNA mutates with evolutionary time and changes in the rRNA can be used to construct phylogenetic trees from which can be calculated evolutionary distances. It is not appropriate to discuss the details of molecular systematics here but the various investigations carried out have revealed few surprises, for example, the nematodes studied, such as *Brugia*, *Onchocerca* and *Nippostrongylus*, all belong to a close-knit group as do the African trypanosomes.

The one unexpected disclosure is that *Plasmodium falciparum* appears to be more closely related to the avian malarias than to the other species in humans and other mammals. *Giardia* turns out to be a very interesting organism having rRNA quite unlike that of any other eukaryote and nearer to that of prokaryotes implying a very early divergence from the main protozoal line. The practical implications of studies such as these are not immediately apparent but might well be used to identify divergent and putative targets for chemotherapy.

6.10 THE FUTURE OF MOLECULAR PARASITOLOGY

Molecular techniques have pervaded all aspects of parasitology and have allowed scientists to design drugs and vaccines targetted against specific macromolecules or metabolic pathways and have also facilitated diagnosis and made it possible to collect vast amounts of epidemiological data with economy of effort. Molecular taxonomy has become a recognized, and in some cases a required, procedure. Computers are already being used in all of these areas and the next stage will be the utilization of the data collected in traditional and novel ways. Within a decade, molecular parasitology has become an established field and over the next decade the fruits of the discoveries in this area should begin to become available.

REFERENCES AND FURTHER READING

Alberts, B., Bray, D., Lewis, J., Raff, M., Roberts, K. & Watson, J.D. (1989) *Molecular Biology of the Cell*, 2nd edn. New York, Garland Publishing.

Darnell, J., Lodish, H. & Baltimore, D. (1990) *Molecular Cell Biology*, 2nd edn. New York, Scientific American Books.

Hyde, J.E. (1990) *Molecular Parasitology*. Milton Keynes, Open University Press.

Sambrook, J., Fritsch, E.J. & Maniatis, T. (1989) *A Laboratory Manual of Molecular Cloning*, 2nd edn. New York, Cold Spring Harbour.

Stryer, L. (1988) *Biochemistry*, 3rd edn. New York, W.H. Freeman.

Walker, J.M. & Gaastra, W. (1983) *Techniques in Molecular Biology*, London, Croom Helm.

Walker, J. M. & Gaastra, W. (1987) *Techniques in Molecular Biology*, Vol. 2. London, Croom Helm.

Watson, J.D., Hopkins, N.H., Roberts, J.W., Steitz, J.A. & Weiner, A.M. (1987) *Molecular Biology of the Gene*, 4th edn. Menlo Park, Benjamin/Cummings.

Chapter 7 / Physiology and Nutrition

L. H. CHAPPELL

7.1 INTRODUCTION

In this chapter, a broad-brush view of parasite physiology is painted, a landscape rather than a portrait, in the school of Cézanne not Gainsborough; the aim is to leave the reader with an impression of this exceedingly broad topic at the inevitable expense of detail. Particular attention will be paid to the ways in which the physiological attributes of parasites impinge on applied aspects of the discipline and which may contribute ultimately to the regulation of parasitic diseases by human intervention.

Parasites possess unique physiological attributes that are a consequence of living in a hostile and ever-changing environment, the body of the host, with the associated necessity of passing through the external environment, possibly on more than one occasion in a single life cycle. It will become apparent to the student of parasitology that an understanding of parasite physiology can also make a fundamental contribution to parasite control.

7.2 SURVIVAL OUTSIDE THE BODY OF THE HOST AND TRANSMISSION

Many parasites release stages of their life cycles into the external environment, these include eggs, cysts and larvae. Such transmission phases are sometimes short-lived but they may possess considerable longevity and the ability to resist environmental degradation.

Parasite transmission between hosts is accomplished in one of three ways:

1 The host ingests eggs, cysts, larvae or an intermediate host containing infective stages of the parasite.

2 The host is inoculated with infective parasites during blood feeding activities of the intermediate (vector) host.

3 The host is actively sought and penetrated or settled by invasive parasite (see Table 7.1).

7.2.1 Cystic stages of parasites

Protozoa parasitic in the alimentary canal, such as *Entamoeba histolytica*, *Giardia duodenalis* and *Balantidium coli*, commonly produce cysts that contain quiescent, infective forms which are passed into the external environment with host faeces and await ingestion by the next individual host. *Naegleria fowleri* is a free-living organism which can cause human amoebic meningitis. It is acquired during bathing in, or by contact with, warm natural freshwaters. This organism, an opportunistic parasite, occurs in three forms, trophozoite, cyst and flagellate but in this case, the trophozoite and not the cyst is the infective stage, entering the human body via the nasal mucosa and migrating to the brain. Coccidian protozoans form resistant, infective cysts. Species of economic importance, such as *Eimeria* in poultry and *Toxoplasma* in cats, produce oocysts containing sporocysts which are voided in host faeces and are acquired by a new host during feeding.

Amongst the helminth parasites, encysted stages concerned with transmission are common. In the Digenea, cysts containing the metacercarial stage may be found either in the external environment (e.g. on vegetation in the liver fluke, *Fasciola hepatica*) or within the body of an intermediate host (e.g. *Opisthorchis sinensis* in freshwater fish). These cysts are complex in structure and serve as resistant hypobiotic stages in the parasite life cycle; the conditions for excystation

Table 7.1 Transmission stages of parasites: a synopsis

	Parasite eaten by host	Arthropod vector transmits parasite	Parasite actively locates host
Protozoa	Coccidian cysts Amoebic cysts	Babesias Trypanosomes Malaria	–
Monogenea	–	–	Oncomiracidium
Digenea	Metacercaria Cercaria (some)	–	Miracidium Cercaria
Cestoda	Egg Cysticercoid Cysticercus Coenurus Hydatid cyst Coracidium Procercoid Plerocercoid	–	–
Acanthocephala	Egg Cystacanth	–	–
Nematoda	Egg Larvae (L_3) (ensheathed or free)	Larvae (microfilariae)	Larvae (e.g. hookworms)

are described later in this chapter (Section 7.3.2).

The encysted larvae of tapeworms are found only within the body of an intermediate host animal. These cysts may contain a single parasite larva (e.g. cysticercus) or contain an increasing number of larval worms produced by asexual internal budding of germinal tissue within the cyst itself (e.g. hydatid cyst of *Echinococcus granulosus*).

Several genera of nematode parasites form cysts involved in transmission. *Trichinella spiralis* larvae encyst in mammalian muscle and the bird gapeworm *Syngamus trachea* may encyst in the haemocoel of the invertebrate transport host. Plant parasitic nematodes produce cysts with remarkable capacity to avoid environmental desiccation: *Heterodera rostochiensis* can survive for up to 8 years within the cyst.

The life cycle of acanthocephalans typically includes an encysted cystacanth stage which resides within the haemocoel of the arthropod intermediate host; a second phase of encystment may take place if this host is ingested by a transport host in which development of the parasite to maturity cannot be achieved.

7.2.2 Helminth eggs

The shelled egg of helminth parasites is ideally suited as a transmission stage and, by virtue of the physico-chemical properties of its resistant shell, is able to withstand the rigours of the external environment for considerable periods of time.

The digenean egg is formed from 30 to 40 vitelline cells and a fertilized ovum surrounded by a protein shell. The phenol oxidase system of enzymes carries out complex cross-linking of these proteins in a process known as tanning. Early studies on *Fasciola* led to the suggestion that all digenean eggshells comprised tanned sclerotin formed by quinone tanning. More recent information, however, indicates that a variety of structural cross-linked proteins may confer upon the egg its rigidity and resistant properties, including sclerotin, keratin and elastin. In schistosomes, use of recombinant DNA techniques has identified female-specific gene products associated with eggshell production; vitelline proteins are rich in certain amino acids (glycine, tyrosine, aspartic acid, lysine and histidine) agreeing with amino acid analysis data for the shell itself. The vitelline peptides of *Fasciola* are rich in dihydroxyphenylalanine (DOPA), tyrosine, aspartate or asparagine, glycine and lysine or arginine. Tyrosine is thought to be converted to DOPA at the post-translation stage before these vitelline proteins are exported from the vitelline glands. Therefore, the mechanical strength of the eggshell derives from tanning of the tyrosine-rich vitelline proteins, involving DOPA formation and subsequent quinone production by phenol oxidase to yield cross-linked proteins. DOPA-rich proteins are common components of biological 'glues' in a variety of free-living animals.

The eggshells of parasitic nematodes are different from those of platyhelminth parasites, being generally more complex in construction and containing non-proteinaceous structural components like chitin. This polymer of *N*-acetyl glucosamine is structurally important in fungi and in the exoskeleton of many invertebrates. Nematode eggshells are typically triple-layered comprising an inner lipid layer, a middle chitinous layer and an outer proteinaceous vitelline layer. Some nematodes, including the filarial nematodes, are ovoviviparous and do not release the egg from the uterus; even in these forms, with a much diminished eggshell, chitin is still present but its role is in doubt.

7.2.3 Mechanisms for locating the host

Active location of the intermediate or definitive host is carried out by a variety of larval helminth parasites including monogenean oncomiracidia, digenean miracidia and cercariae, cestode coracidia and infective L_3 nematode larvae. With the exception of the coracidium, all of these different larvae actively seek out and either attach to or penetrate the next host in the life cycle. The physiological mechanisms underlying the processes of finding a suitable host remain largely mysterious.

Monogenea

Chemotaxis is an essential feature of host finding in the external flukes. Data on a monogenean parasite of European marine flatfish, *Entobdella*

soleae, reveals that the oncomiracidium displays a marked preference for the sole (*Solea solea*) which is mediated, in experimental conditions, by chemical recognition of the fish skin.

Digenea

Two distinct types of larva are involved in active location of the host in the Digenea – the miracidium and the cercaria. The miracidium emerges from the egg in water as a ciliated free-swimming stage of limited life expectancy whose endeavour is to locate and penetrate a suitable mollusc. The cercaria, also short-lived, is released from the snail and may either crawl or swim in water to locate and establish in the next host, as in the schistosomes, or it may encyst on vegetation and await ingestion (e.g. *Fasciola hepatica*).

Experimental studies on these larval stages have failed to reveal the nature of host-finding mechanisms. Many miracidia have sense organs, such as eye-spots and surface papillae, which may help to orientate the larva in its environment so as to bring it in close proximity to its host mollusc. Accordingly, miracidia react in various ways to environmental stimuli such as light, temperature, gravity, water currents and changes in carbon dioxide tension (P_{CO_2}). Chemotactic recognition of the host may occur but is controversial and while some evidence supports miracidial attraction to chemicals released from snails, possibly in mucus or faeces, other data favours the hypothesis that miracidia locate snails by a random, trial-and-error process. Chemoattraction may also be important for host finding by cercariae but, again, data are both limited and controversial.

The ability of cercariae to establish in the next host is clearly affected by water flow, a fact that may have important medical and commercial implications. Cercariae of *Schistosoma mansoni* are adversely affected by water turbulence and they are unable to penetrate even under conditions in which their physical integrity remains unaltered. Whether it is host finding or attachment to the host that is altered remains to be discovered, but the data strongly suggest that one way of reducing the transmission of aquatic parasites with free-swimming larvae is to increase water velocity where practicable.

Cestoda

Pseudophyllidean tapeworms possess a free-swimming, ciliated larva, the coracidium. This larva hatches from the egg in freshwater and is eaten by copepods; transmission here must be regarded as a passive process.

Nematoda

Chemical attraction is probably an important component of host-finding in many plant and animal parasitic nematodes. The infective L_3 larvae are well-provided with sensory structures and reveal a complex behavioural capacity associated with locating their hosts.

7.2.4 Entry mechanisms

Many parasites gain entry to their hosts by active penetration through the epidermis (cercariae, miracidia and some nematode larvae, such as hookworms).

Penetration of a suitable snail by the digenean miracidium is brought about by secretions from the complex of apical glands at the anterior end of the larva. These secretions contain both lubricants and lytic enzymes. During the process of snail penetration some miracidia shed their epidermal ciliated plates (e.g. *F. hepatica*) while in others cilia are retained.

The penetration of mammalian skin by schistosome cercariae is initiated by surface lipids of the host. Some non-essential fatty acids and complex skin lipids will induce the 'penetration response' in cercariae whereby the tail is shed and transformation to the schistosomulum stage commences. Experimentally, these effects can be inhibited by topical application of eicosanoids and eicosanoid-like substances, related to the prostaglandins. Both cercariae and mammalian skin can produce eicosanoids and these compounds may interact during the process of skin penetration (Fig. 7.1). Moreover, it is postulated that cercarial eicosanoids may act as immunomodulators and protect the penetrating larvae from host attack.

The infective larvae of hookworms and related nematodes actively penetrate host skin probably using lytic enzymes to aid in the process.

7.2.5 Heat shock proteins

A great many parasites experience, during the passage of their life cycle, dramatic changes in

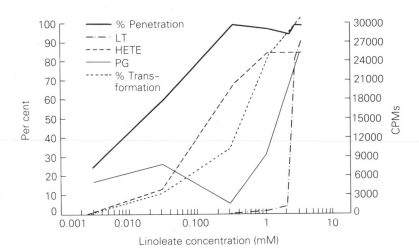

Fig. 7.1 Effects of linoleic acid on penetration through an artificial membrane and on transformation of cercariae of *Schistosoma mansoni*. The left-hand y axis depicts percentage penetration and percentage transformation of cercariae. The right-hand y axis depicts the concentration of cercarial prostaglandin (PG), leukotriene (LT) and hydroxyeicosatetraenoic acid (HETE) in response to elevated linoleic acid levels. (Data from Salafsky & Fusco, 1987.)

ambient temperature. Such changes may occur during the process of host entry, as when an egg or cyst is ingested, an active larva penetrates the skin or when an invertebrate or other cold-blooded animal introduces the infective parasite to the host bird or mammal. Such temperature changes will tend to be metabolically and physiologically harmful to the invading parasite. Accordingly, parasites can synthesize heat shock proteins (HSP) in response to temperature changes in much the same way as do free-living organisms. Heat shock protein production may be prolific, and in schistosomes HSP-70 represents more than 1% of the total protein synthesized by adult worms and is even more abundant in the transforming schistosomulum.

As a group, HSP are highly conserved molecules of diverse function, classified according to molecular weight and encoded by multigene families. In parasites, they are highly immunogenic and may not always be produced in response to temperature change. Increase in temperature induces HSP synthesis in *Leishmania* promastigotes, *Giardia* trophozoites, *Trypanosoma* trypomastigotes, *Naegleria* trophozoites, infective larvae of *Brugia* and schistosomula of *Schistosoma*. By contrast, schistosome cercariae placed in mammalian cell culture media at 23°C produce high levels of HSP when all other protein synthesis has been down-regulated, but they do not respond when placed in water at 37°C. The physiological function of the schistosome HSP-70 is unknown but it may be involved in the reshaping of the macromolecular architecture of the parasite as it moves from one environment to another. Heat shock proteins are also involved in the repair of faulty protein synthesis associated with the ageing process and in this way may be responsible for the considerable longevity typical of some parasites.

7.2.6 Circadian rhythms and parasite transmission

Transmission of some parasites to a new host may be associated with daily or circadian (i.e. around 24 hours) cycles. These can be classified as follows:

1 synchronous cell division (e.g. malaria parasites);
2 release of infective stages from
  (a) final host (e.g. coccidia, pinworms),
  (b) intermediate host (e.g. schistosome, cercariae);
3 migratory patterns (e.g. trypanosomes, microfilariae).

Synchronous cell division

Asexual reproduction, termed schizogony, is profoundly periodic in occurrence in malaria parasites. Cell division within the red blood cell takes place every 24 hours in *Plasmodium knowlesi*, 48 hours in *P. vivax* or 72 hours in *P. malariae*. This periodicity is related to the production of gametocytes in the blood, which are infective to the mosquito vector. Maturation of gametocytes appears to occur at a time of day coincident with the feeding activities of the appropriate species of mosquito. Circadian rhythms of this type are possibly entrained to the daily cycle in body temperature of the homoiothermic hosts; experimentally induced hypothermia in monkeys will disrupt the circadian pattern of malarial development.

Circadian release of infective parasites

There are several examples of the circadian release of infective parasites, timed so as to optimize transmission. Amongst the Protozoa, *Isospora* oocysts are released from the gut of infected birds at roosting time each day (i.e. late afternoon). This may greatly increase the chances of uninfected birds acquiring the oocysts during feeding.

Mammalian pinworms (e.g. *Syphacia muris* in rats and *Enterobius vermicularis* in humans) migrate diurnally from the rectum to the perianal region to lay their eggs. This migration by female worms is related to a lowering of rectal temperature during sleep and it enables the inadvertent 'hand-to-mouth' transmission of parasite eggs to occur without faecal contamination.

The cercariae of many digeneans are released into water from snails at specific times of the day coinciding with the presence of the appropriate new host. This phenomenon has been well-studied for schistosomes, where each species exhibits unique characteristics in the chronology of cercarial shedding. The majority of schistosome species are strictly circadian in their behaviour having a single peak of release each day. *Schistosoma mansoni*, *S. haematobium*, and certain strains of *S. japonicum* and *S. bovis*, all release cercariae during the day, while *S. rodhaini* and other strains of *S. japonicum* release cercariae at night. Of the factors that may influence the periodicity of cercarial shedding from snails, the most important are environmental light and temperature cycles. This has been confirmed by experimental manipulations of light period and temperature: reversal of photoperiod rapidly causes a reversal of the pattern of cercarial shedding and, while alterations of the relative lengths of light and dark periods may be influential, the emission wavelength of light is unimportant. Thermoperiod plays a less significant role than does photoperiod. Human schistosomes tend to shed cercariae during the day, *S. rodhaini*, a parasite of wild rodents, sheds cercariae at night and *S. margrebowiei*, infecting antelope which drink at dawn and dusk, has two peaks of cercarial emission which coincide with the appearance of the potential host animals.

Circadian migratory activities

Some parasites that inhabit deep internal tissues of the host and are transmitted by surface-biting invertebrate vectors are faced with the problem of enabling their infective stages to reach superficial tissues at the time when the vector is present and feeding. As a consequence, we see amongst certain parasites elegant physiological adaptations to facilitate transmission under such circumstances. The microfilarial larvae of the filarial nematodes are well-researched examples of this phenomenon. The adult worms reside in the lymphatic system (e.g. *Wuchereria bancrofti*) or in thick nodules in the skin and subdermal regions (e.g. *Onchocerca volvulus*) while their larvae are transmitted by mosquitoes or simuliid blackflies respectively, both of which feed on peripheral blood. To optimize infection of the vector, many filarial nematodes have developed diurnal rhythms in migration of their larvae. These patterns

can be classified accordingly and are related to the feeding activities of the insect species concerned with transmission:

1 microfilariae numerous in host peripheral blood at night only, absent by day (e.g. *W. bancrofti*, *Brugia malayi*);

2 microfilariae numerous in peripheral blood by day only, absent by night (e.g. the eye worm, *Loa loa*);

3 microfilariae more numerous in peripheral blood during the evening (e.g. the heart worm, *Dirofilaria immitis*);

4 microfilariae present in peripheral blood for entire 24 hour period, but more numerous in the afternoon (e.g. Pacific form of *W. bancrofti*).

When the microfilariae are not in the peripheral blood they accumulate within the pulmonary circulation at the capillary junctions of arterioles and venules. The difference in oxygen tension (P_{O_2}) between arterial and venous blood at these junctions is the initiating trigger for the diurnal migration of the larvae. When the P_{O_2} is in excess of 55 mmHg the larvae of *W. bancrofti* accumulate in the pulmonary circulation; they emerge and migrate to the peripheral circulation when the P_{O_2} decreases to 47 mmHg or below (see Fig. 7.2 & Table 7.2).

7.3 ESTABLISHMENT AND SURVIVAL WITHIN THE HOST

After locating and gaining entry into a suitable host, the parasite has to become established in a physiologically suitable microhabitat in order to grow, either to sexual maturity or to an intermediate stage, such as another larva, whose continued development occurs in the next host. Parasite establishment and growth require that a complex series of physiological conditions are met. These are summarized in Fig. 7.3.

7.3.1 Transformation of digenean cercariae

During attachment and subsequent penetration of the host epidermis, cercariae undergo a physiological transformation in which they rapidly become adapted to the conditions within the host body. Transformation has been studied in some detail in schistosomes but in few other digenean

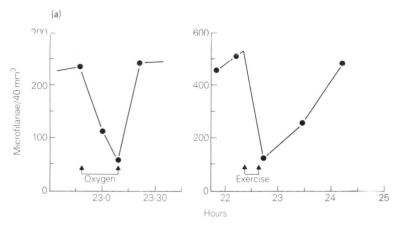

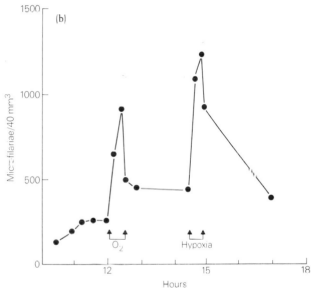

Fig. 7.2 The distribution of microfilariae in the host according to the state of oxygenation. (a) *Wuchereria bancrofti* in humans – both breathing oxygen and vigorous exercise produce a high venous–arterial difference in blood oxygen tension and microfilariae migrate away from the peripheral circulation. (b) *Dirofilaria repens* in dogs. In all graphs, the curves represent the microfilarial count in peripheral blood obtained by conventional skin-shipping and direct counting. (Data from Hawking, 1975).

Table 7.2 The effects of oxygen tension in human venous and arterial blood on the distribution of *Wuchereria bancrofti* microfilariae. (Based on data in Hawking, 1975.)

Activity of host	Arterial P_{O_2} (mmHg)	Venous P_{O_2} (mmHg)	Venous–arterial difference (mmHg)	Microfilarial distribution
Resting (day)	95	40	55	Lungs
Sleeping (night)	85–90	43	42–47	Peripheral blood
Breathing 100% O_2	640	53	587	Lungs
Breathing 14% O_2	51	32	19	Peripheral blood (partial release)
Vigorous exercise	91	20	71	Lungs

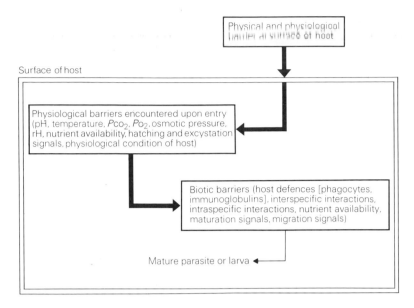

Fig. 7.3 Barriers to establishment encountered by invading parasites.

parasites. The schistosome cercaria is adapted for an existence in freshwater at 25°C in which it swims vigorously, gaining motility from its tail. Once attached to mammalian skin, the tail is lost and penetration commences. It was thought that tail-loss signalled cercarial transformation to the schistosomulum but cercariae with their tails intact will also transform in an isotonic medium at 37°C. On skin contact, the cercaria secretes a powerful protease from its preacetabular glands and morphological changes in its tegument are initiated. Over a period of just a few hours the schistosome cercaria transforms into a migratory larva, the schistosomulum. These two larvae are metabolically distinct, the cercaria is aerobic and the schistosomulum produces lactate anaerobically; they possess morphologically and antigenically diverse surfaces; and they show osmotic and thermal restriction in their tolerance. The dramatic modifications to the worm surface reflect the need for the parasite to defend itself against host immune attack, to which end it can bind host molecules to effect immunological disguise, and also acquire the ability to transport nutrients transtegumentally (see Section 7.8.3). The major features that accompany schistosome transformation are summarized in Table 7.3 and it is probable that these events are common to many species of digenean. Differences that occur

may depend on whether the host invaded is a poikilotherm or a homoiotherm.

7.3.2 Hatching and excystation

Many parasites enter their hosts encapsulated either within cysts or within egg membranes (see Sections 7.2.1 & 7.2.2). Such parasites inevitably enter the host via the alimentary canal, within which they become activated and then liberated from these capsules prior to further development. Not all parasites that gain entry to their host by being eaten are encapsulated; for instance, many helminth larvae reside in the tissues of an intermediate host free from a cyst (e.g. metacercariae of some strigeid digeneans, pseudophyllidean pro- and plerocercoids and many nematode larvae). It is not clear why some intermediate stages of parasites form cysts and others do not; in part, it may reflect the nature of the host response to the parasite since many cysts are made up of contributions from both parasite and host.

Protozoa

Activation and excystation of protozoan cysts has been examined *in vitro* for a relatively small number of species (Table 7.4).

Optimum experimental conditions include tem-

Table 7.3 Transformation of the schistosome cercaria. (Data from Wilson, 1987.)

Condition	Cercaria	Schistosomulum
Habitat	Freshwater	Body of bird or mammal
Temperature	25°C	37–41°C
Motility	Swims using tail	Crawling or burrowing
Glands	Pre-, postacetabular and head glands full: secrete contents on skin contact	Glands empty
Tegument	Trilaminate surface membrane; extensive glycocalyx	Heptalaminate surface membrane: reduced glycocalyx
Permeability	Impermeable to nutrients	Permeable to nutrients
Osmotic tolerance	Survives in water, dies in complex media	Intolerant of freshwater
Energy metabolism	Oxidative, cyanide-sensitive	Anaerobic, cyanide-insensitive
End-products	$CO_2 + H_2O$	Lactate (after 24 hours)
Surface immunochemistry	Antigenically simple	Antigenically complex

Table 7.4 Conditions for *in vitro* excystation of some protozoan parasites. (Data from Lackie, 1975.)

Species	Temperature (°C)	pH	Gas phase	Enzymes added	Bile	Host
Entamoeba histolytica	37	–	Air or anaerobic	Reducing agents	–	Man
E. invadens	8–24	–	–	–	–	Reptiles
Eimeria bovis	40	7.5–8.5	Air or 50% Air/CO_2	Trypsin + reducing agents	1%	Cattle
E. tenella	37–41	7.6	Air or CO_2	Trypsin, HCO_3^-, pancreatin	Present	Poultry
Cystoisospora canis	22–37	–	Air or CO_2	Trypsin	0.5%	Dogs

perature increase, if the parasite is invading a homoiotherm, neutral pH, low P_{O_2}, high P_{CO_2} and the presence of reducing potential. Activation of the parasite within the cyst may be distinct from excystation, the former depending upon high P_{CO_2} and the latter requiring proteolytic enzyme action. In the Coccidia, for example *Eimeria* and *Isospora*, excystation of the sporocyst after its release from the oocyst can involve the breakdown of a localized region of the cyst wall – the Stieda body – by the action of bile salts and trypsin.

Digenea

The eggs of the majority of digeneans hatch in water under suitable environmental conditions of light, salinity and temperature. The eggs of some other digeneans are ingested by molluscs and hatch in the snail gut.

The operculate eggs of *Fasciola hepatica* hatch under the influence of light, and in fact, specific wavelength may be important. Light appears to stimulate activity in the miracidium resulting in permeability changes to the viscous cushion just below the operculum, the hydration of which may force off the operculum allowing the parasite to escape. Schistosome eggs do not have an operculum and the larval parasite emerges on rupture of the eggshell; osmotic pressure is the major physiological effector for the hatching of schistosome eggs, such that a rapid decrease in osmotic pressure, as would be experienced when the egg is passed into freshwater, triggers hatching. The influence of light and ambient temperature are of less significance. In some digeneans, activation may involve the production of lytic enzymes that contribute to the process of excystation.

The metacercariae of many digeneans are enclosed within cysts whose walls vary in specific architecture and dimension. For those species that invade birds or mammals and whose cysts are thin-walled, excystation is initiated by the elevation of ambient temperature alone. Excystation in the laboratory of the more complex metacercarial cysts requires the action of serial treatment with pepsin followed by trypsin as well as temperature changes; bile salts may also play an important role in this process (e.g. *F. hepatica*). Within the Digenea, excystation initiators vary according to species; often certain of these factors activate the encysted larva itself and then a combination of external and internal factors contribute to the final emergence of the parasite. The initiators for digenean excystation include temperature, pH, redox potential, P_{O_2}, P_{CO_2}, osmotic pressure, bile salts and inorganic ions.

Cestoda

The eggs of many tapeworm species hatch in the external environment upon receipt of, and in response to, suitable stimuli. By contrast, the eggs of the Cyclophyllidea hatch in the gut after ingestion by the invertebrate or vertebrate host. The cyclophyllidean egg has a thin outer capsule but the oncosphere larva is enclosed within a thick, protective embryophore. Hatching of the eggs of taeniid tapeworms is a biphasic process whereby the hexacanth larva is first activated, bringing about disruption of the onchospheral membrane, and digestion of the outer capsule is completed by the action of host proteolytic enzymes. In non-taeniid cyclophyllideans, hatch-

ing is largely a mechanical process due to the action of the host mouthparts on the eggshell. Hatching of these eggs can be accomplished *in vitro* in simple physiological saline. In *Hymenolepis* species, however, onchosphere activation requires high P_{CO_2} and the presence of bicarbonate ions and digestive enzymes. Hatching of the taeniid egg requires exogenous pepsin (e.g. *Taenia saginata*) or pancreatin (e.g. *T. pisiformis*) and activation of the encapsulated larva is influenced by host bile salts.

Acanthocephala

Activation of the cystacanth larva takes place on temperature elevation (if a parasite of homoiotherms) and bile salts may also play a role (e.g. *Moniliformis dubius*). Excystation is due either to mechanical (e.g. *Polymorphus minutus*) or enzymic (e.g. *M. dubius*) disruption of the thin outer capsule. The egg (or shelled acanthor larva) hatches in the midgut of arthropods. Hatching *in vitro* depends upon ionic strength of the medium employed, the concentration of bicarbonate ions and pH. Enzymes, such as insect chitinases are without effect.

Nematoda

The eggs of many nematodes hatch in the external environment to release either infective larvae or larvae that develop to the infective stage (e.g. *Ancylostoma, Nippostrongylus*). Upon receipt of the appropriate environmental stimuli, including water, and changes in temperature or oxygen, the enclosed larva liberates lytic hatching enzymes whose action may increase water uptake into the egg. Hatching may be induced therefore by increased turgor pressure within the egg (e.g. *Trichostrongylus*).

The eggs of other parasitic nematodes hatch only after ingestion by a suitable host (e.g. *Ascaris, Toxocara*). *Ascaris* eggs hatch *in vitro* at 37 °C in media with high P_{CO_2}, and neutral pH containing bicarbonate ions and reducing factors. The larva within the egg is activated to produce a hatching fluid comprising enzymes capable of digesting the ascaroside and chitinous layers of the eggshell.

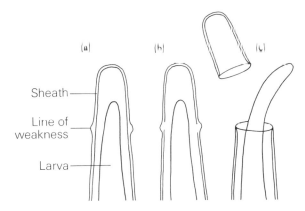

Fig. 7.4 Exsheathment in infective trichostrongyle larvae (e.g. *Haemonchus*). (a) A line of cuticular weakness is developed by localized anterior swelling of sheath. (b) Digestion of inner layers of line of weakness. (c) Rupture of sheath along line of weakness and release of larva. (After Lee & Atkinson, 1976.)

Table 7.5 *In vitro* survival of protoscoleces of *Echinococcus granulosus* in 10% bile from various vertebrates (Data obtained from curves by Smyth & Haslewood, 1965.)

Survival time of protoscoleces (days)	Percentage survival (source of bile)				
	Dog	Sheep	Pig	Ox	Fish
1	100	100	100	100	100
2	100	100	100	100	75
3	100	95	88	92	–
4	100	80	52	36	0

Trichostrongyle L$_3$ larvae are enclosed within a sheath formed from the cuticle of the second larval ecdysis; exsheathment takes place within the gut of a suitable host under the influence of carbon dioxide, bicarbonate ions, reducing agents at neutral pH and at the appropriate temperature. The larval parasite is induced, by these environmental stimuli, to produce an exsheathing fluid that contains enzymes for disruption of the sheath to allow the infective larva to emerge (Fig. 7.4).

7.3.3 The role of bile salts in parasite establishment

Bile, a complex mixture of organic acids, is released into the upper duodenum of vertebrates via the bile duct. Bile contains bile salts, which are steroid-like molecules based on cholic acid, bile acids, which are the degradation products of red blood cells, and bicarbonate ions. Bile salts can be of considerable significance in determining host specificity of many parasites. They affect parasites in a number of ways including: (1) membrane permeability; (2) activation of encysted forms; (3) lysis of parasite surface membranes; (4) synergism with host digestive enzymes; and (5) metabolic action.

One way in which bile salts may act as deter-

minants of host specificity in parasitism is exemplified by studies on the hydatid organism, *Echinococcus granulosus* (Table 7.5). Larval protoscoleces, removed from the hydatid cyst, respond in various ways to bile from different animals *in vitro*: bile rich in deoxycholic acid is lytic to the protoscoleces whereas bile from dogs or other carnivores, the natural hosts for *E. granulosus*, is low in this particular bile acid. The lytic action of bile salts from unsuitable hosts provides a possible mechanism of host specificity at the physiological level.

Additionally, bile salts can affect both establishing and established parasites. Experimental cannulation of the bile duct of rats infected with the tapeworm *Hymenolepis diminuta* brings about a reduction in size and fecundity of the worm whereas cannulation prior to infection inhibits establishment.

Bile salts activate and initiate excystation in many parasites including protozoans (e.g. *Eimeria*), digeneans (e.g. *Fasciola*), cestodes (e.g. *Taenia pisiformis*), and acanthocephalans (e.g. *Moniliformis*, *Polymorphus*). The physiology of these events is not well-understood.

7.3.4 Hypobiosis

The condition under which animals may become quiescent during their development is termed hypobiosis. Periods of arrested development may occur in free-living animals that inhabit arid or otherwise intemperate regions and the hypobiotic state is therefore adopted in response to climatic extremes. In the case of parasites, hypobiosis

occurs not infrequently; immature stages, encapsulated within eggs or cysts or free within host
tissues, are hypobiotic states whereby the parasite
becomes arrested in its development and awaits a
suitable trigger to initiate the continuation of
its development to adulthood. Hypobiosis may
represent a serious problem in commercial terms
when we consider some of the nematode parasites of cattle.

Arrested larval development (ALD) is a feature
of many of the major parasitic diseases of cattle
caused by trichostrongyles including *Dictyocaulus viviparus* (husk or lungworm), *Ostertagia
ostertagi*, *Haemonchus contortus*, *Trichostrongylus axei*, *Oesophagostomum radiatum* and *Co-
operia* species (collectively referred to as the
agents of 'parasitic gastro-enteritis'). Sheep or cattle become infected with these parasites by in-
gesting the L_3 larvae during grazing; the larvae
enter the gut and develop to adults within 21
days, either in the lungs, abomasum or small
intestine, depending upon the species. This pattern of development can become disrupted under
certain conditions and populations of arrested
early L_4 larvae occur, usually within the abomasal
glands of the stomach. The factors that trigger
ALD in these parasites have not been identified
but, host immunity, season of the year and parasite population size may all be involved. Arrested larvae will begin to develop in conditions
of experimental or natural immunodepression
(e.g. during parturition) and following anthelmintic treatment to remove crowded adult worm
infections. There is a wealth of evidence that
ALD in trichostrongyles occurs on a seasonal
basis, being prevalent in autumn in the northern
hemisphere and spring in the southern hemisphere. Arrested larvae resume their development
and become adult worms at the start of the next
grazing season. Artificial induction of ALD has
been accomplished by conditioning parasite L_3
larvae in a climatic chamber; a large proportion of
these 'autumn conditioned' L_3s become hypobiotic in naive calves. Chilling parasite larvae to
4°C for 5–10 weeks achieves similar effects. The
factors that trigger the resumption of development of arrested nematode larvae are unknown.

Arrested larval development (Fig. 7.5) may be
seen, therefore, as a response by L_3 larvae to

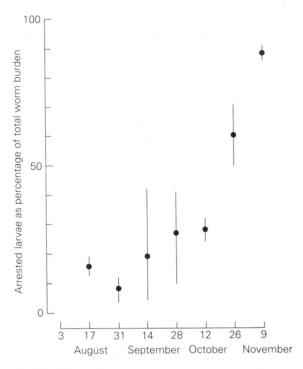

Fig. 7.5 Change in proportion of arrested larvae of
Ostertagia in calves from the west of Scotland between
late summer and autumn. Data are mean and ranges of
percentage of arrested L_4s detected at autopsy 7–14
days after removal from contaminated pasture. (Data
from Armour & Duncan, 1987.)

external environmental stimuli, such as lowering
of ambient temperature or the onset of a hot dry
summer period, or to host-mediated stimuli, such
as immunocompetence. Moreover, these hypobiotic larvae may also be drug resistant and many
anthelmintic drugs are ineffective against them.
Thus, hypobiosis in trichostrongyles represents
a sophisticated adaptation to environmental ad-
versity through physiological responses to complex stimuli.

Nematodes also exhibit a variety of additional
quiescent states that are responses to particular
stress factors of environmental origin. Anabiosis
is an extreme state of arrest which may last for
considerable periods of time: some nematodes
stored dry for over 20 years can then develop
normally. In the anabiotic state the parasite has
no detectable metabolism and the process of ageing is suspended. Desiccation, low temperatures,

osmotic stress and low P_{O_2} can all induce specific anabiotic responses in stages of nematode parasites that occur external to the body of the host.

7.3.5 Migration and site selection

From the point of entry into the host body, the majority of parasites undergo a migration of varying complexity that will take them to their final site of residence. This may be highly specific, as in human schistosomes, cattle lung worms or fish eye- or brain-flukes, or be rather more generalized, as with hydatid cysts that settle in a variety of different sites. Migrations of this type are commonly accompanied by growth and development of the parasite and are thus referred to as *ontogenetic migrations*. These migratory events, which may be of short or long duration, may culminate in a sexually mature parasite or a larval or cystic stage that exhibits hypobiosis, as described above, until ingested by the final host. Usually the migration will take place over a fixed route if normal parasite development is to occur. The physiological triggers and determinants that orchestrate these migrations are not understood but are clearly important factors of host specificity.

Aberrant migratory patterns can occur if a parasite enters an unsuitable host in which normal and complete development cannot be accomplished. This has been widely documented in strongyloid and ascaridoid nematodes where larval worms invading the wrong host, in this case humans, can cause serious damage. These are the migratory *larva migrans* which are typical of hookworms, *Ancylostoma braziliense* and some ascaridoids, such as *Toxocara canis*. On invasion, these parasite larva undertake what is presumably an inadequately signposted migration either in the superficial tissues (e.g. *cutaneous larva migrans*, *Ancylostoma*) or in deeper tissues (e.g. *visceral larva migrans*, *Toxocara*). Since the physiological triggers for normal migration and development are lacking in the incorrect host, these parasites eventually die in an ectopic site; therefore they can cause disease that is often difficult to diagnose. Some common patterns of ontogenetic migration are depicted in Fig. 7.6, but in no case do we understand the physiological mechanisms that are involved.

Some parasites undergo migrations within the host body that are not related to growth and development. These include diurnal migrations associated with transmission (see Section 7.2.6) and diurnal migrations related to the nutritional cycle of the host (e.g. *Hymenolepis diminuta* in the rat intestine).

The complex and varied nature of site selection and pattern of migration seen within parasites, especially helminths, argue compellingly that these parasites possess sophisticated sensory capacity to enable them to recognize and respond to the appropriate stimuli emanating from the tissues of the host body. This aspect of parasite physiology has proved to be a difficult area of research to establish and the sensory biology of parasites remains a topic of ignorance and neglect.

7.3.6 Invasion of tissues

During the process of establishment many parasites invade specific cells in host tissues where they reside either temporarily or for long periods of time. The physiology of cell recognition and adherence in parasitology is of interest since many intracellular parasites are responsible for causing major diseases of both humans and animals. Examples of tissue-invading parasites include malarias and babesias (red blood cells), leishmanias (macrophages), coccidians, cestodes and nematodes (*muscularis mucosa* of the gut), schistosomes (circulatory system), trypanosomes (nervous system) and larval digeneans, cestodes and nematodes (body musculature). The physiology of cell recognition, cell adhesion and penetration is complex and poorly understood and is best described for malaria parasites.

Cell invasion in malaria

Two essentially unrelated features of the cell biology of malaria parasites have attracted considerable attention: red blood cell recognition and invasion and cytoadherence of the parasitized red cell to the endothelial lining of host blood vessels.

Invasion of the red cell by *Plasmodium* is a specific, sequentially-defined process which involves: (1) cell recognition by the merozoite; (2) orientation of the parasite with respect to the red

(a)   Entry to the host via the mouth

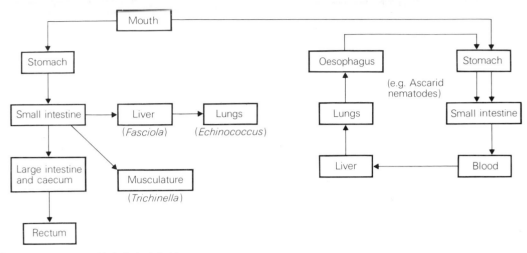

(Many protozoans and helminths inhabit
the gut; the exact site selected varies)

(b)   Entry to the host via the skin

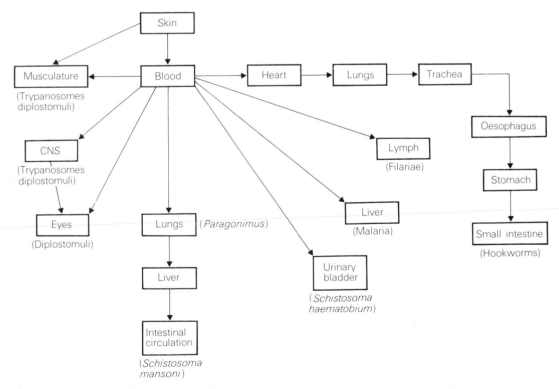

Fig. 7.6 Patterns of ontogenetic migrations carried out by parasites within the body of the vertebrate host.

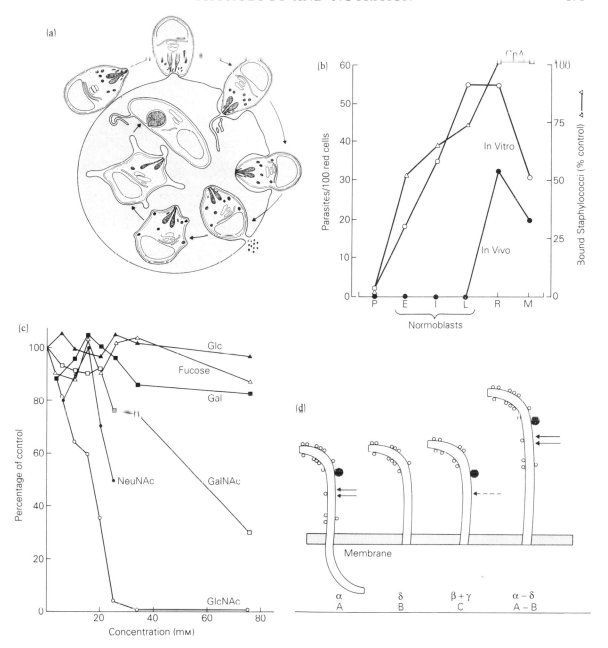

Fig. 7.7 Red cell invasion by *Plasmodium*. (a) Invasion sequence by merozoites of *P. knowlesi* (after Bannister, 1977). (b) Invasion of red blood cells by *P. falciparum in vitro* (○–○) and *in vivo* (●–●) relative to appearance of glycophorin A (GpA): P, pronormoblasts; E, early; I, intermediate; L, late normoblasts; R, reticulocyte; and M, mature erythrocyte. (c) Red cell invasion by *P. falciparum* related to the presence of various sugars. Data expressed as a percentage of control invasion: Glc, glucose; Gal, galactose; GalNAc, *N*-acetyl-D-galactosamine; NeuNAc, *N*-acetyl-neuraminic acid; GlcNAc, *N*-acetyl-D-glucosamine. (d) Glycophorins A, B and C schematically represented on the red cell surface. The small circles are *O*-glycosidically linked oligosaccharides and the large circles are *N*-glycosidically linked oligosaccharides. Arrows indicate sites of tryptic cleavage. A-B is a hybrid molecule. ((b)–(d) from Pasvol & Jungery, 1983.)

cell surface and apposition of the apical complex; (3) formation of a junction between the invading merozoite and red cell surface at the point of contact; (4) induction of invagination of the red cell membrane by secretions from the merozoite; and (5) entry of the parasite through extensive invagination of the red cell surface membrane forming the parasitophorous vacuole (Fig. 7.7a). Recognition of the red cell by the merozoite depends on specific surface receptors and varies according to age of cell, blood group antigens, and host specificity. Invasion of human red cells is diminished in races lacking Duffy antigens (e.g. in *P. vivax*); treatment with *N*-acetyl-D-glucosamine, trypsin or neuraminidase blocks cell invasion by *P. falciparum*. Recent evidence implicates surface glycophorins as major determinants of invasion. Red cell surface glycophorins are sialic acid-rich glycoproteins comprising four subgroups (a, b, c, and d) and their role as surface receptors for *P. falciparum* merozoites is now well-established. Cells lacking either glycophorin a (En(a-)-cells) or glycophorin d (S-s-U-cells) resist merozoite invasion to a significant extent and this resistance can be enhanced by tryptic removal of remaining glycophorin molecules. Glycophorins represent a significant component of the red cell membrane and yet they have no clear role, since they can be absent without any red cell dysfunction. *Plasmodium falciparum* will develop normally within glycophorin-deficient cells and will invade young red cells, in which it cannot develop, relative to the progressive appearance of glycophorins on the cell surface (Fig. 7.7b, c & d).

Although the exact role of red cell glycophorins as receptors for merozoites is unknown and the evidence that they contribute to the initiation of erythrocyte invasion is convincing, other factors are also involved. Once recognition of the cell is accomplished, the invasion continues by the formation of a junction of about 10 nm, containing fine fibrils that extend between the thickened red cell membrane and the apical protruberance of the merozoite: Duffy-negative, invasion-resistant cells do not form this junction but trypsin treatment of these cells renders them permissive to invasion by *P. knowlesi* merozoites and a typical host cell–parasite junction is

formed. The apical organelles of the merozoite, the rhoptries and micronemes, initiate the actual invasion of the red cell by releasing secretions which include a histidine-rich protein. During invasion, the red cell membrane invaginates progressively to enclose the merozoite and the junction moves so that its position is maintained at the mouth of the developing parasitophorous vacuole. This mobility of the junction may be related to membrane fluidity, since treatment of merozoites with cytochalasin B inhibits invasion despite the formation of a junction on attachment of the merozoite; here no junctional migration occurs. On completion of normal invasion, the junction seals up behind the merozoite which now lies completely enclosed within the parasitophorous vacuole. This vacuole is made of original red cell membrane which has undergone molecular reorganization. Not only does the intracellular malarial parasite induce molecular changes within the membrane of the parasitophorous vacuole, but it also affects the red cell surface itself. Under certain conditions, surface electron-dense knobs appear on the red cell membrane and caveola-vesicle complexes may also be formed.

One major feature of *P. falciparum* infections is the sequestration of infected red cells through cytoadherence to endothelial cells of the host circulatory system, mediated through erythrocyte surface knobs. Sequestration is an important adaptation by which the parasite may avoid circulation through the spleen where host defence is active. Pathologically, sequestration may contribute to obstructed blood flow typical of cerebral malaria. Falciparum malaria differs from other human malarias in that only red cells containing young ring stages of the parasite circulate freely, while cells infected with more mature parasites become sequestered.

The surface knob, which is the functional unit of cytoadherence of *P. falciparum*-infected red cells, comprises a cup-shaped membranous structure and associated protrusion of the red cell surface membrane (Fig. 7.8). These knobs contain a unique protein of Mr 80 000, rich in histidine and proline which is lacking in knobless strains of the parasite. Its role in mediating cytoadherence remains unexplained but, since antibody has

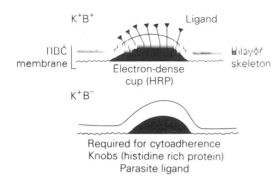

Fig. 7.8 Model for knobs on the red cell surface induced by infection with *Plasmodium falciparum*. K^+B^+, knobby parasite able to bind to blood vessel endothelium; K^+B^-, knobby parasite incapable of endothelial binding due to absence of putative cytoadherence factor. (After Leech *et al.*, 1984.)

been shown to inhibit cytoadherence in a strain specific way, it suggests that additional molecules may also be involved.

Cell invasion in other parasites

Many other protozoan parasites adopt an intracellular habit and therefore invade host cells. Like malaria parasites, these also may reside within a parasitophorous vacuole and enter by invagination of the cell surface rather than by penetration of the membrane itself. This pattern of invasion occurs in the Coccidia (e.g. *Eimeria* invading gut cells and *Toxoplasma* invading macrophages). The parasitophorous vacuole may be temporary (e.g. *Babesia*) and regresses shortly after entry is accomplished. *Toxoplasma*, *Eimeria* and *Leishmania* all invade host cells by inducing their own phagocytosis. Treatment of host cells with drugs that inhibit phagocytosis, such as colchicine or cytochalasin B, alters the pattern of invasion. *In vitro* studies on cell invasion by *Leishmania* parasites suggests a degree of induction and specificity in the mechanisms involved. Promastigotes of various species of *Leishmania* induce phagocytosis by host macrophages and these cells respond to the presence of live parasites by producing pseudopodial whorls or lamellar sheaths; killed parasites are also taken up but at a reduced rate. Cytochalasin B inhibits

macrophage invasion by these promastigotes, which will also invade non-phagocytic cells in culture. Cautious interpretation must be placed on these *in vitro* studies and it is not certain that the same mechanisms necessarily function *in vivo*. After invasion, *Leishmania* parasites reside within a parasitophorous vacuole formed from host cell membranes. Survival of the parasite within a cell, whose role is to ingest and kill invaders, is both of considerable interest and applied significance (see Chapter 8, Section 8.7).

Trypanosoma cruzi, the causative agent of Chagas disease in the Americas, is an obligate intracellular protozoan. In cell culture, trypomastigotes invade fibroblasts by mechanisms that involve interaction between the surface membranes of the host cell and the parasite. The parasite produces a lectin-like protein responsible for cytoadherence aided by a proteolytic activating system; penetration is effected by a tunicamycin-sensitive glycoprotein. The host cell contributes to these events by producing glycoproteins that are active in adherence and also penetration. By contrast, entry of *T. cruzi* into macrophages is by phagocytosis; trypomastigotes survive within the macrophage by escaping the confines of the lysosomal vacuole and replicating within the cytoplasm.

In some cases, host cell invasion by protozoan parasites is the major pathophysiological event. In *Entamoeba histolytica*, secreted products, such as 'amoebapore' (a parasite-derived, pore-forming protein), cytotoxin and proteolytic enzymes all contribute to invasion of the host intestinal cells and to the severe intestinal and liver pathology associated with this phase of amoebiasis.

While several species of helminth parasite invade host cells (e.g. *Trichinella spiralis* in muscle cells) little information on mechanisms of entry is available.

7.4 REPRODUCTIVE PHYSIOLOGY

Many parasites have complex life cycles which involve stages that reproduce by sexual processes in one host and by asexual proliferation in another host. These mechanisms ensure that genetic information is varied and becomes widely disseminated and also that infective parasite stages are produced in sufficient numbers to favour successful transmission in a hostile environment.

7.4.1 Asexual mechanisms

Asexual splitting, or budding, occurs in many parasitic Protozoa, all of the Digenea and in some Cestoda.

Asexual reproduction in the Protozoa involves binary or multiple fission, schizogony, endo-dyogeny or single budding, while in helminth parasites individual numbers are dramatically increased by internal budding (polyembryony). During the life cycle of the Digenea, an asexual phase occurs exclusively within the molluscan intermediate host. Normally two distinct asexual generations are passed within the snail, mother sporocyst to daughter sporocysts or sporocyst to rediae, the net result being that from a single miracidium entering a snail many hundreds of thousands of cercariae may subsequently emerge. Each of these individuals will be genetically identical.

The majority of cestodes produce only a single larva from the egg but proliferative external budding occurs in the urocystis and urocystidium larvae and internal budding occurs within the polycercus, coenurus and hydatid larvae of the Taeniidea. The hydatid cyst of *Echinococcus* can generate several million protoscoleces by asexual budding, while the coenurus rarely produces more than a few hundred larval tapeworms.

7.4.2 Sexual reproduction

Many protozoans reproduce by a form of sexual reproduction but it is not always easy to distinguish between the fusion of individual parasites and the fusion of gametes. In some species the gametes are morphologically distinct, such as male microgametes and female macrogametes, and sexual processes may alternate with asexual reproduction, each taking place in a different host. In malaria parasites, asexual schizogony increases numbers of merozoites in the host blood but sexual gametocytes are also formed which are transmitted to mosquitoes in which host gamete fusion and sexual proliferation takes place. The stimulus which controls the formation of gametocytes has interested malariologists for a long time; current thinking favours the view that trophozoites are directed towards sexual reproduction by environmental factors associated with host cell lysis or degeneration. Here gametocytogenesis can be regarded as an escape mechanism from unfavourable conditions by means of the genomic variation conferred by random fusion of gametes in the mosquito.

The sexuality of trypanosomes is a topic of current interest and the traditional opinion that these parasites reproduce only by asexual binary fission has been challenged. Sexual stages in the life cycle of trypanosomes may occur either within the mammalian host in an extravascular location or within the tsetse fly vector. In support of the latter contention, giant forms of trypanosomes, which are capable of liberating large numbers of new trypanosomes, have been isolated from midgut cells of flies; these giant forms are apparently the product of fusion of two individuals and this represents a mechanism by which genetic interchange might occur. Natural populations of African trypanosomes demonstrate considerable electrophoretic variability which perhaps indicates the importance of sexual mechanisms in increasing genomic variability in the wild.

All monogeneans are hermaphrodites and asexual mechanisms are unknown. Cross-fertilization usually takes place between adjacent individuals but self-fertilization may also occur.

The majority of digeneans are hermaphrodite and both self- and cross-fertilization have been recorded; the schistosomes have separate sexes and cross-fertilization is therefore mandatory. The physiology of egg production is well-understood; eggs are released from the mature ovary and enter the oviduct. Spermatozoa from the partner are stored, after copulation, in a seminal receptacle. These are then released along with a small number of vitelline cells and they make their way to the ootype where the ova are fertilized. During development the eggshell becomes tanned *in utero* and on release the egg is fully protected by the rigid shell (see Section 7.2.2).

Almost all tapeworms are hermaphrodites and each proglottis (segment) contains a full complement of male and female apparatus. Both cross- and self-fertilization occur. Cestodes mature posteriorly such that the terminal segments are the oldest and, when gravid, contain ripe eggs that are either shed independently or within the liberated proglottis itself. The tapeworm egg is not tanned like that of the Digenea, but is surrounded by a capsule comprising various constituent layers.

Acanthocephalans have separate sexes and are sexually dimorphic. Spermatozoa are introduced into the pseudocoelom of the female at copulation; ova within the female are associated with ovarian balls of germinal tissue. Eggs are fertilized and subsequently liberated from the ovarian balls to complete their development free in the pseudocoelom. Female acanthocephalans release large numbers of eggs which are protected by a covering of protein and chitin.

The majority of parasitic nematodes are sexually dimorphic and reproduce sexually; a small number of species reproduce either hermaphroditically or parthenogenetically but no somatic asexual processes have been described. The male nematode generally possesses a single testis and accessory structures, such as a copulatory bursor or paired spicules, which are used during copulation. The female nematode may have one or two sets of gonads; sperm are stored in a seminal receptacle and these fertilize mature oocytes *in situ*. Eggshell formation is initiated by the process of fertilization and continues during egg maturation. In some groups, such as the filarial nematodes, egg hatching takes place *in utero* and the eggshell is accordingly reduced in size and chemical complexity.

Almost all parasites are characterized by their enormous reproductive capacity and they produce, either by sexual or asexual mechanisms, or sometimes both, extremely large numbers of offspring. Physiologically this strategy is demanding in terms of nutrients and energetic commitment to reproduction.

7.4.3 Reproductive synchrony

Reproductive events in a small number of parasite species are synchronized to host sexual cycles and breeding patterns; this relationship serves to liberate infective parasites into the environment simultaneously with susceptible juvenile hosts. The best known examples come from the flagellated protozoans of amphibians and arthropods.

The release of opalinid gametes from the amphibian gut is initiated by host sex hormones. The Hypermastigina that inhabit the gut of arthropods can be stimulated to reproduce sexually under the influence of host moulting hormones. Flagellates of termites are lost with each successive moult since they inhabit the insect hindgut which is lined with cuticle. Here, synchrony of sexual processes in the parasite with moulting in the host ensures reinfection, which in this example is a mandatory phenomenon since the parasite is the source of essential digestive cellulases.

7.5 CHEMICAL COMMUNICATION

Chemical communication between animals of the same species or of different species has long been the subject of intensive research interest. Among the insects, the topic of communication via pheromones has had considerable commercial significance in the field of insect pest control. Surprisingly, little is known about chemical communication between parasites, yet this information could prove to be invaluable in the quest for novel strategies with which to control the world's major parasitic diseases.

Pheromones are probably produced by many helminth parasites and they may serve as sexual attractants: there is considerable indirect, but little direct, evidence to support this contention.

Laboratory identification of parasite pheromones is normally made using *in vitro* bioassays whereby movement of individual worms in aqueous or semi-solid media is assessed within choice chambers of various design. Less frequently, supporting data has been obtained from *in vivo* observations on mate location by parasitic worms, but such information has proved more difficult to interpret unambiguously.

Solubility, enzyme and chromatographic studies on putative pheromones from helminths have all provided indirect evidence for the types of chemical messengers involved in sexual attraction, but no parasite pheromone has yet been identified or characterized chemically. Sterols and peptides have both been implicated as candidate pheromones in parasitic nematodes (e.g. *Nippostrongylus brasiliensis*) and digeneans (e.g. *Echinostoma*). However, detailed biochemical analyses of pheromones of *N. brasiliensis* suggest a more complex picture in which the worms produce a mixture of lipidoidal and hydrophilic substances which can variously attract either the same or the opposite sex. The site of pheromone production appears to be highly varied in helminth parasites. In nematodes, the copulatory organs themselves and the body surface are the apparent source of attractant molecules, while in the digeneans, the tegument, alimentary canal and excretory system all produce unidentified chemoattractants. Studies on sensory receptors of helminths that may detect chemical messengers released into the environment have failed to provide unequivocal evidence on location and precise function of these organelles.

Schistosomes present an interesting and somewhat unusual picture in terms of sexual attraction: they possess separate sexes that show varying degrees of interdependence for growth and development and in which processes chemoattractants must play an important but undisclosed role. Males of *Schistosoma mansoni* will grow and develop to maturity in the absence of females whereas female worms lacking males grow poorly and never reach sexual maturity although they have the potential to do so if males are experimentally introduced even as much as a year later. Other schistosome species show a lesser degree of sexual dependence. Laboratory studies reveal that adult male and female schistosomes attract one another by chemical means involving lipid-based pheromones; homosexual attraction and pairing can occur and under these conditions partial sex-reversal of the smaller male partners has been observed. Females of *S. mansoni* will only grow to sexual maturity after a period of residence within the gynaecophoral groove of the male, implying tactile as well as chemical stimuli as controlling mechanisms. The receptors involved in interpreting this complex array of signals have not been described.

In a parasitic disease whose pathogenesis is directly related to sexual activity of the parasite, pheromones that initiate worm pairing, growth and maturation would represent ideal and novel drug targets.

Aside from chemical aspects of parasite reproductive biology, the general endocrinology of helminths is little understood. Hormones concerned with regulation of parasite growth and development have been examined for only a small number of species and their role remains a matter for some speculation. Parasitic nematodes, like their free-living relatives and arthropods, grow by a series of moults in which the old cuticle is shed and replaced by a new structure. These events are controlled by juvenoid, ecdysteroid and neuropeptide hormones in insects and, by analogy, they should also function similarly in nematodes. Biochemical identification of these hormones from parasite tissues has been confirmed but experimental verification of their biological function has not yet been made, nor has their biosynthesis been demonstrated unequivocally. Therefore, despite the presence of ecdysteroids (ecdysone and 20-hydroxyecdysone) in nematodes and platyhelminths, their function remains elusive, particularly since moulting only occurs in the nematodes. It may be that the complex surface biology of helminth parasites is regulated by conserved families of developmental hormones, but this has yet to be established (Table 7.6).

7.6 NEUROPHYSIOLOGY OF HELMINTH PARASITES

Parasite neurobiology has long been a 'Cinderella' topic with parasitologists but it should be apparent that, with their relatively sophisticated patterns of host-finding, site-location, mate-finding and reproductive biology, parasites are more complex than many textbooks would admit and their neurobiology will be accordingly complicated. Furthermore, a great many antiparasite drugs act on the neuromuscular system of helminths. Thus parasite neurophysiology is gaining in importance as a topic of considerable applied significance.

Table 7.6 Ecdysteroid hormones in helminth parasites. (Data from Mercer, 1985; Rees & Mercer, 1987.)

DIGENEANS	
Schistosoma mansoni	Ecdysone in schistosomula, ecdysone + 20(OH)ecdysone in eggs and adults
Fasciola hepatica	Ecdysone + 20(OH)ecdysone in adults
CESTODES	
Echinococcus granulosus	Ecdysteroids in hydatid cyst fluid
Hymenolepis diminuta	Free and conjugated ecdysteroids in eggs and adults
Moniezia expansa	Ecdysone, 20(OH)ecdysone + 20,26di-(OH)ecdysone in adults; no conjugates
NEMATODES	
Ascaris suum	Free and conjugated ecdysteroids in adults
Anisakis simplex	Ecdysone, 20(OH)ecdysone + 20,26di-(OH)ecdysone in L_3 larvae
Dirofilaria immitis	Ecdysone, 20(OH)ecdysone, free and conjugated ecdysteroids in adults
Oncocerca gibsoni	Ecdysone + 20(OH)ecdysone in adults
Parascaris equorum	Metabolizes exogenous ecdysone

7.6.1 Helminth nervous systems

The nervous system of platyhelminth parasites comprises an anterior complex of cerebral ganglia with, posteriorly, a bilaterally symmetrical series of nerves serving the body. The system is formed of a nerve network containing unmyelinated fibres, the majority of which have motor function. The acanthocephalan nervous system is comparably undeveloped and contains a cerebral ganglion from which arise single and paired nerves; the male worm has a second ganglion associated with the reproductive system.

Information of the nervous system of parasitic nematodes derives from early studies on ascarids and more recent work on free-living forms such as *Caenorhabditis elegans*. A nerve ring, containing the main concentration of neurones, is associated with the oesophagus, from which motor nerves are directed to the head; sensory nerves connect the nerve ring to anterior sense organs. A paired, ganglionated, ventral nerve cord extends posteriorly and connects with a dorsal nerve via a series of commissures. There is a posterior nerve ring which contains ganglia connecting with both motor and sensory neurones. The pattern of the nervous system tends to be consistent throughout the phylum.

7.6.2 Sense organs and sensory biology

The functioning of helminth sense organs has been determined primarily from morphological and behavioural studies and little direct information has been obtained. The major areas of sensory physiology that have been examined are photoreception in free-living larval helminths, chemoreception in host-, site- and mate-location, and sensory recognition of and responses to temperature gradients and gravity.

Many larval helminths respond to light in their environment, utilizing these responses to come into juxtaposition with a suitable host for invasion. Eye-spots are present in some helminth larvae, including most monogenean oncomiracidia and some digenean miracidia. Larval forms lacking such sense organs may nevertheless respond either positively or negatively to light but it is often difficult to distinguish between one stimulus and another in an experimental arena. Photosensitivity is undoubtedly important for orientation of invasive helminth larvae with respect to host finding and often also for the initiation of hatching of helminth eggs.

Response to a thermal gradient may be necessary to accomplish infection of a warm-blooded host (e.g. larval hookworms in a terrestrial environment and schistosomes in an aquatic environment). It is not known what sensory apparatus is involved in these responses. Similarly, parasites may respond to gravity, chemical gradients including oxygen or carbon dioxide, and to the presence of other individual parasites, but the sensory organelles involved here have also yet to be identified. Surface receptors include ciliated sensillae (Monogenea), ciliated pits and papillae (Digenea), tegumental protrusions (Cestoda) and amphids, papillae and ciliated pits in the Nematoda.

7.6.3 Neurotransmission and neurosecretion

Helminth parasites synthesize a number of putative neurotransmitter substances including adrenalin, noradrenalin, acetylcholine, DOPA, dopamine, GABA (gamma-aminobutyric acid) and serotonin (5-HT). Cholinergic synapses are widely distributed throughout the helminths and acetylcholinesterase has been located histochemically in many species. The neuromuscular junctions of nematodes are cholinergic: acetylcholine decreases muscle resting potential while substances like physostigmine increase sensitivity to acetylcholine; piperazine, a widely used nematocidal drug, inhibits the stimulatory effects of acetylcholine. The major inhibitory neurotransmitter in nematodes is GABA (Fig. 7.9) and its action at the synapse is thought to be the focus of activity of the anthelmintic ivermectin. Platyhelminth neurotransmitters include acetylcholine, 5-HT, noradrenalin and dopamine; GABA is

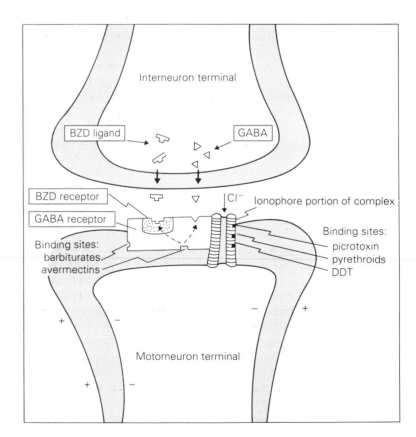

Fig. 7.9 Interaction between the nematode synapse and the drug ivermectin. Potentiation of GABA and benzodiazepine binding (dotted lines) cause Cl^- influx and motorneurone hyperpolarization. Ivermectin (avermectins) also potentiate GABA release, which may explain their anthelmintic mode of action in causing worm paralysis. (After Campbell, 1985.)

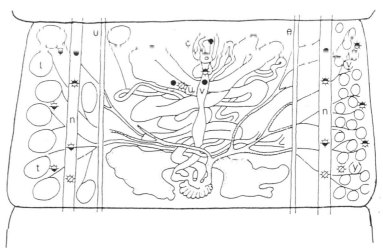

Fig. 7.10 Aminergic and peptidergic nervous elements in the proglottid of a mature cestode (*Diphyllobothrium dendriticum*). All four types of immunoreactivity are present in the two main nerve cords (n). (After Gustafsson, 1985.)

☙, fibres immunoreactive to peptide histidine isoleucine surrounding testicular follicles; ✡, immunoreactivity to FMRF– amide (Phe–Met–Arg–Phe–NH$_2$) in cirrus sac (c); ❀, immunoreactivity to growth hormone releasing factor in vaginal wall (v) and vitelline glands (y); ●, immunoreactivity to serotonin in vaginal wall, uterine pore and cirrus sac.

presumed to be unimportant since ivermectin has no effect on platyhelminth parasites.

Chemical messengers within the nervous system, such as neuropeptides, amines, amino acids and acetylcholine, are the focus of growing attention since they may differ in parasite and host and thus present chemotherapeutic potential. The complexity of this topic has been revealed in recent studies; for instance, in tapeworms, aminergic, cholinergic and peptidergic neurones have been identified (Fig. 7.10). At least 29 neuropeptides have been detected by immunoreactivity with specific antisera to mammalian peptide hormones including bovine pancreatic polypeptide, growth hormone releasing factor, peptide histidine isoleucine, gastrin, gastrin releasing peptide, leu-encephalin, neurotensin, vasotocin, oxytocin and FMRF-amide. Nothing is known of the function of this complex of neuropeptides in cestodes but once disclosed the possibility exists for the development of novel drug targets at the neurophysiological level.

The products of some parasites are capable of causing what appear to be endocrinological lesions in their hosts via the production and release of substances that mimic host hormone action. This occurs in larval helminth infections of snails, such as *Trichobilharzia ocellata*, a bird schistosome which induces host gonadal regression via a substance called 'schistosomin'; this substance is synthesized by the snail itself under the influence of the parasite. Many larval digeneans cause profound growth and sexual changes in molluscs, some of which may be mediated endocrinologically. Plerocercoids of the tapeworm, *Spirometra mansonoides*, release a platelet growth factor (PGF) which is a remarkable mimic of mammalian growth hormone (Table 7.7). The plerocercoid stage shows little host specificity, infecting a wide range of animals, including humans, causing the condition known as sparganosis. The similarities between PGF and human growth hormone have led to speculation that this parasite has acquired the human gene for growth hormone which it is able to express in the plerocercoid stage. Viral transduction is one possible method by which this proposed genetic exchange may have occurred.

7.6.4 The neuromuscular junction in helminths

The muscle cells of nematodes, as revealed by studies on *Ascaris*, are unusual in that they contain both nervous and contractile elements: the

Table 7.7 Comparison of plerocercoid growth factor (PGF) of *Spirometra mansonoides* with mammalian growth hormone (MGH). (Data from Phares, 1987.)

Parameter	PGF	MGH
Weight gain	Increase	Increase
Skeletal growth	Increase	Increase
Somatomedin activity	Increase	Increase
Endogenous GH	Decrease	Decrease
Insulin-like in:		
Normal rats	Yes	No
Hypophysectomized rats	Yes	Yes
Antiinsulin-like in: Diabetogenic	No	Yes
Binding to rabbit and rodent GH and Prolactin receptors	Yes	Yes
Lactogenic in pigeon crop-sac assay	Yes	Yes
Primate GH-activity	Yes	Yes
Molecular weight	24 000	22 000
Isoelectric point	pH 4.7	pH 4.9
Reaction with monoclonal antibody to human GH (%)	61	100

muscle arm synapses with motoneurones of the nerve cord. The contractile portion contains regular arrangements of thick and thin myofilaments, typical of striate muscle, in which H, A and I, but not Z bands are apparent. The neuromuscular junction of other helminth parasites is poorly understood.

7.7 LOCOMOTORY PHYSIOLOGY

Almost all parasites are capable of movement and some stages in the life cycle, especially those concerned with active transmission in the external environment, are highly motile. Even adult parasites can be highly active when viewed after host autopsy, but this may not necessarily reflect their natural state *in vivo*.

Information in locomotory physiology is largely restricted to observations on the effects of various external stimuli upon activity and speed of movement. Amongst the Protozoa, ciliary, flagellar and amoeboid movement have all been described; parasites appear to be no different in this regard from free-living forms. Members of the Apicomplexa, such as *Eimeria* and *Plasmodium*, exhibit gliding movements during which the parasite undergoes no alteration in body shape. The physiological mechanisms underlying this process

are unknown but the notion of a 'linear motor', powered by interactions between actin and myosin filaments in the parasite surface membranes, has been proposed (Fig. 7.11). Ciliary locomotion occurs in some helminth parasites, including oncomiracidia, miracidia and coracidia. In other larval stages and adult helminths, locomotion takes the form of swimming, crawling or burrowing brought about by muscular action. Nematode parasites, like their free-living relatives, move in a characteristic undulatory manner which is the net product of their cylindrical body shape, with opposing dorsal and ventral musculature and a fluid-filled pseudocoelom which acts as a hydrostatic skeleton. Additionally, the pattern of muscle innervation aids the generation of sine waves in the body shape to bring about movement. The resting pseudocoelomic hydrostatic pressure of *Ascaris* is approximately 70 mmHg and it can vary between 16 and 225 mmHg during wave-form production by posteriad contraction of the body musculature; these waves are dorsoventral in plane and the animal lies on its side during locomotion. The three underlying mechanisms associated with nematode locomotion, spontaneous myogenic depolarization, neuromuscular coordination and local changes of hydrostatic pressure, are controlled by serotonin and epinephrine.

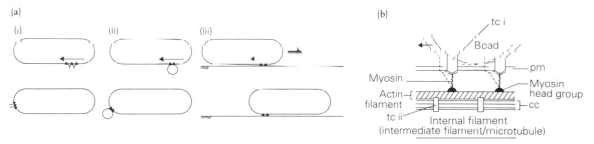

Fig. 7.11 Biomechanics of locomotion in sporozoan (apicomplexan) protozoans. (a) Gliding movement with associated capping and bead translocation; (i) antibody (Y) interacts with cell surface components forming a cluster which may activate the proposed 'linear motor' sweeping complexes to rear of cell; (ii) beads and cell surface interact to initiate 'linear motor' that moves bead to end of cell; (iii) substrate interacts with cell surface to activate motor, since substrate cannot move, cell is moved forwards. (b) Model for proposed 'linear motor', based on interactions of surface actin and myosin: tc i and tc ii, transmembrane components; cc, cortical cytomembranes; pm, plasma membrane. Interaction of myosin head group with ATP and actin filaments leads to conformational changes associated with release of ADP. Addition of further ATP would cause release of myosin head group from actin filament and the bead illustrated would move to the left. (After King, 1988.)

7.7.1 Helminth locomotion for drug assays

One of the most common determinants of the efficacy of antiparasite drugs in laboratory screening tests concerns the action of the compound on parasite motility. While this is perhaps a debatable criterion for *in vitro* drug screening, it does have a proven track record in distinguishing potentially useful drugs from those of little value. Naturally, effects on motility can only be one of many parallel facets of compound evaluation.

The so-called 'micro-motility meter' has been devised as a relatively simple instrument for determining the effects of putative drug substances on helminth movement *in vitro* and it provides an attempt at objectivity in this potentially highly subjective assay procedure (Fig. 7.12). Many drugs will depress parasite motility in this system which is an inexpensive and rapid primary screen. However, the majority of new compounds are still tested on conventional model parasite systems as the motility screen has a predilection for drugs that influence the parasite neuromuscular system and may not identify compounds that are specific for alternative targets.

7.8 NUTRITION OF PARASITES

The popular concept that parasites feed at the expense of the host may have little foundation in

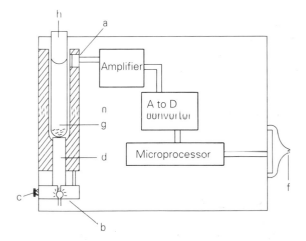

Fig. 7.12 Diagram of a micromotility meter for quantification of helminth movement in the presence of drugs and other substances. Medium is placed in a tube (g) and light from a lamp (b) passes to the meniscus (h) through a plastic light pipe (d); part of the light is deflected to the photodiode (a), the signal amplified, digitized and microprocessed. Addition of a motile worm will result in fluctuations in the digitized signal and worm movement can be compared statistically under different conditions. (After Bennett & Pax, 1987).

reality. There are certain conditions under which the nutritional demands of a parasite may result in physiological stress to the host, such as the acquisition of vitamin B_{12} by *Diphyllobothrium latum* in the gut of humans and anaemia in hookworm disease, but in general there is little evidence to support the notion that parasites cause disease by their nutritional activities, although the act of feeding itself may be physically damaging. In truth, we are rather ignorant of the nutritional physiology of most parasites and are not in a position to define the nutrient requirements of more than the small handful of species investigated. We must be careful not to imply a lack of metabolic dependence of the parasite on its host, but within the established, stable host–parasite relationship, physiological excesses by the parasite tend to be eliminated in favour of activities that lead to a benign association. Moreover, most parasites inhabit sites in the host body where nutrients are themselves often in excess supply.

7.8.1 Nutrient requirements and *in vitro* culture

A widely adopted approach to the study of parasite nutrition involves either short-term maintenance or long-term culture *in vitro*. It is important to distinguish between these two, since the former is used primarily to examine methods of nutrient acquisition by parasites that have been removed from their hosts and held, relatively briefly, in physiologically simple media, while the latter approach employs complex media often to determine the precise nutrient requirements of the parasite. Whichever approach is adopted, the questions about the comparability of data obtained from *in vitro* studies with the events that occur *in vivo* must be addressed, and of necessity, nutritional information obtained from *in vitro* studies must be cautiously interpreted. However, since it is virtually impossible to gather meaningful nutritional facts from *in vivo* studies due to the complexity of the host–parasite interaction, we must continue to rely on *in vitro* cultivation to provide us with answers to fundamental questions on parasite nutrition.

Relatively few parasites have been cultured *in vitro* under axenic and chemically-defined conditions, and many culture media contain complex, undefined additions, such as serum, which are necessary for parasite survival. Defined media have been developed for some protozoans (e.g.

Crithidia fasciculata, *Leishmania donovani* and *L. brasiliensis*, *Trypanosoma cruzi*, *Trichomonas vaginalis*) but this has not yet proved to be possible for helminth parasites.

7.8.2 Nutrient acquisition by parasites

In addition to feeding in the conventional sense, employing mouth and associated alimentary apparatus, many parasites obtain both dissolved and macromolecular nutrients by uptake across the body surface (membrane transport). In some groups of parasite (e.g. Protozoa, Cestoda, Acanthocephala) no mouth or alimentary canal is present so that the body surface forms the major system for molecular exchange.

Alimentary systems

The Monogenea, Digenea and Nematoda all possess a recognizable alimentary system and mouth with which are often associated structures that relate to the particular pattern of feeding. In some protozoan parasites (e.g. sporozoans) a permanent or sometimes temporary oral structure is developed, the cytostome. This is a specialized region concerned with nutrient uptake, such as haemoglobin acquisition by intracellular forms of *Plasmodium*, at which location food vacuoles, surrounded by cytostomal membrane, are formed. Digestion takes place entirely within the vacuole by chronologically distinct acid and alkaline phases.

The Monogenea and Digenea have well-developed alimentary systems comprising a mouth surrounded by a sucker, a muscular and often glandular pharynx, an oesophagus which may have associated glands, and two blind-ending digestive caeca (Fig. 7.13); only rarely is an anus present. The morphological complexity of the platyhelminth gut varies considerably but in a way unrelated to feeding activity. Among the Monogenea, two patterns of feeding predominate: blood feeding and feeding on tissues and mucus. There are physiological adaptations associated with these different types of nutrition: in the majority of blood feeders, the gastrodermal lining of the digestive caeca is shed following each meal, whereas the gastrodermis is non-deciduous

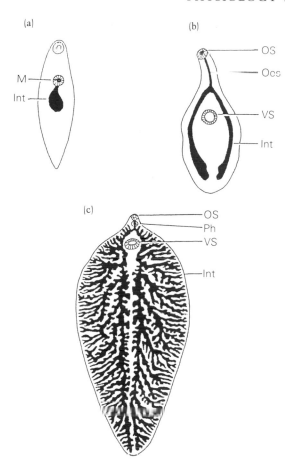

Fig. 7.13 Morphological types of digenean alimentary canal. Int, intestinal caeca, M, mouth; Oes, oesophagus; OS, oral sucker; Ph, pharynx; VS, ventral sucker. (From Chappell, 1980: after Dawes, 1968.)

in tissue feeders. Similarly, amongst the Digenea, blood and tissue feeders are found. Different species of digenean have evolved distinct approaches to the physiological problems associated with haematophagy: schistosomes are exclusively sanguivorous and they digest haemoglobin extracellularly, removing waste metallic iron by periodic regurgitation through the mouth. The liver fluke, *Fasciola hepatica*, by contrast, feeds on both tissue and blood and completes digestion of the blood meal intracellularly in the gastrodermis, passing waste iron to the excretory canals to be voided. In *Fasciola*, a curious gastrodermal cell cycle has been identified, which is related to the various phases of ingestion and digestion of food.

The gut of parasitic nematodes differs little from that of free-living relatives. Significant morphological modifications are developed anteriorly around the mouth and these reflect the diet, taking the form of teeth, jaws and penetrating stylets for engaging host tissues during feeding. The alimentary canal of nematodes comprises an anterior mouth and associated structures, a muscular and sometimes glandular pharynx, oesophagus, intestine in which both digestion and nutrient absorption take place, and a posterior rectum and anus. In a small number of insect–parasitic nematodes, the alimentary canal can be lost completely, under which conditions the exposed hypodermis of the cuticle becomes the site for nutrient acquisition (e.g. *Mermis*, *Bradynema*).

7.8.3 The parasite surface and its role in nutrition

Morphological adaptations

The surface (plasma) membrane and its associated glycoprotein coat forms, in all parasites, at least one facet of the nutritional interface with the host, and in some parasites, where the surface assumes a major nutritional function, there are marked morphological adaptations at this interface. These are most clearly seen in the platyhelminths and acanthocephalans, in which the surface architecture typifies a digestive–absorptive epithelium with its enormous increase in surface area. This is achieved by the development of surface folds and microvilli (e.g. Monogenea, Digenea, Cestoda), tubercles and spines (e.g. Digenea) or surface pores with branching invaginated canals (e.g. Acanthocephala), all of which are illustrated in Fig. 7.14. In the Protozoa, apart from the cytostome referred to above, the surface of parasites differs little from free-living forms in its nutrition-related morphology. However, in the intracellular Protozoa, the nutritional interface additionally includes the surface membrane of the host cell, perhaps in the form of the parasitophorous vacuole, and thus there may be two or more distinct barriers to nutrient acquisition.

Early light microscopy suggested that platy-helminth parasites were enclosed in a protective cuticle, presumed to be a defensive structure against host digestive and immunological attack. Electron microscopy has revealed the true picture and we now know that these parasites possess a metabolically active, non-cuticular surface, termed the tegument. This entire tissue system includes the external glycoprotein glycocalyx, the surface membrane, and beneath it, but above the basement membrane, the anucleate syncytium. The cell bodies, which form the epidermal covering in free-living platyhelminths, are sunken in parasites and are located interior to the basement

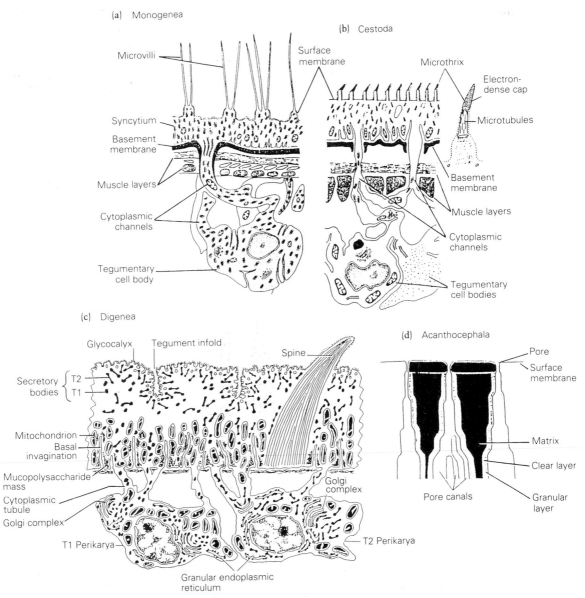

Fig. 7.14 Surface morphology in platyhelminth and acanthocephalan parasites. (From Chappell, 1980: after Lyons, 1970; Smyth & Halton, 1983; Threadgold, 1984; Crompton, 1970.)

membrane within the muscle layers (Fig. 7.14a, b, c). The syncytial layer is replete with organelles of metabolic function and replacement membrane material. Morphological study thus implicates the helminth surface as an active participant in nutrition as well as its apparent role in defence by active surface renewal. The acanthocephalan surface is physiologically comparable although morphologically quite different from the platyhelminths (Fig. 7.14d). Only in the nematodes does a true cuticle occur, but even in this group recent evidence suggests that the cuticle may be permeable to low molecular weight nutrients and to water.

Membrane transport mechanisms in parasites

Transport of nutrient molecules into parasites occurs by one, or a combination of simple diffusion, carrier-mediated transport (facilitated diffusion, active transport, exchange diffusion) or macromolecular transport (endo- or exocytosis). These transport mechanisms can be distinguished kinetically and biochemically and their occurrence may be determined in laboratory studies using *in vitro* methods.

Simple diffusion obeys Fick's law, which states that the rate of solute movement is directly related to the concentration difference of that solute

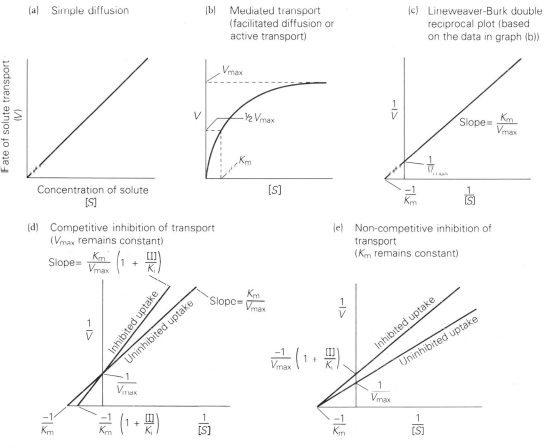

Fig. 7.15 Kinetics of membrane transport mechanisms. Often, nutrient uptake patterns follow a combination of (a) and (b) in which mediated transport is more important at low solute concentrations with diffusion assuming greater significance at higher values. The inhibitor constant (K_i) is determined experimentally as shown in (c) and (d); [I] concentration of inhibition; K_m, transport or Michaelis constant; V_{max}, maximum rate of transport; NB ($K_m = 1/2\ V_{max}$).

on either side of a semi-permeable membrane; it therefore displays linear kinetics (Fig. 7.15a), is independent of metabolic energy and temperature and does not respond to interference by competitive inhibitors. Simple diffusion involves movement of molecules down a concentration gradient. Carrier-mediated transport requires that the nutrient molecules bind specifically to, and complex with, carriers (permeases, transporters, transport sites or loci) in the membrane; these complexes have intramembrane mobility and translocate nutrients to the opposing side of the membrane where complexes are dissociated and nutrients released. This type of transport displays saturation kinetics (Fig. 7.15b, c), is temperature- and energy-dependent (e.g. active transport but not facilitated diffusion), responds to inhibition both competitively and non-competitively (Fig. 7.15d, e) and can be either uphill in terms of solute concentration (e.g. active transport) or down an electrochemical gradient (facilitated diffusion). Saturable transport systems (i.e. carrier-mediated) can be examined kinetically by application of the Michaelis–Menten (enzyme) equations from which can be derived the transport constant, K_t, defining the affinity of the substrate for the carrier, and V_{max}, which is the maximum velocity of the transport process.

Thus transport systems of different parasites may be compared kinetically and their substrate

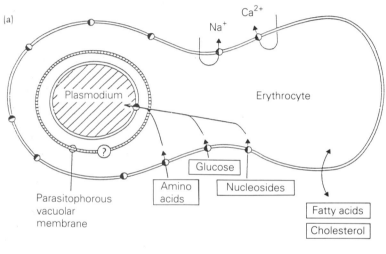

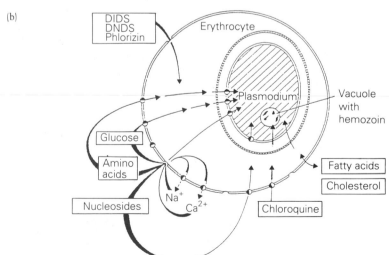

Fig. 7.16 Nutrient transport in the malaria-infected red cell before (a) and after (b) parasite-induced changes in membrane permeability. Carriers are represented as circles, exchange is shown as a double-headed arrow. After induction of permeability changes, the red cell membrane becomes leaky towards the disulphonic stilbenes DIDS and DNDS, and phlorizin. The drug Chloroquine accumulates via a membrane carrier and becomes concentrated within food vacuoles. (After Sherman, 1988.)

specificities determined by inhibitor studies. The applied importance of such work lies in the potential for the development of drugs that target specific transport systems in parasites thereby denying entry of essential nutrients. At present, very few antiparasite drugs operate in this way, but the area is rich with potential for rational chemotherapy.

The nutrient transport systems of relatively few parasites have been examined in critical detail. Below are described, in summary, data collected for four groups of parasite about which a reasonable quantity of information is available.

Plasmodium. Malaria parasites transport carbohydrates, amino acids, purine nucleosides, fatty acids, complex lipids, anions and cations and the presence of the parasite confers upon the infected red cell pathological alterations in nutrient transport that may favour the development of the parasite (Fig. 7.16). The asexual stages of the parasite within the red cell lack stored carbohydrate but require considerable quantities of glucose to fuel their active metabolism and division. Infected erythrocytes use between 10 and 50 times more glucose than uninfected cells and the parasite appears to induce permeability changes in the red cell membrane which facilitate the passage of host glucose and amino acids into the erythrocyte.

The nutritional source of amino acids for intracellular stages of malaria is not fully understood; haemoglobin digestion undoubtedly provides significant amounts, but the infected red cell also shows increased transport of free amino acids in culture. In *P. falciparum*, these changes in amino acid transport rates are first seen 15 hours after invasion and the infected erythrocyte loses energy-coupled transport systems in favour of diffusion; whether these amino acids enter the parasite itself by carrier-mediated transport or by diffusion remains to be determined and awaits the development of methods permitting culture of the asexual stages of the parasite outside the red cell.

Malaria parasites transport exogenous purine nucleosides but not pyrimidines. This may be related to their inability to synthesize the purine ring *de novo*. Parasites liberated from red cells

may accumulate certain purines (e.g. adenosine, guanosine and hypoxanthine) and can incorporate radioactivity from labelled adenosine, AMP and ATP. Once again, however, studies on liberated parasites produce questionable data. Several studies have demonstrated that lipids (i.e. free fatty acids, cholesterol and phospholipid) are readily incorporated into malaria parasites resident within the red cell but the transport processes involved are unknown.

Trypanosomes. The extracellular habit of trypanosomes makes it relatively easy to examine the transport of nutrients. Carbohydrates are transported by specific carriers, some of which are capable of moving more than one species of nutrient molecule. *Trypanosoma lewisi* has two carriers that transport glucose, mannose, fructose, galactose and glucosamine; *T. equiperdum* has three carriers for monosaccharide transport, a glucose transporter, and two distinct systems for glycerol. These carriers have been defined by the use of inhibitors of transport *in vitro*. Amino acid transport into trypanosomes is complicated and several distinct carriers have been described using kinetic and inhibitor studies. *Trypanosoma cruzi* transports basic amino acids (e.g. arginine, lysine) by multiple carriers that possess unusual specificities in terms of substrates carried. Here, both neutral and acidic amino acids inhibit basic amino acid transport, a situation not encountered in mammalian transport systems. By contrast, arginine transport in *T. equiperdum*, a species with at least four distinct amino acid carrier systems, is remarkably substrate specific. *Trypanosoma brucei gambiense* transports amino acids by a mixture of carrier-mediated and non-specific (diffusion) mechanisms. Similarly, in *T. equiperdum* and *T. lewisi*, simple diffusion may assume greater significance than saturable processes when amino acids are present at higher concentrations, although it is not easy to extrapolate from these *in vitro* studies to what may happen in the blood stream of a mammal. Lipid acquisition by trypanosomes is complex and is associated with membrane-bound enzymes (acetyltransferase and phospholipase A$_1$), while 3'-nucleotidase/nuclease has been implicated in the uptake of purines.

Tapeworms. Much of the pioneering work on membrane transport of low molecular weight nutrient molecules in parasites was carried out on tapeworms, in particular *Hymenolepis diminuta*, the rat tapeworm, at Rice University (Texas) (see Pappas & Read, 1975). This parasite has proved to be an ideal model since it can be obtained readily in large quantities in the laboratory, is non-pathogenic both to the rat and to humans, and having no gut, absorbs all of its nutrients across the tegument. The tapeworm transports carbohydrates by both carrier-mediated systems and by diffusion – glycerol and glucose enter by separate carriers but both depend on sodium ion concentration. Amino acid and purine/pyrimidine transport are complex processes: there are six separate amino acid carriers; four transporting neutral amino acids, one for acidic and one for basic amino acids; and at least three purine/pyrimidine carriers with multiple-binding capacity. Fatty acid transport is similarly complicated and separate systems, transporting short-chain and long-chain moieties, have been described.

Digenea. Transport of nutrients across the digenean tegument is complicated by the presence of an alimentary canal and undoubtedly, *in vivo*, the gut plays a major role in nutrition, aided to an unknown extent by tegumental transport systems. Study of the latter can be carried out *in vitro* either by ligation of the pharynx or by using short-term studies in which oral ingestion of nutrients assumes an insignificant role. Both *Schistosoma mansoni* and *Fasciola hepatica* transport monosaccharides by carrier-mediated mechanisms of varying substrate specificities and while schistosomes transport amino acids also by tegumental carriers, *Fasciola* appears to lack these, absorbing amino acids by simple diffusion only. No explanation for this difference is available but it might relate to the differential dependence on the worm gut as a source of nutrients.

Digestive enzymes

The digestive enzymes of most parasites have been little studied, thus there is a paucity of information on their characteristics, substrate specificities, pH optima, secretory regulation and location. Digestive enzymes occur in the food vacuoles of protozoan parasites and in both the alimentary canal and the body surface of helminths.

The alimentary protease of *S. mansoni* has become a topic of considerable research interest, ironically primarily because of its serodiagnostic potential rather than because of its digestive role. Schistosomes ingest large quantities of host blood via the mouth and digest haemoglobin readily; the schistosome gut is typically delineated by the presence of the black pigment, haematin, which is the result of this digestion. A single proteolytic enzyme (haemoglobinase) occurs in the gut of *S. mansoni*, this being a thiol-protease with a pH optimum of around 3 and which hydrolyses globin to peptides but not to individual amino acids. It is not known whether these peptides are further hydrolysed intracellularly within the gastrodermis or if the tegumental carriers provide the major source of free amino acids from host serum. Other alimentary peptidases have recently been located histochemically, suggesting that haemoglobin digestion could be completed in the schistosome gut. The proteases of schistosomes vary in their occurrence during the life cycle and may thus have stage-specific functions; the haemoglobinase itself is only expressed in the developing schistosomulum and adult worm, in which stages haematophagy becomes physiologically important (Fig. 7.17). This adult protease is highly antigenic and is useful in the diagnosis of schistosomes in subclinical human cases as a prelude to chemotherapy.

Several helminths possess surface enzymes, either of parasite origin or derived from the host, that may have a digestive function. *Hymenolepis diminuta* synthesizes digestive phosphohydrolases, hydrolysing phosphate esters, monoglyceride hydrolases and ribonucleases, all of which function in a digestive capacity at the tegumental surface. The tapeworm can also bind host digestive enzymes, such as amylases, whereupon enzyme activity may become enhanced, although the mechanism of this so called 'contact digestion' is open to interpretation. Conversely, tapeworms can bind and inhibit host enzymes (e.g. trypsin, chymotrypsin) and this is possibly one adaptation for parasite survi-

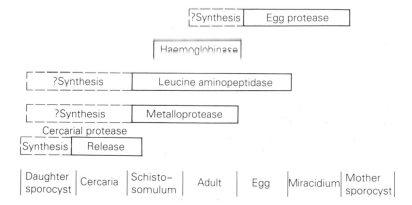

Fig. 7.17 Schistosome proteases may be expressed in a stage-specific manner. (After McKerrow & Doenhoff, 1988.)

val in an enzymatically hostile environment.

Surface membrane-bound enzymes have also been described in some protozoans. *Leishmania* spp, and other trypanosomatids, for instance, possess a 3'-nucleotidase/nuclease complex in their surface membranes which can hydrolyse 3'-nucleotides and nucleic acids. The hydrolytic activity of this complex is implicated in the acquisition of purines associated, perhaps, with the inability of the parasite to synthesize the purine ring *de novo*. The membrane bound acyltransferase of African trypanosomes may also play a part in nutrient acquisition, in this case in the uptake and internalization of lipid, particularly phopholipids.

7.8.4 Excretory physiology

Since the life histories of many parasites are complex, involving alternation of free-living and parasitic stages, it might be expected that the processes of regulation of water and ion content and removal of toxic excretory products would reflect such complexity. Little information, however, is available.

In parasites, two types of excretory system are found: (1) the contractile vacuole of protozoans; and (2) the protonephridial system of platyhelminths. Contractile vacuoles are present in many ciliates, but are absent from amoebae and sporozoans. It seems likely that these vacuoles are involved in both osmoregulation and excretion of nitrogenous waste.

The protonephridial system of platyhelminth parasites comprises numerous blind-ending tubules that interconnect and open to the outside at a single nephridiopore. Each tubule has at its terminus a flame cell or cluster of cells, so called because the wave-like beating of the flagella is reminiscent of a flickering candle. Each 'flame' contains between 50 and 100 flagella whose beat regulates fluid flow in the excretory tubule and possibly draws solutes into the terminal organ of the protonephridial system from the surrounding parenchyma. Ultrafiltration may occur at this stage and it has been demonstrated experimentally that a sufficiently high filtration pressure could be developed by the flagellar beat of the flame cell.

Excretion of nitrogenous waste (e.g. ammonia) may take place via diffusion or via the protonephridial system. Detailed analysis of the protonephridial canal fluid of the tapeworm *H. diminuta* reveals the occurrence of inorganic ions (sodium, potassium, chloride, carbonate), ammonia, amino acids, urea and lactic acid. This analysis accounts for 90% of dry matter and the pH of the canal fluid is 4.5 with a $P\text{CO}_2$ of 120 mmHg. These data imply an excretory role but the key elements of ultrafiltration and active transport within the system have yet to be demonstrated.

The excretory systems in acanthocephalans and nematodes are poorly understood and there is little physiological evidence for either group to substantiate what is inferred on morphological grounds alone.

7.9 PARASITE PHYSIOLOGY IN A WIDER CONTEXT

There are many areas of physiology which clearly reflect the unique and often remarkably complex life styles adopted by parasites. The inherent multi-faceted nature of the parasitic life cycle suggests an enormous adaptability, ultimately residing within the genome, which must surely convince the student of the sophisticated status of the parasitic animal. As more information accrues, we come to realize that there are more targets for chemical, immunological or environmental attack upon which we may effect parasite control, but at the same time we discover that the subtleties of parasite evolution have conferred upon these organisms a considerable buffer against the types of onslaught we can currently mobilize.

REFERENCES AND FURTHER READING

Arme, C. (1988) Ontogenetic changes in helminth membrane function. In C. Arme & L.H. Chappell (eds) *Molecular Transfer Across Parasite Membranes. Parasitology*, **96**, (supplement), S83–S104.

Armour, J. & Duncan, M. (1987) Arrested larval development in cattle nematodes. *Parasitology Today*, **3**, 171–176.

Bannister, L.H. (1977) The invasion of red cells by *Plasmodium*. In A.E.R. Taylor & R. Muller (eds) *Parasite Invasion*, pp. 27–55. Symposia of the British Society for Parasitology, **15**, Oxford, Blackwell Scientific Publications.

Barker, G.C. & Rees, H.H. (1990) Ecdysteroids in nematodes. *Parasitology Today*, **6**, 384–387.

Bennett, J.L. & Pax, R.A. (1987) Micromotility meter: instrumentation to analyse helminth motility. *Parasitology Today*, **3**, 159–160.

Berendt, A.R., Ferguson, D.J.P. & Newbold, C.I. (1990) Sequestration in *Plasmodium falciparum* malaria: sticky cells and sticky problems. *Parasitology Today*, **6**, 247–254.

Campbell, W.C. (1985) Ivermectin: an update. *Parasitology Today*, **1**, 10–16.

Chappell, L.H. (1980) *Physiology of Parasites*. Glasgow, Blackie.

Chappell, L.H. (1988) The interactions between drugs and the parasite surface. In C. Arme & L.H. Chappell (eds) *Molecular Transfer Across Parasite Membranes. Parasitology*, **96**, (supplement), S167–S193.

Cordingley, J.S. (1987) Trematode eggshells: novel protein biopolymers. *Parasitology Today*, **3**, 341–344.

Crompton, D.W.T. (1970) *An Ecological Approach to Acanthocephalan Physiology*. Cambridge, Cambridge University Press.

Dawes, B. (1968) *The Trematoda*. Cambridge, Cambridge University Press.

De Jong-Brink, M., Elsaadany, M.M. & Boer, H.H. (1988) *Trichobilharzia ocellata*: interference with endocrine control of female reproduction of *Lymnaea stagnalis*. *Experimental Parasitology*, **65**, 91–100.

Elford, B.C. (1986) L-Glutamine influx in malaria-infected erythrocytes: a target for antimalarials? *Parasitology Today*, **2**, 309–312.

Ellis, D.S. (1986) Sex and the single trypanosome. *Parasitology Today*, **2**, 184–185.

Fairweather, I. Halton, D.W. (1991) Neuropeptides in platyhelminths. *Parasitology*, **102**, (supplement), S77–S92.

Furlong, S.T. (1991) Unique role for lipids in *Schistosoma mansoni*. *Parasitology Today*, **7**, 59–62.

Geary, T.C., Klein, R.D., Vanover, L., Bowman, J.W. & Thompson, D.P. (1992) The nervous systems of helminths as targets for drugs. *Journal of Parasitology*, **78**, 2125–2130.

Gibbs, H.C. (1986) Hypobiosis in parasitic nematodes – an update. *Advances in Parasitology*, **25**, 129–174.

Gottlieb, M. (1989) The surface membrane 3'-nucleotidase/nuclease of trypanosomatid protozoa. *Parasitology Today*, **5**, 257–260.

Gottlieb, M. & Dwyer, D.M. (1988) Plasma membrane functions: a biochemical approach to understanding the biology of *Leishmania*. In P.T. Engelund & A. Sher (eds) *The Biology of Parasitism*, pp. 449–465. New York, Alan R. Liss.

Gustafsson, M.K.S. (1985) Cestode neurotransmitters. *Parasitology Today*, **1**, 72–75.

Hadley, T.J. & Miller, L.H. (1988) Invasion of erythrocytes by malaria parasites: erythrocyte ligands and parasite receptors. *Progress in Allergy*, **41**, 49–71.

Halton, D.W., Fairweather, I., Shaw, C. & Johnston, C.F. (1990) Regulatory peptides in parasitic platyhelminths. *Parasitology Today*, **6**, 284–290.

Haseeb, M.A. & Fried, B. (1988) Chemical communication in helminths. *Advances in Parasitology*, **27**, 169–207.

Hawking, F. (1975) Circadian and other rhythms of parasites. *Advances in Parasitology*, **13**, 123–182.

Howard, R.J. (1988) Malarial proteins at the membrane of *Plasmodium falciparum* – infected erythrocytes and their involvement in cytoadherence to endothelial cells. *Progress in Allergy*, **41**, 98–147.

Jewsbury, J.M. (1985) Effects of water velocity on snails and cercariae. *Parasitology Today*, **1**, 116–117.

Joiner, K.A. (1991) Rhoptry lipids and parasitophorous

vacuole formation: a slippery issue. *Parasitology Today*, 7, 226–227.

Kazacos, K.R. (1986) Racoon ascarids as a cause of larva migrans. *Parasitology Today*, 2, 233–255.

Kearn, G.C. (1986) The eggs of monogeneans. *Advances in Parasitology*, 25, 175–273.

King, C.A. (1988) Cell motility of sporozoan Protozoa. *Parasitology Today*, 4, 315–319.

Lackie, A.M. (1975) The activation of infective stages of endoparasites of vertebrates. *Biological Reviews*, 50, 285–323.

Lee, D.L. & Atkinson, H.J. (1976) *The Physiology of Nematodes*, 2nd edn. London, Macmillan.

Leech, J.H., Aley, S.B., Miller, L.H. & Howard, R.J. (1984) *Plasmodium falciparum* malaria: cytoadherence of infected erythrocytes to endothelial cells and associated changes in the erythrocyte membrane. In J.W. Eaton & G.J. Brewer (eds) *Malaria and The Red Cell*, pp. 63–77. New York, Alan R. Liss.

Lo Verde, P.T. & Chen, L. (1991) Schistosome female reproductive development. *Parasitology Today*, 7, 303–308.

Lyons, K.M. (1970) The fine structure and function of the adult epidermis of two skin parasitic monogeneans, *Entobdella soleae* and *Acanthocotyle elegans*. *Parasitology*, 60, 39–52.

McKerrow, J.H. & Doenhoff, M.J. (1988) Schistosome proteases. *Parasitology Today*, 4, 334–340.

MacKinnon, B.M. (1987) Sex attractants in nematodes. *Parasitology Today*, 3, 156–158.

Maresca, B. & Carratu, L. (1992) The biology of the heat shock response in parasites. *Parasitology Today*, 8, 140–166.

Martin, R.J., Pennington, A.J., Duittoz, A.H., Robertson, S. & Kusel, J.R. (1991) The physiology and pharmacology of neuromuscular transmission in the nematode parasite, *Ascaris suum*. *Parasitology*, 102, (supplement), S41–S58.

Mellors, A. & Samad, A. (1989) The acquisition of lipids by African trypanosomes. *Parasitology Today*, 5, 239–244.

Mercer, J.G. (1985) Developmental hormones in parasitic helminths. *Parasitology Today*, 1, 96–100.

Miles, M.A. & Cibulskis, R.E. (1986) Zymodeme characterization of *Trypanosoma cruzi*. *Parasitology Today*, 2, 94–97.

Mons, B. (1985) Induction of sexual differentiation in malaria. *Parasitology Today*, 1, 87–89.

Newport, G., Culpepper, J. & Agabian, N. (1988) Parasite heat-shock proteins. *Parasitology Today*, 4, 306–312.

Oppenheimer, S.J. (1989) Iron and malaria. *Parasitology Today*, 5, 77–79.

Ouaissi, M.A. (1988) Role of the RGD sequence in parasite adhesion to host cells. *Parasitology Today*, 4, 169–173.

Pappas, P.W. (1983) Host–parasite interface. In C. Arme & P.W. Pappas (eds) *Biology of the Eucestoda*, 2, 297–334. London, Academic Press.

Pappas, P.W. (1988) The relative roles of the intestines and external surfaces in the nutrition of monogeneans, digeneans and nematodes. In C. Arme & L.H. Chappell (eds) *Molecular Transfer Across Parasite Membranes*. *Parasitology*, 96, (supplement), S105–S121.

Pappas, P.W. & Read, C.P. (1975) Membrane transport in helminth parasites – a review. *Experimental Parasitology*, 33, 469–530.

Pasvol, G. & Jungery, M. (1983) Glycophorins and red cell invasion by *Plasmodium falciparum*. In D. Evered & J. Whelan (eds) *Malaria and the Red Cell*, pp. 174–186. London, Pitman.

Pasvol, G. & Wilson, R.J.M. (1989) Red cell deformability and invasion by malaria parasites. *Parasitology Today*, 5, 218–221.

Pax, R.A. & Bennett, J.L. (1991) Neurobiology of parasitic platyhelminths: possible solutions to the problems of correlating structure with function. *Parasitology*, 102, (supplement), S31–S39.

Peattie, D.A. (1990) The giardins of *Giardia lamblia*: genes and proteins with promise. *Parasitology Today*, 6, 52–56.

Perkins, M.E. (1992) Rhoptry organelles of apicomplexan parasites. *Parasitology Today*, 8, 28–32.

Perry, R.N. (1989) Dormancy and hatching of nematode eggs. *Parasitology Today*, 5, 377–383.

Phares, C.K. (1987) Plerocercoid growth factor: a homologue of human growth hormone. *Parasitology Today*, 3, 346–349.

Polla, B.S. (1991) Heat-shock proteins in host–parasite interactions. In C. Ash & R.B. Gallagher (eds) *Immunoparasitology Today*, A38–A41. Cambridge, Elsevier Trends Journals.

Popiel, I. (1986) The reproductive biology of schistosomes. *Parasitology Today*, 2, 10–15.

Rees, H.A. & Mercer, J.G. (1986) Occurrence and fate of parasitic helminth ecdysteroids. *Advances in Invertebrate Reproduction*, 4, 173–186.

Rumjanek, F.D. (1987) Biochemistry and Physiology. In D. Rollinson & A.J.G. Simpson (eds) *The Biology of Schistosomes: from Genes to Latrines*, pp. 163–183. London, Academic Press.

Russell, D.G. & Dubremetz, J.F. (1986) Microtubular cytoskeletons of parasitic Protozoa. *Parasitology Today*, 2, 177–179.

Salafsky, B. & Fusco, A. (1987) Eicosanoids as immunomodulators of penetration by schistosome cercariae. *Parasitology Today*, 3, 279–281.

Schulz-Key, H. & Karam, M. (1986) Periodic reproduction of *Onchocerca volvulus*. *Parasitology Today*, 2, 284–286.

Seebeck, T., Hemphill, A. & Lawson, D. (1990) The

cytoskeleton of trypanosomes. *Parasitology Today*, **6**, 49–52.

Shaw, C. & Johnson, C.F. (1991) Role of regulatory peptides in parasitic helminths and their vertebrate hosts: possible novel factors in host–parasitic interactions. *Parasitology*, **102**, (supplement), S93–S105.

Sherman, I.W. (1983) Metabolism and surface transport of parasitised erythrocytes in malaria. In D. Evered & J. Whelan (eds) *Malaria and the Red Cell*, pp. 206–217. London, Pitman.

Sherman, I.W. (1988) Mechanisms of molecular trafficking in malaria. In C. Arme & L.H. Chappell (eds) *Molecular Transfer Across Parasite Membranes. Parasitology*, **96**, (supplement), S57–S81.

Sherman, I.W. & Zidovetski, R. (1992) A parasitophorous duct in *Plasmodium*-infected red blood cells. *Parasitology Today*, **8**, 2–3.

Sinden, R.E. (1983) Sexual development of malaria parasites. *Advances in Parasitology*, **22**, 154–216.

Smyth, J.D. & Halton, D.W. (1983) *The Physiology of Trematodes*, 2nd edn. Cambridge, Cambridge University Press.

Smyth, J.D. & Haslewood, G.A.D. (1963) The biochemistry of bile as a factor determining host specificity in intestinal parasites, with particular reference to *Echinococcus granulosus*. *Annuals of the New York Academy of Science*, **113**, 234–260.

Smyth, J.D. & McManus, D.P. (1989) *The Physiology and Biochemistry of Cestodes*. Cambridge, Cambridge University Press.

Sommerville, R.I. & Rogers, W.P. (1987) The nature and action of host signals. *Advances in Parasitology*, **26**, 240–294.

Stretton, A.O.W., Cowden, C., Sithigorngul, P. & Davis, R.E. (1991) Neuropeptides in the nematode *Ascaris suum*. *Parasitology*, **102**, (supplement), S107–S116.

Sukhdeo, M.V.K. & Mettrick, D.F. (1987) Parasite behaviour: understanding platyhelminth responses. *Advances in Parasitology*, **26**, 73–144.

Tait, A. & Sacks, D.L. (1988) The cell biology of parasite invasion and survival. *Parasitology Today*, **4**, 228–234.

Tanabe, K. (1990) Glucose transport in malaria infected erythrocytes. *Parasitology Today*, **6**, 225–229.

Theron, A. (1986) Chronobiology of schistosome development in the snail host. *Parasitology Today*, **2**, 192–194.

Threadgold, L.T. (1984) Parasitic Platyhelminthes. In J. Bereiter-Hahn, A.G. Matoltsy & K.S. Richards (eds) *Biology of the Integument* Vol. 1 *Invertebrates*, pp. 132–191. Heidelberg, Springer-Verlag.

Tinsley, R.C. (1989) The effects of host sex on transmission success. *Parasitology Today*, **5**, 190–195.

Trager, W. (1986) *Living Together: The Biology of Animal Parasitism*. New York, Plenum.

Wahlgren, M., Carlson, J., Udomsangpetch, R. & Perlman, P. (1989) Why do *Plasmodium falciparum*-infected erythrocytes form spontaneous erythrocyte rosettes? *Parasitology Today*, **5**, 183–185.

Webster, L.A. & Wilson, R.A. (1970) The chemical composition of protonephridial canal fluid from the cestode *Hymenolepis diminuta*. *Comparative Biochemistry and Physiology*, **35**, 201–209.

Wharton, D.A. (1980) Nematode egg shells. *Parasitology*, **81**, 447–463.

Wharton, D.A. (1986) *A Functional Biology of Nematodes*. London, Croom Helm.

Wilson, R.A. (1987) Cercariae to liver worms: development and migration in the mammalian host. In D. Rollinson & A.J.G. Simpson (eds) *The Biology of Schistosomes: from Genes to Latrines*, pp. 115–146. London, Academic Press.

Young, A.S. & Morzaria, S.P. (1986) *Biology of Babesia*. *Parasitology Today*, **2**, 184–185.

Chapter 8 / Immunology

F. E. G. COX

8.1 INTRODUCTION

The immune system of any animal is concerned with defence against invading organisms and the removal of potentially malignant cells. Over the last decade, the immune response has been studied in the minutest detail, with the aid of the latest techniques of molecular biology, to such an extent that the principles of all the processes involved are now clearly understood. This understanding of the immune response has been applied to infectious diseases and it is now clear that immunity to any infection is consistent with the central dogmas of immunology which involve a small number of well-defined processes operating sequentially or concurrently. Essentially, the role of the immune system is to deliver an appropriate immune response which must be delivered at the right place and at the right time. Immunity to infections is complicated by the facts that the various agents involved are very different from each other, they occupy different sites in their hosts and many have evolved methods of evading the immune response. These complications often result in inappropriate immune responses which bring about pathological changes instead of protection.

In order to understand the nature of immunity to parasites it is necessary to distinguish between microparasites and macroparasites (see Chapter 4, Section 4.2.2). Protozoa, like viruses and bacteria, are microparasites that multiply within their vertebrate hosts posing an immediate threat unless constrained. Helminths, on the other hand, are macroparasites that do not normally multiply within their vertebrate hosts and only present a real threat when large numbers have been built up following repeated exposure to infection. The host defence strategies must therefore be different in response to protozoa and helminths, but there are several principles that apply to parasites of all kinds.

In classical acquired immunity, the host becomes infected, the invading organisms are recognized as foreign, an appropriate immune response is elicited and the organisms are eliminated within about 14 days, after which the host is resistant to reinfection with the same organism. This is what happens in the majority of bacterial and viral infections and the importance of the immune response is clearly seen if the host has some deficiency in the immune system, in which case a rapidly fulminating and usually fatal infection ensues. In most parasitic infections caused by protozoa, the parasite becomes established and the host begins to overcome the infection but is frustrated by the ability of the parasite to evade the immune response or to move out of range. The net result is that the infection is partially controlled but the immune response may become misdirected towards, for example, host molecules deposited in various tissues, resulting in immunologically mediated damage compounded by the overrelease of toxic molecules by cells of the immune system. In helminth infections, the situation is similar and immune responses elicited against, for example, migrating larval stages may cause damage to host tissues while the adult worms escape attack either by moving out of range or disguising themselves in some way. A long lasting infection accompanied by some immunopathology is characteristic of most parasitic infections including those in humans (Table 8.1). This kind of uneasy compromise situation is, by its very nature, unstable and can easily change to one in which the host is overwhelmed or one in which the host recovers and becomes immune to reinfection.

Table 8.1 Some important parasitic infections of humans showing the duration of the infection, evidence for clinical immunity and the occurrence of gross immunopathology. In all cases, there is evidence of clinical immunity but not necessarily in all infected individuals. Gross immunopathology can be correlated with the duration of the infection

	Duration of infection (years)	Clinical immunity	Gross immunopathology
Plasmodium vivax	1–2	+	−
Leishmania tropica	1–2	+	−
L. donovani	Life	+	+
Trypanosoma cruzi	Life	+	+
Ascaris lumbricoides	1–2	+	−
Necator americanus	2–4	+	−
Schistosoma mansoni	3–4	+	+
Wuchereria bancrofti	Life	+	+

The prevalence of long chronic infections, which often became fulminating, led scientists at one time to believe that there were no effective immune responses against the majority of parasites. It is now clear, however, that immunity to parasitic infections is the rule, the evidence coming from three sources: (1) in endemic areas the prevalence of infections declines with age whereas signs of immunity, such as raised levels of antibodies, increases; (2) immunodepressed individuals, such as those with *AIDS*, quickly succumb to infection; and (3) in every experimental model investigated, acquired immunity can be demonstrated under certain circumstances. It is clear then that acquired immunity does occur and parasite immunologists are now largely concerned with the nature of acquired immunity with the specific aims of developing vaccines and eliminating or moderating immunopathological side-effects. Such investigations have led to a fairly clear understanding of the ways in which parasites manage to evade the immune response but, before this can be discussed, it is necessary to consider the nature of acquired immunity itself.

8.2 THE IMMUNE RESPONSE

An immune response is elicited whenever an animal is injected with a foreign protein or polysaccharide or is infected with any organism that gains access to the cells of the body. Much of what we know about the immune response has been derived from experiments involving the in-jection of well-defined proteins or peptides into laboratory mice but the general principles apply to infectious agents in humans and domesticated animals. An outline of what happens is shown in Fig. 8.1 but the actual processes involved are much more complex than this. In summary, the invading organism, or part of it, is taken up by a macrophage as part of the normal process of phagocytosis, broken down in the phagocytic vacuole and fragments are transported to the surface accompanied by specific molecules called class II MHC molecules. The fragments constitute the antigen and the whole complex, consisting of the antigen and the class II molecule, is recognized by receptors, called T cell receptors, on T helper lymphocytes and binds tightly to them. The macrophage, which acts as an antigen presenting cell (APC), then secretes a lymphocyte activating factor called interleukin 1 (IL-1) which stimulates the T helper lymphocytes to secrete other cytokines which activate other sets of cells. In mice there are two kinds of helper T cells, T_H1 and T_H2 and similar cells exist in humans and other animals. T_H1 cells secrete IL-2, T cell growth factor, which activates cytotoxic cells known as Tc cells, or gamma-interferon, IFN-γ, which activates macrophages. T_H2 cells secrete IL-4 and IL-5 which activate B lymphocytes to produce antibodies. Immunological specificity is conferred at the stage of antigen processing and presentation and is maintained by the production of memory T and B cells which continue to circulate long after the antigen has disappeared.

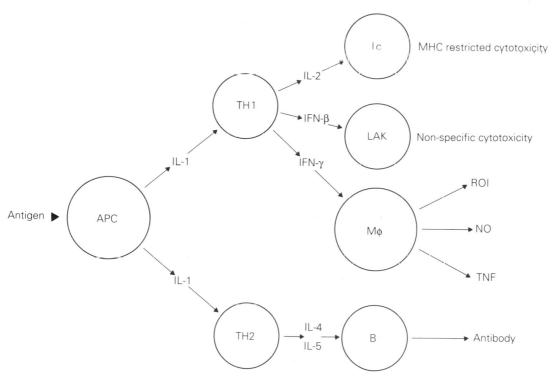

Fig. 8.1 The main events in the immune response. (For further details see Cox & Liew, 1992, and the text.)

This general pattern varies in detail, for example, antigen presentation can be exogenous, as described above, or endogenous in which the antigen is synthesized by the infected cell and presented in the context of Class I molecules. The overall result is the activation of B cells, cytotoxic T cells and macrophages which together constitute the three arms of the immune response in which the effector mechanisms are antibodies, cytolysis and macrophage-mediated killing. These three arms, acting either independently or in combination, are instrumental in the destruction of parasites and the acquisition of immunity. Once immunity has been established, resistance to reinfection is brought about by recalling the original immune response through the activation of the circulating memory cells.

8.2.1 Antibodies

Antibodies are immunoglobulins which consist of basic units of four polypeptide chains, two heavy and two light (Fig. 8.2). The antigen binding part, the Fab portion, is formed from one heavy and one light chain, thus there are two chains on each basic unit. The opposite end of the molecule is the Fc portion which is capable of binding to any cell with an appropriate Fc receptor. There are five classes of immunoglobulin, IgG, IgM, IgA, IgE and IgD of which only the first four are involved in immune responses to parasites. Immunoglobulin G and IgA are further divided into subclasses, each of which has its own specific function.

Immunoglobulin M, a pentameric form of the basic molecule, possesses ten antigen-binding sites. It is the first immunoglobulin produced during an immune response and its structure is ideally suited to binding to organisms with repeated antigenic sites as in many bacteria. Immunoglobulin G, the most abundant immunoglobulin, is a monomeric form which is produced after, and replaces, IgM and persists after the disappearance of the antigen. Immunoglobulin A,

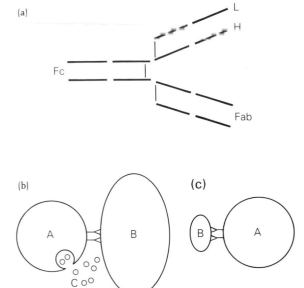

Fig. 8.2 Antibodies and some of their effects. (a) Diagrammatic representation of an immunoglobulin molecule. One end of the molecule (Fc) binds to cells of various kinds while the other (Fab) binds antigen in a cup-like structure formed by a heavy and a light chain. (b) Immunoglobulin E binds to the target cell by its Fab portion and to a cell such as a mast cell, basophil or eosinophil by its Fc portion causing the host cell to release pharmacologically active or toxic substances. (c) Antibodies bind to a parasite by their Fab portions and to cells such as macrophages by the Fc portion, a process known as antibody dependent cell mediated cytotoxicity (ADCC). (A, host cell; B, parasite; C, host cell products.)

which occurs either as a monomer or a dimer combined with a secretory piece, occurs in secretions in the gut and other mucous surfaces and also in the serum. Immunoglobulin E is a cell-bound monomer which does not occur free in the serum.

Immunoglobulin M and IgG can act in various ways. They can cause microorganisms to agglutinate, in the same way as mismatched blood, or cause them to be phagocytosed by binding to the antigens by the Fab portion and to the phagocytic cell by the Fc portion. They can also fix complement, a series of factors in the blood which, when activated in order, result in the lysis

of the target cell (Fig. 8.3). The final steps in the activation of the complement cascade are not specific and cells in the vicinity of the target cell can also be destroyed, a process known as by-stander lysis. Complement components are also chemotactic for white cells, enhance phago-cytosis and increase vascular permeability. Complement can also be activated in the absence of antigen–antibody reactions by various substances, including bacterial lipopolysaccharides, through what is known as the alternative pathway (Fig. 8.3). The complement system is, therefore, ideally suited for the destruction of microorganisms and is also potentially dangerous for macro-parasites, but can also cause inflammation and damage to the host.

Immunoglobulin A is also able to bind to microorganisms preventing them from attaching to and invading mucosal surfaces and, as it is not destroyed by enzymes in the gut, IgA is an ideal immunoglobulin as a first line of defence against intestinal pathogens. Immunoglobulin E is a cell-bound immunoglobulin which binds to cells, including mast cells, eosinophils and basophils, by its Fc portion and, if the Fab portion binds antigen, the cell releases various pharmacologically active substances (Fig. 8.2). This is the basis of anaphylaxis, asthma and hay fever but also plays an important role in the immune response to helminth worms.

8.2.2 Cell-mediated immunity

Cell-mediated immunity is a term that has out-lived its usefulness and the processes previously grouped together under this heading comprise both cytotoxic T cell activity and macrophage activation which have in common independence from antibody but little else. Provided that this is realized, the term is a convenient one and is still retained here and in a number of immunological textbooks. Cytotoxic T cells are able to recognize cell surface antigens in the context of MHC class I molecules and then lyse the cells carrying the antigen, a process known as class I restricted cell killing. This is the main method of defence against virally infected cells and tumour cells but, except in the case of theileriasis, is not

(a)

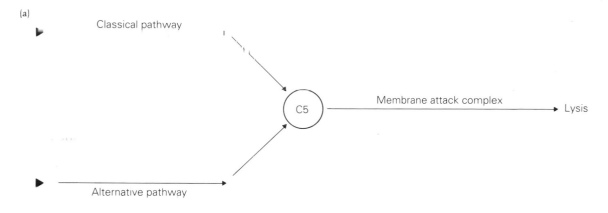

(b)

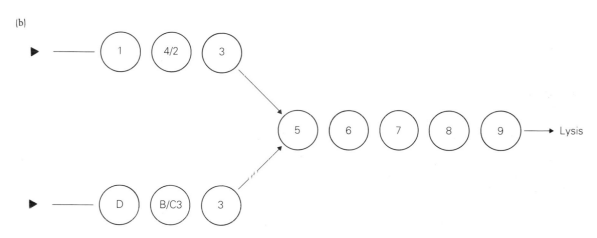

Fig. 8.3 Complement activation. Complement consists of serum components which, when activated in sequence, bring about the lysis of a target cell. (a) In the classical pathway, activation is initiated by an antigen–antibody complex. In the alternative pathway, activation is brought about by a number of factors including microbial products. (b) In the classical pathway, the first component, C1, binds to an immunoglobulin molecule and activates, in turn C4, C2 and C3. In the alternative pathway the complement components activated in order are D, B/C3 and C3. C3 plays a central role activating C5 which initiates the membrane attack complex components C6, C7, C8 and C9 forming a pore in the target cell membrane through which ions are lost and water enters the cell resulting in cell lysis. Control mechanisms are not shown in these diagrams.

the primary defence mechanism in any parasitic infection.

Macrophages are phagocytic cells which take up microorganisms and destroy them within phagolysosomes. Normally this occurs at a low level but, under certain circumstances, macrophages can become activated and exhibit enhanced phagocytosis and destruction of ingested microorganisms. In addition, macrophages carry receptors for the Fc portion of the immunoglobulin molecule which forms a bridge between a parasite and the macrophage and thus activated macrophages can damage or destroy even relatively large parasites, a process known as antibody dependent cell mediated cytotoxicity (ADCC) (Fig. 8.2). Macrophages can be activated in several ways, the most important of which is by IFN-γ although complement components may

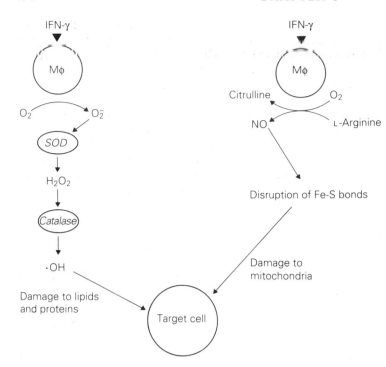

Fig 8 4 Activation of macrophages and parasite killing through reactive oxygen intermediates and nitric oxide. (a) Macrophages activated by interferon-gamma (IFN-γ) produce superoxide (O_2^-) which is converted to hydrogen peroxide, H_2O_2, by the enzyme superoxide dismutase (SOD). Hydrogen peroxide is converted to water and the hydroxyl radical $\cdot$OH. Superoxides, peroxides and hydroxyl radicals are all cytotoxic mainly by damaging lipids and proteins especially those of the membrane. (b) Interferon-gamma activated macrophages also produce nitric oxide (NO) from L-arginine. Nitric oxide disrupts iron–sulphur bonds and causes damage to mitochondria. (For further details, see the text.)

also be involved. Activated macrophages possess higher levels of proteolytic enzymes than resting cells and also produce more reactive oxygen intermediates and nitric oxide. Reactive oxygen intermediates include superoxide, peroxides and hydroxyl radicals all of which are highly toxic to living organisms or cells (Fig. 8.4). Other important products of activated macrophages are nitric oxide and tumour necrosis factor (TNF), both also highly toxic to living cells (Fig. 8.4). Activated macrophages can, therefore, destroy ingested microorganisms and also any organisms within close proximity but can also contribute to the damage to host cells and tissues. Macrophages are also under the influence of other cytokines including those that enable these cells to remain in the vicinity of antigens and sensitized lymphocytes. Such accumulations of cells can, therefore, encapsulate a potentially dangerous pathogen but can also contribute to pathology by causing local inflammatory reactions.

Several other kinds of cells are also involved in immunity to parasites. Neutrophils, also called polymorphs, are phagocytic cells that occur in the circulation in large numbers and constitute the first line of defence against microorganisms but usually die at the site of infection. Neutrophils possess receptors for IgG and complement but cannot act as antigen presenting cells. Eosinophils also occur in the circulation and, although they have the capacity to ingest foreign particles, are not actively phagocytic nor do they produce reactive oxygen intermediates. Eosinophils possess receptors for complement and for IgG and, most importantly, IgE. On stimulation, which can occur if surface immunoglobulins are cross-linked, they extrude granules containing potentially toxic molecules. Basophils also possess receptors for complement and IgG and IgE and degranulate when stimulated, releasing histamine and other pharmacologically active substances as do mast cells, which usually occur in the mucosa and connective tissues, whose action is more localized. Platelets, normally involved in blood clotting, possess receptors for IgG and IgE. All these cells with immunoglobulin or complement receptors can participate in ADCC and can contribute to immunopathological processes.

There are also a number of less clearly defined cells which include natural killer (NK) cells, large granular lymphocytes without T cell receptors, and lymphokine activated killer (LAK) cells. Both of these cell types, which may represent different forms of the same kind of cell, are cytotoxic to tumour and other target cells, including parasites, but do not possess the characteristics of Tc cells and do not exhibit class I restricted killing.

8.2.3 Hypersensitivity

'Hypersensitivity' is a term applied to an enhanced state of responsiveness following sensitization to a particular antigen and represents the unacceptable face of the immune response. Hypersensitivity states are important in parasite immunology and four types are generally recognized, three of these involve preformed antibody and occur soon after exposure to a particular antigen, causing immediate type hypersensitivity whereas the last is antibody independent and is slow to develop. Type I involves the binding of antibody, particularly IgE, to mast cells or basophils which degranulate releasing pharmacologically active substances such as histamine. This kind of reaction results in smooth muscle contraction and increases in vascular permeability and causes asthma or hay fever if the antigen is applied locally or anaphylaxis if it is given systemically. In parasitic infections, Type I reactions are important in defence against helminth worms, which may be expelled if the environment in which they live is disturbed by the release of pharmacologically active substances, but can also contribute to immunopathology. Type II hypersensitivity involves antibodies binding to cell surfaces via their antibody binding sites and activating complement bringing about the lysis of the cell. Alternatively, antibody bound to the cell surface can bind, by the Fc portion, to phagocytic cells facilitating phagocytosis which can also occur if the phagocytic cell binds to complement already fixed on the cell surface by antibody. This is the basis of ADCC which plays a defensive role in many parasitic infections but can also contribute to pathology if parasite antigens have bound to host cells. Type III hypersensitivity occurs when antibody and antigen are both in solution and combine to form complexes which activate complement or lodge in various organs which they physically block. There is no apparent defensive role for type III hypersensitivity reactions in parasitic infections but they are frequently involved in pathological processes. Type IV hypersensitivity, a form of cell-mediated immunity also called delayed type hypersensitivity (DTH), involves cytokine-activated cells particularly macrophages, plays an important role in defence against many parasitic infections but can also contribute to pathology.

8.2.4 Regulation of the immune response

The very nature of the immune response, particularly the dangers inherent in any malfunctioning, demands that it must be carefully controlled and regulated. This regulation has been the subject of a great deal of research during the past few years and it is now clear that regulation acts at all stages of the immune response. The nature of the antigen and the way in which it is processed and recognized by the T helper cells determines the eventual outcome of the immune response. Some antigens preferentially activate T_H1 cells leading to the production of IL-2 and IFN γ. On the other hand, small amounts of IL-4 can synergize with IFN-γ to enhance macrophage activation. Cytokines are, therefore, central to the regulation of the immune response and an understanding of their properties has replaced earlier concepts of suppressor T lymphocytes. A number of parasites are able to interfere with this fine regulation (see Section 8.7) and this can lead to the activation of an inappropriate pathway resulting in pathology instead of protection. In order to understand the nature of protective immune responses to parasites and any concurrent pathology it is essential to know a considerable amount about parasite antigens and the ways in which they are processed and much current research is directed towards finding ways of inducing protective immunity while ameliorating pathological processes.

8.3 IMMUNITY TO MICROORGANISMS

Acquired immunity is the rule in viral and bacterial infections and such infections usually follow a characteristic pattern of an incubation period, a period of patent infection and a crisis followed by relatively rapid recovery. On subsequent exposure to the same pathogen, the resultant infection is usually either asymptomatic or mild. There are a number of exceptions to this general rule but in most cases the infection is rapidly overcome and the ensuing immunity lasts for a considerable period of time.

Immunity to viruses involves a number of different components. Viruses are obligate intracellular pathogens and the main defence mechanism is the production of non-specific interferon-alpha (IFN-α) which inhibits the entry of viral particles into the host cells. Once within the cell, viral antigens located on the cell surface can be recognized in the context of class I MHC molecules and destroyed by cytotoxic T lymphocytes. Viral particles so released are neutralized by appropriate antibodies, IgM and IgG in the serum and IgA in secretions, which provide the main line of defence against reinfection. However, in many viral infections the immune response is not as simple as this and it is becoming clear that other mechanisms including IFN-γ and TNF may also be involved and are particularly effective when working in combination with one another and with other components of the immune system.

Bacteria are considerably larger than viruses and tend to be extracellular rather than intracellular. The first lines of defence are non-specific barriers, such as the skin, gastric juices or lysozyme. If the bacteria succeed in entering the body they are immediately subject to non-specific phagocytosis by neutrophils or macrophages. This phagocytosis may be reinforced when antibodies coating the bacteria bind to Fc receptors on the phagocytic cells thus conferring a degree of specificity. Macrophages and neutrophils destroy bacteria within the phagolysosome through a combination of oxygen-independent and oxygen-dependent processes and the latter are enhanced by IFN-γ secreted by specifically activated T lymphocytes. Obligate intracellular bacteria, such as those that cause leprosy and tuberculosis, are also killed by IFN-γ activated macrophages. Complement is also involved in immunity to bacterial infections and can be activated through the classical pathway following the binding of antibody,

particularly IgM, or through the alternative pathway triggered by bacterial cell wall components. Complement also facilitates phagocytosis by binding to complement receptors on phagocytic cells. As is the case for viral infections, there is increasing evidence that other components of the immune system may be involved in immunity to bacteria, e.g. cytotoxic T lymphocytes can destroy cells harbouring bacteria releasing them for subsequent destruction by IFN-γ activated phagocytic cells.

8.4 ANIMAL MODELS AND *IN VITRO* SYSTEMS

Because of their nature it is almost impossible to unravel the intricacies of the human immune responses to parasites and considerable reliance has had to be placed on animal models and *in vitro* systems. Although it is not so difficult to study parasitic infections in domesticated animals, the convenience of laboratory models has made these attractive to laboratory workers and, for all those working on parasites, the availability of sophisticated laboratory techniques has led to the accumulation of massive amounts of data which might or might not be relevant to natural infections. Animal models range from those in which the actual parasite of economic importance can be maintained in a convenient laboratory animal, to those in which a natural parasite of such an animal is considered as an analogue of the one of real interest. For example, at one end of the spectrum, the human malaria parasite, *Plasmodium falciparum*, can be maintained in squirrel monkeys and *Schistosoma mansoni* in baboons while at the other end, the rodent malaria parasite, *P. berghei*, and the rodent nematode, *Nippostrongylus brasiliensis*, have been much studied as models of malaria and intestinal nematodes. Laboratory mice are the most favoured model hosts and a number of inbred strains are available in which both the infections and the immune responses can be reproduced with a great deal of precision. Certain strains are resistant to particular infections whereas others are susceptible. C57/B1 mice, for example, are resistant to infection with *Leishmania major* and Balb/c mice are susceptible while the reverse is true for the nematode, *Trichinella spiralis*. The best studied immune responses are those to microorganisms that live within macrophages, *L. donovani*, *Salmonella typhimurium* and various mycobacteria, in which immunity is conferred by a single gene and this has facilitated the most detailed dissection of the processes involved in immunity to these and related organisms. Although experiments with laboratory models have provided a vast amount of information about immune responses to parasites the data collected must be treated with caution and extrapolations to humans and domesticated animals made with great care. Experiments in one animal can only provide clues as to what is happening in another and, as in any kind of detective work, it is essential that conclusions should not be drawn unless corroborated by evidence from other sources. In this context, it is important to realize that the immune responses of an animal like a mouse to a large parasite such as a helminth worm must inevitably be very different in scale from those in a much larger host.

A number of well-tried and well-characterized biochemical, molecular and cellular techniques are available for the study of various aspects of the immune response *in vitro*. Essentially these are concerned with the identification and characterization of antigens and the use of these to elicit the production of antibodies and to activate cytotoxic T lymphocytes or macrophages. With such systems it is possible to investigate the roles of a range of molecules in the initiation, activation or inhibition of immune responses. It is also possible to maintain many parasites *in vitro* thus eliminating the use of animals altogether. Various caveats apply to *in vitro* experiments which have the inherent disadvantage that, given the right conditions, some kind of immune response can be detected in any infection but whether or not it has any relevance in the whole body may be uncertain. For example, macrophages can always be activated by IFN-γ *in vitro* but in animals infected with *L. major* this activation is inhibited by IL-4 and IL-10 and numerous cells bind to schistosome larvae in culture but there is no way of determining whether or not this happens in an animal. Nevertheless, the use of animal models and *in vitro* systems has provided a vast amount

of invaluable information which, provided that it is treated with caution, can contribute to our understanding of immunity to parasites in humans and domesticated animals. In the examples given below, the emphasis is on what is known about immunity to parasitic infections in their natural hosts but reference is also made to relevant information drawn from animal models and *in vitro* systems.

8.5 IMMUNITY TO PROTOZOA

There is evidence for the existence of immunity in the majority of protozoan infections although, in many cases, it is not fully understood. Protozoa, being microparasites, multiply within their hosts and, unless this multiplication is contained, the host will inevitably and quickly die. The fact that most infected individuals do not succumb to fulminating infections indicates that some kind of immunity has intervened. It is convenient to consider immune responses with respect to three separate functions: (1) bringing the initial infection under control; (2) eliminating residual parasites; and (3) preventing reinfection, and the actual immune mechanisms involved in each of these may be quite different from one another. It is also clear that each kind of immune response is multifactorial and involves different effector mechanisms operating in a predetermined temporal sequence and several mechanisms may be operating at any one time. All protozoa have ways of evading the immune response and all generate a myriad of immunological reactions unrelated to protection. All these factors, coupled with the fact that different species or strains of hosts may react in quite distinct ways to the same protozoan infection, have combined to make it difficult for any investigator to identify *the* actual immune mechanism(s) involved in any particular infection. Nevertheless, it is now possible to recognize many of the protective mechanisms and to unravel them from the counterprotective, immunopathological and irrelevant ones.

8.5.1 Intestinal protozoa

The majority of amoebae and flagellates that inhabit the lumen of the gut are harmless commensals that elicit no immune response. *Entamoeba histolytica* falls into this category until it invades the intestinal wall and, possibly later, the liver and other organs. Invasion is accompanied by the production of complement-activating antibodies capable of lysing the trophozoites. However, there is no correlation between antibody production and protection and repeated infections involving the invasion of the gut wall are relatively common even in the presence of high levels of antibody. In hepatic amoebiasis, reinfection is rare. In experimental animals, spontaneous recovery and resistance, which appear to have a genetic basis, have been demonstrated and can, in some cases, be correlated with the acquisition of cell-mediated immunity. *In vitro* studies have shown that the amoebae are sensitive to oxygen radical killing suggesting that macrophage or neutrophil activation may be involved in amoebiasis, and neutrophils that cannot normally kill trophozoites are able to do so if pretreated with IFN-γ particularly if combined with TNF. Amoebiasis is accompanied by a range of immunological dysfunctions, including immunodepression, and the amoebae are able to resist immune attack in several ways, e.g. trophozoites coated with antibody are resistant to complement lysis and are also able to attract and kill non-activated neutrophils. This direct killing on the part of the parasite also contributes to hepatic lesions. Human amoebiasis is, therefore, a complicated disease and one in which relatively little is known about the actual mechanisms involved in either protection or immunopathology.

Human giardiasis, which does not involve tissue invasion, occurs in various forms, some individuals quickly overcome the infection whereas others develop chronic disease. This suggests that immunity does occur and that it is genetically controlled and this is supported by evidence from murine models using *Giardia muris*. Serum antibodies, including IgA which could act at the mucosal level, are produced during human infections and ADCC and complement-mediated lysis have been demonstrated *in vitro*. However, there is no real evidence of acquired immunity or for any mechanism for such immunity in humans. The pathology involves both mechanical damage and toxins produced by the parasites themselves.

8.5.2 Leishmaniasis

Leishmaniasis exists as a complex of diseases that, from an immunological as well as a clinical point of view, fall into three major categories: cutaneous, mucocutaneous and visceral, but even this simple classification is confused by the existence of a spectrum of intermediate forms. In the cutaneous forms, as exemplified by *L. major* and *L. tropica* in the Old World and *L. mexicana* in the New World, recovery, albeit slow, is the rule and this is characterized by increased DTH reactions, little antibody production and resistance to reinfection. In mucocutaneous leishmaniasis, caused by *L. braziliensis*, recovery does occur eventually, although it may be too slow to save some victims, and both DTH and antibodies are present. In visceral leishmaniasis, caused by *L. donovani*, recovery is rare and infections are characterized by poor DTH responses but considerable antibody production. All this evidence indicates a central role for cell-mediated immunity and no requirement for antibody. All forms of leishmaniasis, particularly that caused by *L. donovani*, are accompanied by prolonged and profound immunodepression.

Experiments in laboratory animals, particularly mice which can easily be infected with most *Leishmania* species have thrown considerable light on immunity to leishmaniasis and, in the case of *L. major* and *L. donovani*, resistance is genetically determined and is controlled by alleles at a single locus unrelated to the major histocompatibility complex. Immunity is dominated by cell-mediated immune responses involving the activation of macrophages by IFN-γ. The parasites live in resting macrophages, where they are able to survive within the phagolysosome, but cannot survive in activated macrophages within which they are killed by reactive oxygen intermediates and nitric oxide. In infected animals, this immunity can be counteracted by cytokines, such as IL-4, which are involved in activating B lymphocytes for antibody production. There is evidence that similar mechanisms operate in leishmaniasis in humans and this would explain the apparent paradox concerning the role of antibody in this disease.

8.5.3 American trypanosomiasis (Chagas disease)

American trypanosomiasis, caused by *Trypanosoma cruzi*, progresses in a series of well defined stages, an incubation period of 2–3 weeks followed, in children but rarely in adults, by an acute phase lasting 4–5 weeks and a chronic phase that lasts indefinitely. Some individuals die early on in the infection but most keep it under control for a variable period. In immunosuppressed patients, the infection is usually acute and fatal. This evidence suggests the existence of an effective but incomplete immune response and this has been well-characterized in laboratory animals, particularly mice which can easily be infected with *T. cruzi*. After infection via the faeces of the vector, the parasites enter macrophages, where they survive by escaping from the phagosome before lysosome fusion, and other cells including muscle and nerve cells. Within these cells, the parasites multiply as amastigote forms but periodically produce trypomastigote forms which enter the blood. The trypomastigote forms, which cannot undergo antigenic variation as do their African counterparts, are subject to attack by antibody and complement and most, but not all, are killed. Some survive because the parasites are able to cleave immunoglobulin molecules and also inhibit complement lysis. *Trypanosoma cruzi* can also inhibit the production of IL-2 and down-regulate the expression of IL-2 receptors which contributes to the generalized immunodepression that accompanies the infection. Amastigotes in macrophages are killed by reactive oxygen intermediates if their host cells are activated by IFN-γ, but those in muscle and nerve cells are relatively safe from attack. However, parasite antigens on the outside of infected cells can be recognized by antibodies and such cells are destroyed by complement-mediated lysis as are uninfected cells passively coated with such antigens. There is some controversy about whether these antigens are parasite-derived or represent shared epitopes with host-cell components. However, the outcome is the same, an apparently autoimmune disease in which both infected and uninfected cells are destroyed (see Section 8.8).

8.5.4 African trypanosomiasis

Unlike *T. cruzi*, the trypanosomes that cause sleeping sickness in humans, *T. brucei gambiense* and *T. b. rhodesiense*, and nagana in cattle and other animals, *T. b. brucei*, *T. congolense* and *T. vivax*, have no intracellular stages but live and multiply as trypomastigotes in the blood and other tissue fluids where they are easily susceptible to immune attack. The parasite's antigens are extremely immunogenic and induce an antibody response which agglutinates the trypanosomes and initiates complement-mediated lysis but no other immune response seems to be relevant to protection. The trypanosomes, however, are able to evade this immune attack by undergoing a process of antigenic variation. Each trypanosome is completely enveloped in a thick glycoprotein coat, the variant surface glycoprotein, VSG (Fig. 8.5). In *T. brucei*, the surface coat is produced while in the salivary glands of the vector tsetse fly. Most trypanosomes injected by the tsetse fly possess an identical surface coat but a few display variations in the polypeptide structure resulting in coats that are antigenically distinct. An immune response to the major variable antigen type (VAT) soon destroys all trypanosomes carrying this particular antigen but the minor variants survive to produce successive parasitaemias which in turn become dominant and are destroyed and replaced (Fig. 8.6). This variation is almost limitless thus the parasite is always one-step ahead of the immune response of the host. Infections with African trypanosomes are, therefore, characterized by successive waves of parasitaemia in which the predominant surface antigens during each wave are different from those of the preceding ones. When the trypanosomes are taken up by a tsetse fly they lose the coat and acquire it again in the salivary glands but there is no reversion to a single basic anti-

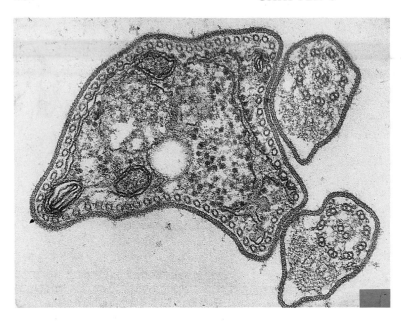

Fig. 8.5 Electronmicrograph of bloodstream *Trypanosoma brucei* showing the thick glycoprotein coat consisting of variant antigen. (Photograph by Professor K. Vickerman.)

genic type. Antigenic variation also occurs in *T. congolense* and *T. vivax* but, in this case, surface antigen is apparently acquired while in the blood of the mammalian host.

Despite this antigenic variation, many infections in humans become chronic and cattle that have been continuously exposed to infection, and drug-treated when ill, eventually acquire a considerable degree of immunity to reinfection. This indicates that, in the field, a host may eventually be able to build up an immunity to all the important variants in a particular area. Immunity, however, is acquired at some cost because repeated immunological responses to a moving antigenic target eventually leads to gross immunopathological changes and prolonged immunodepression.

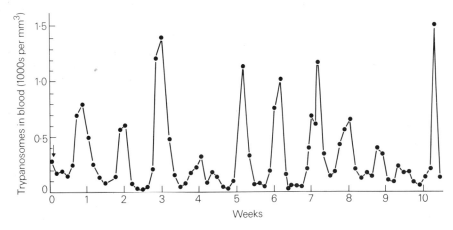

Fig. 8.6 Antigenic variation in trypanosomiasis. This graph shows the fluctuations in parasitaemia in a patient with trypanosomiasis. Such fluctuations are characteristic of human and animal trypanosomiasis. After each peak, the trypanosome population is antigenically different from that of earlier or later peaks. (After R. Ross and D. Thompson, *Proceedings of the Royal Society of London*, **B82**, 411–415, 1910.)

8.5.5 Coccidiosis

Coccidiosis embraces a wide spectrum of infections caused by members of the genus *Eimeria* mainly in chickens but also *E. bovis* in cattle. Infections are normally self-limiting and are governed by the genetically controlled number of asexual schizogonies. Immune responses are generated during the asexual phases of the infection but play little part in the control of these initial stages although they are extremely effective against reinfection with homologous parasites. The actual mechanisms of immunity involved have not been clearly defined but *in vitro* experiments have implicated IFN-γ during the intracellular stages and neutrophil activation and antibodies later. Immunity to reinfection seems to involve IgA antibodies in the gut.

8.5.6 Toxoplasmosis

Toxoplasma gondii, in hosts other than cats, multiplies indefinitely in various kinds of cells including resting macrophages in which the parasites are able to survive by preventing phagosome lysosome fusion. The parasite induces both IgM and IgG antibodies and the latter, which are produced early during the infection, persist for the lifetime of the infected individual and can be correlated with acquired immunity and resistance to reinfection. Immunity involves the killing of antibody-coated parasites by IFN-γ activated macrophages. Parasites coated with antibody seem to have reduced capacity to inhibit phagosome–lysosome fusion and, despite possessing considerable amounts of antioxidants, superoxide dismutase, catalase and glutathione peroxidase, are killed by reactive oxygen intermediates. The immune response does not, however, result in sterile immunity and dormant encysted forms of the parasite can be found in muscle and nervous tissue long after the acute phase of the infection and in the presence of high levels of circulating antibodies. The importance of the immune response can be appreciated from the fact that immunodepressed individuals suffer from fulminating infections and disseminated toxoplasmosis is one of the results of infection with the AIDS virus. Similarly, if the parasite passes across the placenta from an infected mother, the immunologically incompetent fetus will be infected and may be aborted or suffer neurological damage.

8.5.7 Malaria

Despite the fact that more effort has been put into the investigation of immunity to malaria than any other parasitic infection, the mechanisms involved are still not well-understood. Malaria infections may be acute and rapidly fatal, particularly in young children and strangers to an endemic area, but the course of the disease is usually long and chronic, about 2–4 years in the cases of *Plasmodium vivax*, *P. ovale* and *P. falciparum* and probably lifelong in *P. malariae*. This implies that immunity does occur and, although it is able to bring the potentially fulminating infection under control, it is incomplete and unable to bring about the total elimination of all parasites. The evidence for acquired immunity comes from the observations that individuals can recover and are resistant to reinfection with the homologous strain, there is a correlation between prevalence and age (see Chapter 4, Section 4.2.3) and mothers can pass on protective antibodies in milk to their offspring. The important questions, therefore, are what is the nature of the immune response involved, how do the parasites evade these responses and to what extent are they involved in the pathology of the disease?

Malaria parasites are relatively host-specific and cannot be studied in laboratory animals so much of the experimental work has involved the use of *P. falciparum* which can be maintained in squirrel and owl monkeys, which are expensive and difficult to obtain, and laboratory models, the most intensively studied being *P. knowlesi* and *P. cynomolgi*, in monkeys and *P. berghei*, *P. yoelii*, *P. vinckei* and *P. chabaudi* in rodents. From such studies, a vast amount of data has been accumulated and it is apparent that there are major differences between the immune responses that occur in different parasite and host combinations and even between different strains of parasite in the same host. It is not at all clear how these various findings relate to human malaria and current investigations are more concerned with what actually happens in humans and this work is facilitated by the fact that *P. falciparum* can be easily maintained in culture.

Many of the important antigens of the malaria parasites have been characterized and it is apparent that each stage in the life cycle has its own repertoire although some may be shared. The sporozoite possesses a dominant surface coat, the circumsporozoite (CS) protein, which is characterized by tandem repeats of as few as four amino acids and is highly immunogenic eliciting a strong antibody response that causes a precipitation around the parasite and inhibits invasion of the liver cells. However, even if only one parasite reaches the liver, an infection will be initiated. The mechanisms of immunity to the liver stages are unclear but seem to involve cell-mediated responses apparently directed at both the residual sporozoite antigen and other antigens unique to this stage. Immunity to the liver stages only occurs early in the infection and there are few signs of cellular infiltration or any other immune reactions against the late schizonts. The blood stages possess a large number of antigens distinct from those of the sporozoite or liver stages; some are characteristic of the early rings, mature schizonts or merozoites, some are associated with the membranes of the infected erythrocytes and some pass into the plasma when the infected cell ruptures. Most of these antigens possess tandem repeats of amino acids and many have been implicated in protective immune responses by eliciting specific antibodies that inhibit invasion of fresh erythrocytes or agglutinate merozoites, or participate in intracellular killing. However, none is able to induce total immunity. An alternative possibility is that the products of activated macrophages, such as reactive oxygen intermediates or TNF, can kill intra-erythrocytic forms and there is evidence that this does occur. The gametocytes possess their own characteristic antigens, again containing tandem repeats, and are highly immunogenic. Normally, however, these antigens are not exposed and the intracellular gametocytes cannot be killed in the mammalian host but infected individuals mount antibody responses that are effective against microgametes, macrogametes and zygotes within the mosquito if the immune serum is taken up with these stages.

The evidence for a particular immune mechanism in malaria is confusing and conflicting but a consensus view is that antibody responses limit the invasion of the liver by sporozoites, cell-mediated mechanisms operate in the liver but are not totally effective, a variety of antibody and non-antibody mechanisms are involved in containing the erythrocytic stages and antibodies are involved in the killing of sexual stages in the mosquito. This consensus is consistent with the fact that immune responses control the infection but do not eliminate it.

The mechanisms whereby malaria parasites evade the immune response are not at all clear and the various explanations are controversial. It may be that the dormant stages of *P. vivax* and *P. ovale* (see Chapter 1, Section 1.8.1) are able to evade the immune response by remaining in the liver but this cannot be true for the recrudescing forms *P. falciparum* and *P. malariae*. There is some evidence for antigenic variation and malaria parasites do exhibit a considerable degree of diversity, e.g. between geographical isolates and even within particular populations of parasites. It is possible that this diversity prevents the rapid acquisition of immunity and could explain why it builds up so slowly and why immunity to heterologous infections is so difficult to achieve.

This inherent antigenic diversity, the expression of novel antigens at each stage of the infection and the hide-and-seek nature of the infection as the parasite moves in and out of the relative safety of a host cell, results in a complex of immunological responses which together slowly control the infection but at the same time contribute to the pathology and eventually the immunodepression associated with the disease.

8.5.8 Babesiosis

Babesiosis in cattle, other domesticated animals and occasionally humans, is caused by infection with *Babesia* species which multiply within erythrocytes causing diseases similar to malaria. Babesial infections may be fatal but are characteristically long-lasting and parasites usually persist in the blood even after clinical recovery. Animals that have recovered are resistant to reinfection providing evidence for acquired immunity. Much of what is known about immunity to babesiosis has been derived from studies on *B. rodhaini* and *B. microti* in rodents. Little is known about immune responses to the sporozoites except that they elicit an antibody response but it is not known if this is effective against reinfection. The actual mechanisms of immunity to the erythrocytic stages are not at all clear but appear to be, in part at least, antibody-based presumably involving the inhibition of erythrocyte invasion as there are no other relevant stages in the life cycle. However, serum transfer experiments and *in vitro* inhibition experiments have been equivocal suggesting the involvement of other mechanisms and there is some evidence for cell-mediated killing involving activated macrophage products. A number of babesial antigens have been identified but not yet implicated in protective immunity. On the other hand, immunization against babesiosis in cattle and dogs has been relatively easy to achieve and can be correlated with the acquisition of antibodies. (See Section 8.9.)

8.5.9 Theileriosis

Theileria species, of which the most important species are *T. parva* and *T. annulata*, that infect cattle are the only parasites that live in lymphocytes. The parasites, injected by a tick, enter lymphoid cells and multiply at a rate that outpaces the immune responses of the host causing considerable mortality in the case of *T. parva* and morbidity in the case of *T. annulata*. *Theileria* species cannot be maintained in laboratory animals so what is known about immunity to these infections has been acquired from experiments with cattle so is directly relevant to the field situation. Some animals recover naturally and others recover after drug treatment, and in both cases, subsequently resist reinfection with the homologous parasite implying the acquisition of immunity. Little is known about immunity to the sporozoites except that they elicit antibodies to a cluster of surface antigens which neutralize infectivity *in vitro* and a recombinant form of the dominant surface antigen has been used to immunize cattle with some success. Infected lymphocytes are the targets of immune attack and the single most important aspect of the immune response is that these are destroyed by specific T-cells in a functional class I MHC restricted manner and that later non-specific cytotoxic cells are involved. These mechanisms have been exploited in the development of effective vaccines (see Section 8.9).

8.6 IMMUNITY TO HELMINTH WORMS

With the exception of larval cestodes, helminths do not multiply within their vertebrate hosts. Immune responses are, therefore, not concerned with the immediate problem of bringing a life-threatening infection under control, as happens in protozoan and microbial infections, but with curtailing the lives of established worms and preventing reinfection. Most helminths have life cycles in which larval stages migrate around the body before the adults settle in their final sites and, although some kind of immunity to reinfection is a general rule, existing worms are not always destroyed. Immunity elicited by larval stages is effective against subsequent invading larvae giving rise to a situation in which the host harbours a population of adult worms but is immune to reinfection – a state called concomitant immunity.

Adult and larval worms are too large to be destroyed by antibodies, with or without complement, and the universal effector mechanism seems to be some form of ADCC in which the worm becomes coated with antibody which binds to the macrophages, neutrophils, eosinophils or mast cells that actually destroy the parasite by secreting various toxic substances onto its surface. As in protozoan infections, the immune mechanisms involved are multifactorial and differ according to the stage of development of the invading worm and the site of infection. Similarly, helminths are capable of evading the immune response and these responses may become misdirected and contribute to the pathology of the infection.

8.6.1 Intestinal nematodes

Many intestinal nematodes have life cycles with larval stages that migrate around the body of the host until the adult finally settles in the intestine (see Chapter 2, Section 2.5). Immunity is elicited by, and is effective against, the migrating larvae, the best examples being *Dictyocaulus viviparus* and *Ancylostoma caninum* against which vaccines have been developed. In addition, many nematode infections are characterized by 'self-cure' reactions in which the adults in the intestine are spontaneously eliminated, for example, *Nippostrongylus brasiliensis* in rats or, after infection with a new batch of larvae, as in *Haemonchus contortus* in sheep.

The immune responses to nematodes are therefore capable of reducing or terminating existing infections and also preventing reinfection. The mechanisms of the two processes are different as, in the first, the target is the adult worm and, in the second, it is a larval stage. Immunity to nematodes is further complicated by the fact that the various immune responses are stage specific. The acquisition of immunity and the elimination of adult worms is linked with the physiology of the gut and inflammatory reactions elicited by the worms themselves. T lymphocytes in the gut-associated lymphoid tissues play a central role in immunity by first recognizing the worm antigens and then producing a range of cytokines that activate other cells to release mediators of inflammation, such as leukotrienes, prostaglandins and platelet activating factor, which transform the gut into a hostile environment for the worm causing it to migrate away from its favoured site and rendering it liable to expulsion. Mucosal mast cells also play a major role in acquired resistance being particularly involved in anaphylactic reactions involving IgE. High levels of IgE are characteristic of helminth infections and both IgE and activated mast cells are found during the acquisition of immunity to a number of nematodes including *H. contortus* and *Ostertagia circumscripta* in sheep. Experimental evidence also supports the concept of a role for such a system, e.g. mice depleted of IgE exhibit diminished resistance to several nematode worms. Immunoglobulin E is also involved in immunity to

larval stages where it acts directly in ADCC reactions forming a bridge between the worm and eosinophils, bound *via* their Fc receptors, causing these cells to release their contents onto the surface of the larval worm bringing about irreversible damage. This kind of damage occurs when the larval stages leave the gut and begin to migrate around the body and, in different kinds of infections, different patterns of ADCC, involving other antibodies and other kinds of cells, are seen. The general principle, exemplified by *Ascaris lumbricoides*, seems to be that the migrating larvae elicit the production of a range of antibodies that take part in ADCC mechanisms after which the larvae escape back into the gut, where they are relatively safe from attack, having left in train a series of immune responses that are effective against subsequent larval invasion thus preventing overcrowding. On reinfection, the larvae released into the gut are coated with secreted IgA and destroyed on their way through the gut wall. Similar patterns of events occur in infections, such as those caused by the hookworms *Ancylostoma duodenale* and *Necator americanus*, in which the larvae enter the body through the skin. Immune responses such as these are very effective but the consequent inflammatory reactions and the death of larvae in various tissues can contribute to the pathology associated with helminth infections.

8.6.2 Filariasis

The filarial worms that infect humans develop slowly from the infective larval stage for about 3–12 months before the adult worms eventually settle in the lymphatics or subcutaneous tisues. The microfilariae produced then circulate in the blood or remain in the skin for many years. Epidemiological studies indicate that in any individual there is a gradual accumulation of worms but that eventually a plateau is reached suggesting that the worms are either able to avoid eliciting an immune response or that they can evade it but, in either case, some immunity eventually develops. Serological evidence suggests that both the adult and larval worms are recognized but it is still not clear whether or not there is any acquired immunity or, if there is, what mechanisms are involved. There is little evidence for any immunity to reinfection. One of the problems inherent in studying filariasis is that there are two cultivable laboratory models and the information from monkeys, cats and rodents that has accumulated has been fragmentary and contradictory, e.g. in some models irradiated larvae can induce protective immunity but not in others. In *in vitro* experiments, it has been shown that the microfilariae can bind antibodies and that these can participate in ADCC reactions involving eosinophils. It is not at all clear if this also happens in the host or, if it does, whether it contributes to the pathology of the disease. Both lymphatic and ocular filariasis are accompanied by gross pathological changes, elephantiasis and blindness, but it is not clear if these have any immunological basis and current opinion favours simple obstruction.

8.6.3 Schistosomiasis

Our understanding of immunity to human schistosomiasis has been both aided and confused by a plethora of experimental work using laboratory animals. The species of *Schistosoma* that infect humans can be maintained in baboons, monkeys, rats, guinea pigs, hamsters and mice and, in each of these hosts, some degree of immunity can quickly be achieved. This contrasts with the situation in human populations in which the actual acquisition of any immunity has only recently been convincingly demonstrated. In most human populations, the prevalence of schistosomiasis can be correlated with age. In both individuals and in populations the parasite load increases over the first 15–20 years of life and then declines suggesting the acquisition of some degree of immunity to reinfection. Schistosome infections in various hosts exhibit certain common features and, by identifying a number of immunological parameters in humans, it is possible to postulate what actually happens in the best studied species, *S. mansoni* and, by extrapolation, other species.

Schistosome infections begin when cercariae released from infected snails penetrate the skin where they transform into schistosomula and, during the process, incorporate a number of host antigens, including red blood and histocompatibility antigens, into their outer surface. The schistosomula, effectively disguised as host, and therefore immunologically invisible, migrate around the body and eventually enter blood vessels associated with the gut or bladder where they mature into adults. Adults live for about 5 years during which time they evade immunological attack but produce thousands of eggs some of which lodge in tissues including the liver. In these sites, the eggs are recognized as foreign and elicit delayed hypersensitivity reactions which lead to the formation of granulomas, fibrosis and, eventually, liver failure. However, the antigens involved are shared with the young schistosomulum larvae and are instrumental in the destruction of these stages in the skin or lungs.

In laboratory and *in vitro* studies, two possible patterns of immunity can be recognized: (1)

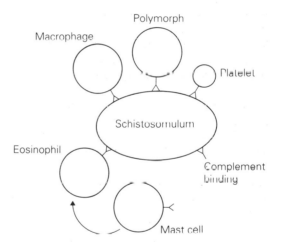

Fig. 8.7 The possible involvement of cells and antibodies in the killing of schistosomula. Macrophages, polymorphs, eosinophils and platelets have all been implicated. Mast cells provide a signal for the activation of eosinophils. Antibodies are indicated as Y shaped molecules. In one kind of mechanism antibody is predominantly involved and binds macrophages, eosinophils, polymorphs and platelets to the surface of the schistosomulum. The other mechanism of killing is antibody independent and the schistosomula are killed directly by IFN-γ activated macrophages. (For further details, see the text and Vignali, Bickle & Taylor, 1989.)

ADCC involving antibodies and eosinophils, neutrophils, macrophages, platelets and mast cells, either individually or in combination (Fig. 8.7); or (2) antibody-independent mechanisms involving activated macrophages. In rats, the primary mechanism involves IgG2a and eosinophils but the protective effect of IgG2a is blocked by IgG2c which recognizes the same antigenic epitopes. In humans, protective immunity develops very slowly and, although children may become infected and recognize the parasite antigens within the first 5 years of life, it takes about 10 more years before it becomes effective. In *S. mansoni* infections, protection involves IgE and eosinophils but early in life the IgE antibody responses are blocked by other immunoglobulins, in a similar way to that in rats. In *S. haematobium* infections, the protective antibody is IgE and the blocking antibody is IgG4. However, schistosomiasis is a complex disease and, when eventually the whole immune response is understood, it is unlikely to be as simple as this.

Schistosomiasis is not a serious problem in domesticated animals but *S. bovis* can cause morbidity and some deaths in cattle. Immunity does occur and can be artificially induced with irradiated larval forms (see Section 8.9) but the mechanisms involved are not at all clear.

8.6.4 Cestodiasis

There is no evidence of any acquired immunity to the adult stages of the most important tapeworms of humans, *Taenia saginata* and *T. solium*. However, immunity does develop to the larval stages of taeniid tapeworms and it is possible to induce immunity to these stages with crude or defined antigens. Immunity is stage specific and oncosphere antigens induce protection against the earliest larval stages but, if these survive, there is no protection against later stages. Conversely, antigens derived from metacestode larvae induce immune responses against later stages but not the early ones. The mechanisms of immunity are not at all clear but there is evidence that both eosinophils and antibodies are involved suggesting ADCC. Despite this lack of knowledge, however, it has been possible to develop a recombinant vaccine against *T. ovis* (see Section 8.9).

8.7 EVASION OF THE IMMUNE RESPONSE

From the moment a parasite enters its host it is recognized as foreign and is subjected to a range of immune responses which transform the host into a hostile environment. Many microorganisms are able to counteract the immune response, to some extent, by rapid multiplication, which allows them to build up large numbers and infect new hosts before being brought under control. Parasites, partly because of their complex life cycles, require a considerable period during which to complete their development and have evolved numerous ways of evading the immune response. These methods have been so successful that many parasites can survive for the lifetime of their host. The best known examples are the African trypanosomes, which undergo antigenic variation (see Section 8.5.4) and the schistosomes (see Section 8.6.3), which disguise themselves by taking up host antigens. There are, however, many more examples and virtually every parasite has evolved some way of evading the immune response or its consequences.

There are three essentially different aspects of evasion and these relate to the site of infection, specific methods for evading the consequences of the immune response and interference with the response as a whole. The simplest way of evading the immune response is by not stimulating it and many parasites can do this by entering immunologically inert cells where they are safe from attack while they remain there. However, few parasites can afford to stay hidden for long; protozoa have to infect new cells and helminths have to grow so it is inevitable that most, if not all, will elicit an immune response at some time. Parasites have, therefore, had to evolve specific ways of evading the consequences of such immune responses and there are many ways in which this can be done. Finally, parasites can escape from the immune response by interfering with the response as a whole, e.g. by suppressing it or stimulating it non-specifically, which not only protects the parasite but also leaves the host open to attack from other organisms.

The various methods of evasion are shown diagrammatically in Fig. 8.8. Initially a parasite can

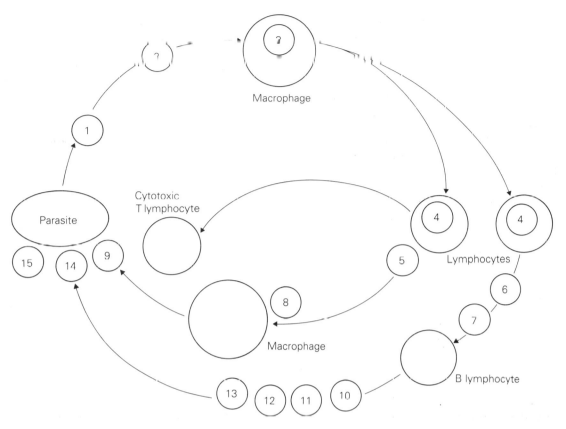

Fig. 8.8 Parasites can evade the immune response by: (1) becoming intracellular or entering an immunologically privileged site soon after invading the host; (2) becoming disguised with host antigens; (3) surviving in a macrophage; (4) living in a lymphocyte; (5 & 6) causing lymphocytes to produce the wrong kind of cytokines; (7) causing polyclonal B cell stimulation; (8) inhibiting the activation of macrophages; (9) producing antioxidants; (10) undergoing antigenic variation; (11) shedding antigens; (12) cleaving antibody molecules; (13) inactivating complement; (14) by blocking antibody binding; and (15) moving out of range of immune attack. (For more details, see the text and Parkhouse, 1984.)

avoid eliciting an immune response by becoming intracellular or entering an immunologically privileged site. This method is suitable for helminths, such as metacercariae, which do not grow within their hosts, and is even more efficient if the site occupied is out of contact with the immune system, e.g. the eye or nervous system. The immune response can also be avoided if the parasite becomes disguised with host antigens as soon as it enters the host, as happens with the schistosomes (see Section 8.6.3), although this method of evasion has not been incontrovertibly demonstrated in other helminth infections.

Some parasites evade the immune response by invading cells of the immune system itself. *Leishmania* species, *Trypanosoma cruzi*, and *Toxoplasma gondii*, can all live within phagocytic cells and circumvent the activities of these cells in various ways. Normally, such cells take in particulate material by endocytosis and form a phagosome with which lysosomes fuse to produce a phagolysosome within which digestion occurs. *Leishmania* species can survive within the phagolysosome. *T. cruzi* is able to lyse the phagosome before lysosome fusion and lives in the cell cytoplasm while *T. gondii* prevents phagosome–lysosome fusion. *Theileria* species are the only parasites that infect lymphocytes, they invade lymphoblasts and transform them into continuously dividing cell lines in which each of the daughter cells receives some of the dividing parasites, thus the cells of the immune system are used to perpetuate the proliferation of the parasite.

Parasites can also interfere with the control of the immune response either by inducing T helper cells to produce inappropriate signals or by suppressing the production of particular cytokines or receptors for them. The induction of inappropriate signals is clearly seen in leishmaniasis in which immunity is mediated, via T_H1 lymphocytes, by IFN-γ activated macrophages but during infection the production of IFN-γ is suppressed by IL-4 and IL-10 produced by T_H2 cells. Disturbances in the balance between T_H1 and T_H2 cells have also been demonstrated in *S. mansoni*, *Trichinella spiralis* and *Trichuris muris* infections in mice. *Trypanosoma cruzi* can inhibit the production of IL-1 and IL-2 and *Plasmodium falciparum* infections are accompanied by reduced levels of IL-2. Inappropriate stimulation of TH2 cells also leads to the production of nonfunctional antibodies either because they are of the wrong isotype or because they lack any specificity. Polyclonal B lymphocyte activation, leading to the exhaustion of antibody producing cells, which occurs in many parasitic infections, probably also results from misdirected control mechanisms. Among other signs of a disturbed immune system are raised antibody levels and cellular responses such as eosinophilia. Perturbations in the control of the immune response do not only permit the survival of parasites but can also contribute to the pathology of the disease caused (see Section 8.8).

Mention has already been made of the fact that *Leishmania* species, *T. cruzi* and *T. gondii* can survive in resting macrophages. The products of activated macrophages can also be counteracted and virtually all parasites possess one or more antioxidant systems such as superoxide dismutase, catalase or glutathione peroxidase. *Toxoplasma gondii* possesses high levels of all three but *Leishmania* species are rich only in superoxide dismutase. High levels of superoxide dismutase also occur in *T. spiralis*, taeniid larvae and schistosomes but it is not at all clear whether

or how these are involved in the evasion of the immune response.

The best known example of evasion involving escape from an active antibody response is the antigenic variation shown by the African trypanosomes (see Section 8.5.4). A distinction must be made between the different antigens specific to particular stages of a life cycle, which remain constant during the course of an infection, and true variant antigens, such as those of the African trypanosomes, which represent switches in the expression of an antigen by individual parasites during the course of an infection. Malaria parasites exhibit a very high degree of antigenic diversity and there is also evidence of antigenic variation.

Antibody-dependent immune responses can also be evaded by shedding surface antigens and this has been demonstrated for a number of parasites including *Entamoeba histolytica*, *Ancylostoma caninum* and *T. spiralis* and many other species including the African trypanosomes. In malaria, the circumsporozoite antigen can be shed and there is convincing evidence that certain of the tandemly repeated antigens of the erythrocytic stages can also be released. Shedding antigens leads to the production of antigen–antibody complexes which direct the immune response away from the invader and constitute what is effectively a smoke-screen effect.

The binding of antibody can be inhibited by specific proteases that cleave the immunoglobulin molecule thus inhibiting ADCC or complement activation. Schistosomula, for example, produce a schistosome derived inhibitory factor (SDIF) which inactivates IgG and *T. cruzi* can also inactivate antibody molecules. Other parasites, including larval taeniid tapeworms and *E. histolytica*, possess anticomplementary factors that protect them from complement-mediated damage.

Finally, many parasites simply avoid the consequences of the immune response they themselves have elicited. The simplest examples are the intestinal nematodes, such as *Ascaris lumbricoides* and hookworms that migrate around the body, stimulating various immune responses as they do so, but escape their consequences by moving into the gut. This is similar to what happens in schistosomiasis but the nematode worms do not need to disguise themselves, although some of them change their antigens during their life cycles. The immunological response is, therefore, effective against reinfection and these patterns of migration probably evolved to prevent overcrowding in the gut.

8.8 IMMUNOPATHOLOGY

The pathology associated with parasitic infections can be considered under three headings: (1) damage caused directly by the parasites; (2) damage caused as a result of the immune response; and (3) damage resulting from a combination of both. With our increasing understanding of the role of cytokines in the immune response, it is becoming apparent that most of the pathology seen in parasitic infections does have some immunological basis. The obstruction caused by the presence of massive larval stages of tapeworms, the blocking of lymphatics by filarial worms, the ulcers and abscesses caused by E. histolytica and the anaemia caused by trypanosome haemolytic fatty acids are examples of pathology which do not seem to have any immediate immunological basis.

The main factor responsible for pathological changes is the immune response itself and the combination of persisting antigens and active immune responses results in immunopathological effects, that are, in many cases, more damaging than the infection itself. The complex interplay between parasites and the immune system, during which parasite antigens are recognized but the parasites are not eliminated, results in long, chronic infections accompanied by a breakdown in overall control as the host attempts to turn off abortive immune responses. The net result is that the response is not switched off as normally happens when the source of stimulation is removed. Instead, circulating antigens either coat cells or lodge in vessels or tissues causing hypersensitivity reactions of all types; immediate type I (anaphylaxis), type II (complement activation and cell lysis), type III (immune complex formation) and delayed type IV.

Among the protozoan infections immunopathological changes are particularly obvious in malaria and damage is considerable, largely because of the release of irrelevant smoke screen antigens, associated with the schizogonic stages, which bind to uninfected red blood cells, resulting in anaemia, or form antigen—antibody complexes which lodge in blood vessels, such as those of the kidney, causing blockage and complement-mediated damage. Overall, the pathology of malaria resembles a generalized inflammatory response in which the various mediators, released as part of a complex of cell-mediated responses, play a part in the destructive process, e.g. reactive oxygen intermediates cause damage to endothelial cells, TNF contributes to red cell adherence in the capillaries of the brain, eventually leading

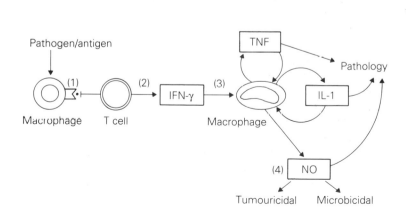

Fig. 8.9 Some of the interrelationships between specific and non-specific effector mechanisms and between protective and pathological effects in parasitic infections. (After Liew & Cox, 1991. Reprinted with permission from *Immunoparasitology Today* (Ash, C. & Gallagher, R.B. eds), p. A20, Elsevier Trends Journals, Cambridge.)

to blockage and cerebral malaria, and to cachexia or wasting, and IL-1 is responsible in part for the fevers associated with the disease. The situation in babesiosis is similar suggesting that it is the nature of the infection, in these cases blood-dwelling, rather than the species involved that determine the nature of the pathology seen. Virtually all parasitic diseases can be considered as inflammatory ones each with its own particular characteristics associated with the nature of the parasite itself and the ways in which it evades the immune response. Infections with the African trypanosomes are characterized by successive waves of parasitaemia as parasites of a particular antigenic type are destroyed by complement-mediated lysis and replaced by parasites with a different surface coat. Internal and surface antigens released by the lysed trypanosomes bind to erythrocytes, causing anaemia, form antigen–antibody complexes which lodge in blood vessels, such as those of the kidney, causing blockage and complement-mediated damage and also initiate various inflammatory reactions including the kinin and coagulation cascades. Chagas disease is complicated by the fact that *T. cruzi* possesses antigens that cross-react with those of the host muscle and nerve cells. These initiate long-term autoimmune reactions which eventually result in the destruction of both infected and uninfected host cells and if, as is often the case, the heart is involved, cardiac failure. Cutaneous leishmaniasis is a spectral disease with strong immunological responses at one pole and weak ones at the other. Where the immune responses are weak, the parasites cause the damage and where they are strong they destroy the parasites but in between the pathology is dominated by the immune response and involves the formation of granulomas at or near the site of the lesion. In kala-azar, most of the pathology results from the direct destruction of infected cells and the formation of antigen–antibody complexes that initiate a number of generalized inflammatory reactions. In toxoplasmosis, most of the pathology is associated with inflammatory responses to disintegrating cysts and the destruction of surrounding cells.

The best-studied helminth infection is schistosomiasis in which the adult worms, disguised with host antigens and therefore immunologically invisible, produce eggs that lodge in the liver and elsewhere releasing their antigens which initiate typical type IV granulomatous responses. These granulomas eventually shut off the portal blood flow, causing portal hypertension, and become fibrotic resulting in liver failure in the case of *S. masoni* infections and calcification of the bladder in the case of *S. haematobium*. Filarial worms initiate inflammatory reactions when they die and microfilariae cause immediate hypersensitivity type I reactions associated with increased eosinophilia. Many helminth worms cause this kind of reaction at their site of entry into the body and elsewhere during their migrations particularly in the wrong host. Avian schistosomes cause dermatitis in humans and larvae of *Toxocara canis*, from dogs, may even cause blindness.

It is becoming increasingly apparent that the immunopathology associated with parasitic diseases can, to a large extent, be explained in terms of the various cytokines involved and the interplay between them. For example, IL-4 is involved in the production of IgE and IL-5 stimulates eosinophilia so together these cytokines could contribute to immediate hypersensitivity type I reactions. In cutaneous leishmaniasis, IFN-γ contributes to protection but IL-4 and IL-10 inhibit IFN-γ production and are effectively disease promoting. Tumour necrosis factor is involved in immunity to parasites but also causes damage to neighbouring cells. Interleukin-1, important in macrophage activation, also causes fever (Fig. 8.9). Observations such as these are beginning to pinpoint the molecules that contribute to pathology and, when their roles are fully understood, it should be possible to ameliorate or reverse some of the pathological damage. For example, patients with cerebral malaria might be treated with anti-TNF monoclonal antibodies and those with leishmaniasis with IFN-γ.

8.9 IMMUNIZATION AGAINST PARASITIC INFECTIONS

Much of the research effort expended on increasing our understanding of immunity to parasitic infections has been devoted to the possibility of developing vaccines. Vaccination has many advantages over chemotherapy or the prevention of transmission as a method for controlling any disease and forms an essential part of any integrated control programme. Vaccination against a number of viral and bacterial diseases has been remarkably successful, including the eradication of smallpox and the virtual elimination of poliomyelitis in many countries, and currently about half the children in the world are routinely vaccinated against the major diseases of childhood. These successes have not been emulated among the parasitic diseases partly because the immune responses to parasites are multifactorial, partly because parasites have developed diverse ways of evading the immune response and partly because the antigens involved have been complex and difficult to identify. However, the techniques of molecular biology have now been used with great success to characterize potentially protective antigens and progress towards a number of vaccines is now being made.

Vaccines can be categorized under five headings: (1) live wild types with low virulence; (2) attenuated organisms; (3) inactivated or killed organisms; (4) subunits, including extracts or metabolic products; and (5) synthetic and recombinant antigens. Anti-idiotype vaccines have been useful experimentally but are not realistic alternatives to more conventional vaccines. Immunization has been relatively easy to achieve in laboratory models and drug-controlled infections attenuated or irradiated parasites, homogenates, extracts and purified antigens have all been used with varying degrees of success. However, the best immunity can normally only be achieved with living parasites and vaccines based on these are unlikely to be acceptable for use in humans. Despite all the problems involved, considerable progress has been made particularly in the field of veterinary medicine. The first antiparasitic vaccine was that developed against the cattle lungworm, *Dictyocaulus viviparus*. This vaccine, which consists of irradiated larvae that live sufficiently long to stimulate an immune response but not long enough to cause an infection, has been used successfully for over 30 years in Europe and one against *D. filariae* in sheep is in use in India. However, the same kind of vaccine developed against the dog hookworm, *Ancylostoma caninum*, which had been marketed in the United States and was very successful, has been discontinued in favour of chemotherapy. Similar vaccines against gapeworm in cattle, caused by *Syngamus trachea* and *Haemonchus contortus* in sheep have not been developed but an irradiated larval vaccine, based on cercariae and schistosomula, has been shown to be effective against

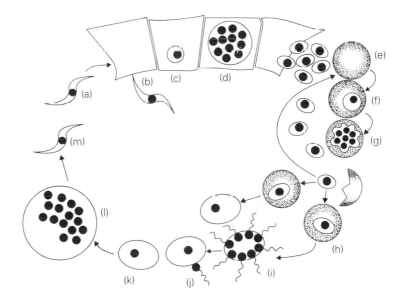

Fig. 8.10 The life cycle of the malaria parasite showing possible vaccine targets. Possible targets are (a) the sporozoites injected by the mosquito (b–d) developmental stages in the liver, (e–g) stages in the blood and (h–k) sexual stages. There is no immunity aginst the oocyst (l) or the sporozoites in the salivary glands of the mosquito (m). (After Cox, 1992. Reprinted with permission from *Nature*, **360**, p. 417. Copyright Macmillan Magazines Limited.) See also Chapter 1, Section 1.8.

Schistosoma bovis and *S. mattheei* in cattle but has not been developed commercially.

There are also a few successul vaccines against parasitic protozoa of veterinary importance. Simple vaccines consisting of whole blood containing parasites with reduced virulence are routinely used against *Babesia bovis* and *B. bigemina* in cattle. Protection against *Theileria annulata* can be achieved with culture-attenuated infected lymphoid cells and such a vaccine is widely used in the Middle East and India. Attenuated cell lines are less effective against *T. parva* but immunization can be achieved by injecting cryo-preserved sporozoites followed by appropriate chemotherapy and, experimentally, some success has been achieved using a recombinant sporozoite surface antigen. Immunization of chickens against coccidiosis can be achieved by infecting the birds with small numbers of oocysts.

This limited success has encouraged the use of parasite extracts as antigens and dogs can be protected against *B. canis* using a vaccine based on culture supernatants. Metabolic products of larval tapeworms, particularly the oncospheres, have also been used to immunize sheep against the larval stages of *Taenia ovis, T. hydatigena, T. multiceps* and *Echinococcus granulosus*, cattle against *T. saginata* and pigs against *T. solium* but this kind of vaccine has been overtaken by the development of a recombinant vaccine based on a 45 kDa *T. ovis* oncosphere antigen, the first effective recombinant vaccine against any parasitic disease. Recombinant vaccines have also been developed against *Haemonchus contortus* in sheep.

There are no immediate prospects for a vaccine against any human parasitic disease and, apart from the actual technical problems involved there are many major obstacles to overcome including those relating to safety, cost and delivery. Pragmatic arguments suggest that some vaccines might be possible; there is evidence of acquired immunity and resistance to reinfection in virtually all parasitic infections. In leishmaniasis, for example, self-infection, somewhat reminiscent of the procedures once used for smallpox, have been practised for centuries and there have been a number of partially successful field trials involving the use of living or killed promastigotes one of which is proceeding in the Middle East under the auspices of the World Health Organization (WHO).

Developments in the field of molecular biology have increased the prospects of vaccines against the major parasitic infections of humans for which there is no possibility of a middle course based on avirulent or attenuated organisms or their metabolic products. Most progress has been made in the development of a vaccine against malaria (Fig. 8.10) and there have been human trials based on the use of the circumsporozoite antigen, both recombinant and synthetic, and also a synthetic one based on a mixture of sporozoite and schizont antigens. The results obtained have been equivocal with some of the volunteers tested showing good protection, some none and some open to either interpretation. Nevertheless, the synthetic peptide vaccine has been extensively used in South America where over 27 000 people have been vaccinated with a considerable degree of success. An alternative approach has been the development of vaccines that block the development of sexual stages in the mosquito but these are still at the experimental stage and, in any case, they are unlikely to be acceptable on their own and would have to be used as part of a cocktail vaccine. The development of vaccines against cutaneous and visceral leishmaniasis remains a WHO priority but research is still at the experimental stage while the factors leading to protection on one hand and pathology on the other are sorted out. Currently the antigens that stimulate T lymphocyte and B lymphocyte responses are being investigated. Vaccines against Chagas disease are unlikely because of the auto-immune components of this infection and antigenic variation makes the possibility of a vaccine against sleeping sickness unlikely.

A vaccine against schistosomiasis has recently become a high priority and considerable progress has been made towards the development of synthetic and recombinant vaccines including one based on schistosome glutathione s-transferase and another based on schistosome paramyosin. As in all the other parasitic infections, however, the full implications of possibly disrupting the cytokine control of the immune response will have to be carefully considered. No other vaccine against a helminth in humans is currently underway but successes in the veterinary field particularly vaccines against larval cestodes will inevitably be transferred to the field of human disease.

The development of vaccines involves not only immunological considerations but must also take into account the role of vaccines in integrated control programmes, the feasibility of delivering vaccines and their likely success and their interaction with chemotherapy.

REFERENCES AND FURTHER READING

Ash, C. & Gallagher, R.B. (eds) (1991) *Immunoparasitology Today*. Cambridge, Elsevier Trends Journals. (Also published as *Parasitology Today*, 7 (No. 69) and *Immunology Today*, 12 (No. 129).

Behnke, J.M. (1987) Evasion of immunity by nematode parasites causing chronic infections. *Advances in Parasitology*, 26, 1–71.

Behnke, J.M. (ed.) (1990) *Parasites: Immunity and Pathology*. London, Taylor and Francis.

Clark, I.A., Hunt, N.H. & Cowden, W.B. (1986) Oxygen-derived free radicals in the pathogenesis of parasitic disease. *Advances in Parasitology*, 25, 1–44.

Cohen, S. & Warren, K. (eds) (1982) *Immunology of Parasitic Infections*, 2nd edn. Oxford, Blackwell Scientific Publications.

Cox, F.E.G. (1991) Malaria vaccines – progress and problems. *Trends in Biotechnology*, 9, 389–394.

Cox, F.E.G. (1992) Malaria: another route to a vaccine. *Nature*, 360, 417.

Cox, F.E.G. & Liew, F.Y. (1992) T cell subsets and cytokines in parasitic infections. *Parasitology Today*, 8, 371–374.

den Hollander, N., Riley, D. & Befus, D. (1988) Immunology of giardiasis, *Parasitology Today*, 4, 124–131.

Denis, M. & Chadee, K. (1988) Immunopathology of *Entamoeba histolytica* infections. *Parasitology Today*, 4, 247–252.

Grau, G.E. & Modlin, R.L. (1991) Immune mechanisms in bacterial and parasitic diseases: protective immunity versus pathology. *Current Opinion in Immunology*, 3, 400–485.

Heath, A.W. (1990) Cytokines and infection. *Current Opinion in Immunology*, 2, 380–384.

Hughes, H.P.A. (1988) Oxidative killing of intracellular parasites mediated by macrophages. *Parasitology Today*, 4, 340–347.

Immunology Letters (1990) Cellular Mechanisms in Malaria Immunity. *Immunology Letters*, 25, 1–293.

Kierszenbaum, F. & Sztein, M.B. (1990) Mechanisms underlying immunosuppression induced by *Trypanosoma cruzi*. *Parasitology Today*, 6, 261–264.

Liew, F.Y. (1989) Immunity to protozoa. *Current Opinion in Immunology*, 1, 441–447.

Liew, F.Y. (ed.) (1989) *Vaccination Strategies of Tropical Diseases*. Boca Raton, CRC Press.

Liew, F.Y. & Cox, F.E.G. (1991) Nonspecific defence mechanisms: the role of nitric acid. In C. Ash & R.B. Gallagher (eds) *Immunoparasitology Today*, p. A20. Cambridge, Elsevier Trends Journals.

McLaren, D.J. (ed.) (1989) *Vaccines and Vaccination Strategies*. Symposia of the British Society for Parasitology, Vol. 26. *Parasitology*, 98, (supplement), S1–S100.

Miller, L.H. & Scott, P. (1990) Immunity to protozoa. *Current Opinion in Immunology*, 2, 368–374.

Mitchell, G.F. (1989) Problems specific to parasite vaccines. *Parasitology*, 98, (supplement), S19–S28.

Mitchell, G.F., Tiu, W.U. & Garcia, E.G. (1991) Infection characteristics of *Schistosoma japonicum* in mice and relevance to the assessment of schistosome vaccines. *Advances in Parasitology*, 30, 167–200.

Moddaber, F. (1989) Experiences with vaccines against cutaneous leishmaniasis of men and mice. *Parasitology*, 98, (supplement), S49–S60.

Parkhouse, R.M.E. (ed.) (1984) *Parasite Evasion of the Immune Response*. Symposia of the British Society for Parasitology, Vol. 21. *Parasitology*, 88, (supplement), S571–S682.

Parkhouse, R.M.E. (ed.) (1985) *Parasite Antigens in Protection, Diagnosis and Escape*. Current Topics in Microbiology and Immunology, 20. Berlin, Springer-Verlag.

Parkhouse, R.M.E. & Harrison, L.J.S. (1989) Antigens of parasitic helminths in diagnosis, protection and pathology. *Parasitology*, 88, (supplement), S5–S19.

Pearce, E.J. & Sher, A. (1990) Immunity to helminths. *Current Opinion in Immunology*, 2, 375–379.

Playfair, J.H.L., Blackwell, J.M. & Miller, H.R.P. (1990) Parasitic diseases. In E.R. Moxon (ed.) *Modern Vaccines. A Lancet Review*, pp. 129–136. London, Edward Arnold.

Roitt, I., Brostoff, J. & Male, D. (1989) *Immunology*, 2nd edn. London, Gower Medical Publishers.

Ross, R. & Thompson, D. (1910) A case of sleeping sickness studied by precise enumeration methods: regular periodical increase of parasite detailed. *Proceedings of the Royal Society of London*, B82, 411–415.

Sher, A., James, S.L., Correa-Oliveira, R. & Hieny, S. (1989) Schistosome vaccines: current progress and future prospects. *Parasitology*, 98, (supplement), S61–S68.

Tizard, I. (1989) *Immunology and Pathogenesis of Trypanosomiasis*. Boca Raton, CRC Press.

Vignali, D.A.A., Bickle, Q.D. & Taylor, M.G. (1989) Immunity to *Schistosoma mansoni. Immunology Today,* **10,** 410–416.

Wakelin, D. (1978) Genetic control of susceptibility and resistance to parasitic infections. *Advances in Parasitology,* **16,** 219–308.

Wakelin, D. (1989) Immunity to helminths. *Current Opinion in Immunology,* **1,** 448–453.

Wakelin, D. (1989) Nature and nurture: overcoming constraints on immunity. *Parasitology,* **98,** (supplement), S21–S35.

Wakelin, D. & Blackwell, J.M. (1988) *Genetics of Resistance to Bacterial and Parasitic Infections.* London, Taylor and Francis.

World Health Organization (1990) Malaria vaccine development: pre-erythrocytic stages. *Bulletin of the World Health Organization,* **68,** (supplement), 7–196.

Chapter 9 / Chemotherapy

W. E. GUTTERIDGE

9.1 INTRODUCTION

Chemotherapy involves the use of small, usually organic, molecules (i.e. *chemicals*) to treat (= *therapy*) diseases of humans and domestic animals.

The approach was developed by Ehrlich and his colleagues in Germany towards the end of the nineteenth century. Ehrlich was at that time studying the staining properties of a number of dyes for protozoa such as trypanosomes and malaria parasites. He conceived the idea that it ought to be possible to find dyes (i.e. drugs) which would selectively destroy pathogens but leave host cells undamaged. His approach was most successful initially with diseases caused by parasites. By 1930, as a result of careful selection and chemical modification, first of dyes and later of other chemicals, in order to improve their activity against parasites and to reduce their toxicity to hosts, a number of drugs had already been developed which were useful in the control of parasitic diseases. These included organic arsenicals and suramin for the trypanosomiases, plasmoquine for malaria and organic antimonials for schistosomiasis.

It is not always appreciated that initially Ehrlich's approach was far less successful with antibacterial drugs. It needed the discovery in the 1930s of antimetabolites such as the sulphonamides and the development in the 1940s of antibiotics (natural products with antibacterial effects) such as the penicillins to open up this area of chemotherapy. It is an interesting and so far unexplained observation that only a tiny minority of antiparasitic drugs (e.g. **amphotericin B, ivermectin, lasalocid, metronidazole, monensin, quinine, salinomycin, tetracycline***) began life as natural products (note that metronidazole is now manufactured synthetically).

Today, all the common parasitic diseases of humans and domestic animals, with the exception of cryptosporidiosis, can be treated at least in part by drugs. For some of them, chemoprophylactic agents are also available. In the absence, in most instances, of vaccines, they, together with measures to reduce transmission (improved public health; vector control), are the mainstays of our current efforts to control such diseases (see Chapter 10). If and when vaccines are ever marketed on any scale, they are unlikely to replace drugs. Rather, they will be added to the armoury of control measures needed to contain parasitic diseases.

9.2 KEY CURRENT DRUGS AND THEIR LIMITATIONS

These are best reviewed on the basis of their activity against the six major groups of parasites: the kinetoplastid, 'anaerobic' and sporozoan protozoa; and the digenean, cestode and nematode worms.

Four key factors can constrain our ability to control an infectious disease, of medical or veter-

---

*Note that all such drugs have at least three names: a systematic chemical name (e.g. 2,4-diamino-5-*p*-chlorophenyl-6-ethylpyrimidine); a simpler generic name to avoid cumbersome repetition of the systematic one (e.g. the generic name of the compound mentioned above is pyrimethamine); a trade name used to describe a marketed formulation of a particular compound (e.g. Wellcome market pyrimethamine in most territories as Daraprim). In this chapter, all the drugs mentioned are referred to by their generic names, unless the contrary is clearly indicated. The chemical formulae of those in **bold type** (**key current drugs**) or *italics (drugs in development)* are given in alphabetical order in the Appendix located at the end of the chapter.

inary significance, with a drug: (1) the intrinsic efficacy of the drug can be limiting; or (2) it can be compromised by drug resistance; (3) adverse reactions can prevent its use for the period of time required to effect a cure; and (4) its route of administration may not be convenient (as a general rule, orally-active formulations tend to be preferred for human use; injectible or pour-on formulations for animals). Most of the currently available antiparasitic drugs suffer from one or more of these limitations; some from all of them. One further constraint applies to drugs intended for use in farm animals: the latter cannot be treated with drugs during the so-called 'withdrawal' period, that is the time required (usually 1–3 weeks) for drug residues to be reduced to insignificant levels (usually <1 part per million) before animals or animal products can be marketed for human use.

9.2.1 Kinetoplastid protozoa

Early cases of human sleeping sickness are usually treated with **suramin**, though occasionally **pentamidine** is used. Late stage cases, where trypanosomes are present in the CNS, are treated with **melarsaprol** or **eflornithine** (*Trypanasoma brucei gambiense* infections only) since neither suramin nor pentamidine cross the blood–brain barrier. An intramuscular injection of pentamidine will give up to 6 months chemoprophylaxis. Only eflornithine is active by the oral route. A wide range of adverse reactions is shown to all of them except eflornithine. Melarsaprol, an arsenical, is particularly toxic and is best administered only to hospitalized patients. Eflornithine is expensive to synthesize, has to be given in large doses over long periods of time and does not work against *T. b. rhodesiense* infections.

Suramin is also used to treat trypanosome infections in equines and camels. **Quinapyramine** works well in camels, pigs and cattle, but **homidium bromide**, **isometamidium** and **diminazine aceturate** are the principal drugs used to control such diseases in sheep and cattle. Quinapyramine and isometamidium are also used chemoprophylactically. Injectable formulations of all of these drugs are available, but severe reactions at the injection site are common. The biggest limitation here, however, is resistance: intensive use of any of these drugs leads rapidly to problems of acquired drug resistance, much of which is cross-resistant to the other trypanocides.

Short, orally-administered courses of **nifurtimox** or **benznidazole** are effective and safe in the control of acute infections of Chagas disease. Attempts to eradicate the chronic stage of the disease by long-term therapy, however, usually run into problems with adverse reactions and often give disappointing results in terms of efficacy, especially in the northern half of Latin America.

The mainstay of chemotherapy for leishmaniasis remains the antimonials, **sodium stibogluconate** and **meglumine antimonate**. **Pentamidine**, however, is sometimes used for visceral cases and **amphotericin B** for the mucocutaneous disease. All of these drugs are usually administered by injection, though for the cutaneous disease, topical formulations are coming into fashion in some regions. Oral administration is not possible. Levels of adverse reactions following systemic administration can be severe.

9.2.2 'Anaerobic' protozoa

Trichomoniasis, giardiasis and amoebiasis in both humans and domestic animals can all be controlled by **metronidazole**. This drug is orally active, very efficacious and relatively free of side-effects. It is, however, mutagenic in some *in vitro* test systems and this at one time led to concern about putative carcinogenicity in humans. However, since there was, and indeed still is, no widely available alternative to it for trichomoniasis, the drug was not withdrawn; the alternative in giardiasis is mepacrine and the alternatives in amoebiasis are diloxanide, which is only effective in non-invasive cases, and **tinidazole**, which is another potentially mutagenic 5-nitroimidazole. Later, retrospective studies of patients given metronidazole gave no indications of enhanced rates of cancer, so that such a possibility is no longer a matter of concern. Subsequently, **satranidazole**, yet another 5-nitroimidazole, was marketed in some territories for giardiasis and amoebiasis.

9.2.3 Sporozoan protozoa

Coccidiosis, especially in broiler chickens, is controlled by the continuous administration of drugs in the diet. Currently, the ionophores such as **lasalocid**, **salinomycin** and especially **monensin** are the coccidiostats of choice. Others, such as **amprolium**, **clopidol**, **decoquinate** and **robenidine**, are however still used. Drug resistance is a major problem in this area. One approach to overcoming it is the use of mixtures of drugs (e.g. **sulphaquinoxaline + amprolium + ethopabate + pyrimethamine**, marketed as Supracox). In another approach, since drug resistant strains tend to be selected out in the absence of drug pressure, diet-formulating companies often rotate the coccidiostat they incorporate into their diets to minimize such problems. A third approach is to develop new chemical entities (e.g. the recently introduced diclazuril) which are not cross-resistant to existing drugs.

It is as well, even in this age of AIDS, that toxoplasmosis is a relatively rare disease in humans, because it can be controlled with difficulty only with a **pyrimethamine + sulphadiazine** combination. The drugs are given orally but the treatment period is much longer and the dosing level higher than that required for malaria (see below), leading to serious side-effects, notably bone marrow depression and skin rashes. No drug or drug combination has yet been marketed that can clear cysts from the brain.

Despite many claims to the contrary, no marketed chemotherapeutic agent, including spiromycin, has yet been shown to have genuine activity in cryptosporidiosis.

Fortunately, malaria, which is the most common of the parasitic diseases of humans, can be treated by a wide range of drugs, all active by the oral route. Until recently, the standard drugs were **chloroquine** for treatment, **primaquine** to prevent relapse (*Plasmodium vivax* malaria) and **chloroquine** or **pyrimethamine + sulphadoxine** (Fansidar) or **pyrimethamine + dapsone** (Maloprim) for prophylaxis. However, drug resistance, especially chloroquine resistance and multi-drug resistance, is now rife in *P. falciparum*. Thus, **amodiaquine**, Fansidar, **quinine**, quinine together with **tetracycline** or one of two new drugs, **meflo-**quine (sometimes in combination with pyrimethamine + sulphadoxine as Fansimef) and **halofantrine**, are frequently now required for effective treatment. Similarly, chloroquine or amodiaquine together with Fansidar, Maloprim, **proguanil** or mefloquine, are now often used for chemoprophylaxis. This in turn has highlighted problems with adverse reactions (especially with amodiaquine, Fansidar and Maloprim and more recently with mefloquine and halofantrine) and the speed with which resistance can develop with the newer antimalarials (both mefloquine and halofantrine).

Babesiosis in cattle can be controlled by **diminazine aceturate**, and theileriosis by **parvaquone** (East Coast Fever only), **buparvaquone** and possibly **halofuginone**. Injectable formulations are available in all cases.

Oral cotrimoxazole (**trimethoprim + sulphamethoxazole**) or injectable formulations of **pentamidine** are highly efficacious against *Pneumocystis carinii* pneumonia (PCP) in humans. However, for reasons which are still not clear, rates of adverse reactions are much higher ($\geq 60\%$) in AIDS patients, where this disease commonly occurs, than is normal with such drugs (e.g. 1–2% with cotrimoxazole). Aerosolized formulations of pentamidine are tolerated very much better and can be used for long-term prophylaxis in AIDS patients. However, significant numbers of such patients then develop disseminated PCP. Furthermore, such a treatment regimen has not yet been used sufficiently widely and for long enough for it to be known if drug resistant strains will rapidly develop, as in the trypanosomiases. Preliminary indications are that they will. Standard prophylaxis against PCP in AIDS patients is, therefore, moving strongly back towards the use of cotrimoxazole.

9.2.4 Digenean worms

Sheep and cattle infected with liver fluke are likely to contain all three developmental stages: (1) early immature stages; (2) immature stages; and (3) adults. Treatment of animals harbouring such mixed-stage infections will be ineffective unless all three stages are eliminated. Unfortunately, most flukicides tend to be stage specific.

Most drugs are active against the adults, including **rafoxanide**, **closantel**, **nitroxynil**, **niclofolan**, **bromophos**, **bithinol sulphoxide**, **oxyclozanide** and **albendazole**. A few of these compounds show some activity against late immature stages, including rafoxanide, closantel and nitroxynil, but none of them have activity against early immature stages. **Diamphenethide** is very effective against early immature and immature stages, but is not against adults. **Triclabendazole**, however, works well against all stages. Fluke infections in humans are controlled by **praziquantel** (note activity against *Fasciola hepatica* itself is poor).

Now that the older drugs (e.g. the antimonials, niridazole, lucanthone and hycanthone) have been phased out, the treatment of schistosomiasis in humans relies on three drugs, **oxamniquine**, **metrifonate** and **praziquantel**. Oxamniquine works well against *Schistosoma mansoni*, metrifonate against *S. haematobium* and praziquantel against both species and also against *S. japonicum*. All are active by the oral route and are generally well-tolerated. These drugs, especially praziquantel, which is usually effective in a single dose, have revolutionized the treatment of schistosomiasis in recent years. Resistance to them has not so far been a clinical problem.

9.2.5 Cestode worms

A number of chemotherapies are available to control these infections in both farm and companion animals, including **dichlorophen** (mainly companion animals), **niclosamide**, **resorantel**, **bunamidine**, **mebendazole** (sheep and cattle only), **nitroscanate** (companion animals only), **pyrantel** (horses only, at double dosage) and **praziquantel** (companion animals only).

Key drugs for human use are **niclosamide**, **albendazole** and **praziquantel**. All three drugs can be administered via the oral route, work well against adult stages in the intestine and are generally well-tolerated. The limitation of human chemotherapy in this area relates to the difficulties in eradicating tissue stages. Niclosamide works only against adult worms in the gut; albendazole and praziquantel can kill *Taenia solium* cysticerci, offering the prospects of a chemotherapeutic cure of neurocysticercosis, but only if

higher doses and longer treatment periods are used. **Mebendazole** and especially albendazole are now being used with some success in cases of hydatid disease due to *Echinococcus* species, often as an adjunct to surgery, since they can prevent metastatic spread.

9.2.6 Nematode worms

A large number of drugs are available to control the gastro-intestinal nematode infestations of domestic animals. Included here are **piperazine** (small animals), **haloxon** (horses), **dichlorvos** (especially pigs), **naphthalophos** (sheep), the benzimidazoles, the benzimidazole carbamates and their prodrugs (e.g. **thiabendazole**, **albendazole**, **oxfendazole**, **fenbendazole**), **morantel** (cattle), **pyrantel** (horses and dogs), **levamisole** and **ivermectin**.

In the 1960s and especially in the 1970s, the principal drugs used were the benzimidazoles and the benzimidazole carbamates, levamisole and morantel and pyrantel. Once ivermectin became available in 1981, however, the situation changed rapidly. Today, the predominant chemotherapy used in veterinary medicine for gastro-intestinal nematodes is **ivermectin**. The market is huge: sales of ivermectin are now of the order of US$ 1 billion per annum!

The drugs used today are highly efficacious and generally well-tolerated. Resistance has been a problem with the benzimidazoles and benzimidazole carbamates, especially in Australasia where they have been used extensively in the past, and some breeds of dog have been badly affected by ivermectin. The possible environmental impact of ivermectin, since much of it is excreted in the faeces in which it has potent cidal activity on dung beetles, is a matter of current debate. The biggest problem with chemotherapy in this area, however, is the rapidity with which reinfection occurs. None of the compounds listed above have useful chemoprophylactic properties. To get round this problem various devices are now marketed which can be inserted into the rumen of sheep and cattle and which release small but carefully controlled amounts of anthelmintic (e.g. oxfendazole, pyrantel) over long periods of time. So far, the control obtained using these controlled release devices has been good. A major concern about them, however, should be the implications for the development of drug-resistant strains.

A large number of drugs are also available for the treatment of human gastro-intestinal infestations, including **piperazine**, **thiabendazole**, **albendazole**, **mebendazole**, **levamisole**, **pyrantel** and **bephenium**. All are active via the oral route. The most commonly used is mebendazole, which probably covers the broadest spectrum, with moderate or high activity extending to all the most prevalent infestations. Adverse reactions to it are rare and there have so far been no indications of problems of drug resistance.

The situation with the human, tissue-dwelling nematodes, the filariae, is far less satisfactory. At present, the chemotherapy of choice for the lymphatic filariases is **diethylcarbamazine**, which until recently was thought to require multiple dosing over several weeks to be effective, provokes severe adverse reactions and has only limited activity against adult worms. Recent WHO-sponsored trials suggest that single-dose therapy can be equally efficacious. Its distribution in cooking salt is even being reconsidered. Adult worms can in theory be eliminated with **suramin**, but the degree of efficacy obtained with this compound is very variable and adverse reactions are very common. Until recently, diethylcarbamazine was used also for onchoceriasis, with all the limitations mentioned above, including a very severe so-called Mazzotti reaction where microfilariae were killed in the eye. **Ivermectin**, however, has recently been registered for this indication. It is active in a single, amazingly low (μg per kg!) oral dose, is well-tolerated, does not usually induce a Mazzotti reaction and gives protection for about 6 months. It is revolutionizing the control of this most debilitating disease. There is even speculation that ivermectin might be macrofilaricidal.

9.3 CHEMICAL STRUCTURES OF KEY CURRENT DRUGS

The chemical structures of key current antiparasitic drugs are given in the Appendix at the end of this chapter. A number of general points can be made about them.

1 Antiparasitic drugs contain a very restricted range of elements: carbon, hydrogen, oxygen and nitrogen are the only almost universal elements. Sulphur occurs in some, often as part of a ring structure (e.g. **nifurtimox**, **levamisole**). Fluorine, chlorine, iodine and phosphorus also occur occasionally, commonly in the anthelmintic phenols (e.g. **nitroxynil**) and organophosphorous compounds (e.g. **metrifonate**), but not with the overall frequency seen in herbicides and insecticides. Inorganic elements are rare, the only ones in the Appendix being arsenic and antimony in two of the most toxic series of drugs on the list, the arsenicals (e.g. **melarsaprol**) and the antimonials (e.g. **sodium stibogluconate**).

2 Ring structures are very common. The benzene ring itself is present in almost half of the key antiparasitic drugs in use in the clinic (e.g. **pentamidine**, **dapsone**, **dichlorophen**, **oxfendazole**). Many of the others have nitrogen-containing rings, such as the pyrimidine ring (e.g. **pyrimethamine**), the imidazole ring (e.g. **metronidazole**), the quinoline ring (e.g. **chloroquine**) and the piperazine ring (e.g. **diethylcarbamazine**). Some such rings have the nitrogen atom in the quaternary configuration; at physiological pH, they are positively charged (e.g. **quinapyramine**, **bephenium**).

3 Common substituents on the rings are methyl (e.g. **clopidol**, **morantel**), methoxy (e.g. **trimethoprim**), hydroxymethyl (e.g. **monensin**), hydroxy (e.g. **parvaquone**) and amino (e.g. **homidium bromide**) groups, though more bulky substituents do occur (e.g. the trifluoromethyl groups in **mefloquine**). Nitro-groups are also quite common (e.g. **metronidazole**, **oxamniquine**), though at one time they were viewed with suspicion, since many nitro-compounds were found to be mutagenic (see Section 9.2.2). Sulphydryl groups are not found: these are too biologically active to be selectively toxic.

The overall conclusion from this analysis is that drugs which have sufficient selectivity to be useful in the clinic are made up of the same elements combined in similar chemical combinations to the chemicals which generally occur naturally in

Table 9.1 Summary of possible mechanisms of action of antiparasite drugs

Cofactor synthesis	Nucleic acid synthesis	Protein synthesis	Membrane function	Microtubule function	Energy metabolism	Neuromuscular function
Antiprotozoals						
Amprolium	Amodiaquine	Eflornithine	Amphotericin B		Buparvaquone	
Dapsone	Benznidazole	Quinapyramine	Lasalocid		Clopidol	
Ethopabate	Chloroquine	Tetracycline	Monensin		Decoquinate	
Proguanil	Diminazine		Salinomycin		Meglumine	
Pyrimethamine	aceturate				antimonate	
Sulphadiazine	Halofantrine				Melarsaprol	
Sulphadoxine	Homodium				Parvaquone	
Sulphamethoxazole	bromide				Primaquine	
Sulphaquinazoline	Isometamidium				Robenidine	
Trimethoprim	Mefloquine				Sodium	
	Metronidazole				stibogluconate	
	Nifurtimox				Suramin	
	Pentamidine					
	Quinine					
	Satranidazole					
	Tinidazole					
Anthelmintics						
				Albendazole	Dichlorophen	Bephenium
				Fenbendazole	Niclosamide	Bromophos
				Mebendazole	Nitroxynil	Dichloros
				Oxfendazole	Oxyclozanide	Diethylcarbamazine
				Thiabendazole	Resorantel	Haloxon
				Triclabendazole		Ivermectin
						Levamisole
						Metrifonate
						Morantel
						Naphthalophos
						Oxamniquine
						Piperazine
						Praziquantel
						Pyrantel

Unknown (antiprotozoals): diclazuril; halofuginone. Unknown (anthelmintics): bithinol sulphoxide; bunamidine; closantel; diamphenethide; niclofolan; nitrosconate; rafoxanide.

cells. They owe their selective effects to the subtlety of their differences from natural chemicals. They are not exotic chemical structures made up of unusual elements and chemical groups: molecules of these types are much more likely to be totally antilife.

9.4 BIOCHEMISTRY OF DRUG ACTION

This topic will be considered under three headings: (1) mechanisms of action; (2) mechanisms of selective toxicity; and (3) mechanisms of acquired resistance to drugs.

9.4.1 Mechanisms of action

The proper establishment of the biochemical mechanism of action of a drug requires a systematic study of its effects on the various metabolic processes of the pathogen at the lowest concentration that stops its growth, kills it or leads to its expulsion from the host. Only when such a survey is complete can it be concluded confidently that a particular pathway or reaction is most sensitive to inhibition by a drug and can therefore be described as the primary target of that drug. Systematic studies of this type have been carried out with many antibacterial and antifungal drugs but only rarely with those active against parasites. Thus in most cases it is only possible to indicate what is the *most likely* mechanism of action of an antiparasite drug.

Extensive studies of the biochemical mechanisms of action of antibacterial drugs over many years has led to the identification of four key areas of metabolism as targets for drug action: (1) cofactor synthesis; (2) nucleic acid synthesis; (3) protein synthesis; and (4) wall synthesis. Note the absence of lipid and carbohydrate metabolism from this list. Presumably, these processes are too similar in bacteria and mammalian cells for *selective* inhibition to be possible.

Bacteria have the prokaryotic type of cellular organization. In contrast, all parasites are eukaryotes and in addition, helminths are multicellular organisms. Not surprisingly, therefore, studies on the mechanisms of action of antiparasitic drugs have produced a different list of the

areas of metabolism vulnerable to chemotherapeutic attack. In particular, there is no real equivalent of the bacterial cell wall in parasites, but they do contain important structural lipids (sterols) and proteins (microtubules), which are not present in bacteria. In addition, helminths contain a neuromuscular system, which is not found at all in unicellular microorganisms.

Thus the mechanisms of action of antiparasitic drugs can be grouped under seven headings: (1) cofactor synthesis; (2) nucleic acid synthesis; (3) protein synthesis; (4) membrane function; (5) microtubule function; (6) energy metabolism; and (7) neuromuscular function. The distribution of key currently used drugs between these seven headings is given in Table 9.1.

Dealing first with drugs interfering with cofactor synthesis, the anticoccidial drug, **amprolium**, is a structural analogue of the dietary vitamin, thiamine, the phosphorylated derivative of which is involved in many decarboxylation reactions. It appears to work by blocking the uptake of this compound from the gut. In contrast, the antisporozoan sulphonamides (e.g. **sulphadiazine**, **sulphadoxine**, **sulphamethoxazole**, **sulphaquinazoline**), sulphones (**dapsone**) and 2-substituted p-aminobenzoic acids (**ethopabate**) block the biosynthesis of tetrahydrofolate at the dihydropteroate synthase step. Tetrahydrofolate is an important cofactor in many active one carbon transfer reactions, especially the synthesis of deoxythymidylate, required for DNA synthesis. The antisporozoan 2,4-diaminopyrimidines (**pyrimethamine**, **trimethoprim**) and **proguanil** (after cyclization) block the same pathway at a later reaction, that catalysed by dihydrofolate reductase. Thus the two groups of compounds tend to potentiate each other.

Antiprotozoal drugs interfering with nucleic acid synthesis all appear to work by interaction with the DNA primer, rather than with the polymerase proteins. The antimalarials, **amodiaquine**, **chloroquine**, **halofantrine**, **mefloquine** and **quinine** and the antitrypanosomals, **homidium bromide** and **isometamidium**, are all able, *in vitro*, to intercalate between the base pairs of the DNA and thus probably, *in vivo*, disrupt its functioning. The antiprotozoal diamidines (**diminazine aceturate**, **pentamidine**) are un-

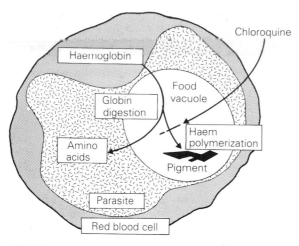

Fig. 9.1 Diagrammatic representation of the action of chloroquine. Inhibition of haem polymerase blocks the detoxification of the haem product of haemoglobin digestion. (After Wellems, 1992. Reprinted with permission from *Nature*, **355**, p. 108. Copyright Macmillan Magazines Limited.)

able to intercalate and probably interact ionically. The Chagas drugs **benznidazole** and **nifurtimox** and the anti-anaerobic protozoal 5-nitroimidazoles (**metronidazole**, **satranidazole**, **tinidazole**) probably alkylate the DNA after suitable activation of the nitro-group in a one electron reduction.

The trypanocide **quinapyramine** appears to block protein synthesis by displacement of magnesium ions and polyamines from the ribosomes. **Tetracycline** blocks protein synthesis in bacteria at the chain elongation and peptide bond formation steps and probably works the same way in malaria parasites. There is a school of thought that suggests that the 4-aminoquinoline, quinolinemethanol and phenanthrinemethanol antimalarials (**amodiaquine**, **chloroquine**, **halofantrine**, **mefloquine**, **quinine**) also work in this way (but, see the preceeding paragraph). The idea originally was that the drugs accumulate in the food vacuole (lysosome), thereby raising the pH, interfere with haemoglobin digestion and thus starve the parasite of amino acids. The problem with this hypothesis is that other sources of amino acids are open to the parasite (e.g. plasma, biosynthesis *de novo*). A more recent suggestion is that the haem

generated during this digestion, which is cytotoxic, is detoxified by an enzyme, haem polymerase, which is inhibited by chloroquine (Fig. 9.1). The trypanocide, **eflornithine**, interferes with the metabolism of the amino acid, ornithine. It works as a suicide substrate of the enzyme ornithine decarboxylase and thus blocks polyamine biosynthesis.

Amphotericin B appears to interact with leishmanial sterol (ergosterol), thus rendering the plasma membrane leaky to ions and small molecules (e.g. amino acids). In contrast, the anticoccidial ionophores (**lasalocid**, **monensin**, **salinomycin**), on the basis of their effects on other cells, probably interact with alkali metal cations and specifically inhibit the transport of potassium ions into mitochondria.

The anthelmintic benzimidazole carbamates (**albendazole**, **fenbendazole**, **mebendazole**, **oxfendazole**, **triclabendazole**) bind *in vitro* to tubulin isolated from worms and thus block its assembly into microtubular proteins. Electron microscopic studies on worms from treated animals indicate gradual loss of cytoplasmic microtubules, suggesting that such binding occurs *in vivo*. This agrees with biochemical studies which indicate impairment of glucose uptake and acetylcholine esterase secretion in worms from drug-treated animals; both of these processes require functional microtubules.

The antiprotozoal arsenicals (e.g. **melarsaprol**) and antimonials (**sodium stibogluconate**, **meglumine antimonate**) most likely work by blocking the kinases of glycolysis, especially the cytoplasmic pyruvate kinase, since they readily react with enzymes which have sulphydryl groups at their active sites. An alternative hypothesis, based on disruption of trypanothione reduction (the latter also contains biologically-active sulphydryl groups, see Chapter 6, Section 6.7.2) has recently been suggested. These two hypotheses may not be mutually exclusive. In contrast the anti-trypanosomal effects of **suramin** are probably mediated via the alpha-glycerophospate oxidase and dehydrogenase involved in NADH oxidation. However, it is interesting to note that the two hot-spots of positive charge on trypanosomal glycosome-located glycolytic enzymes are the same distance apart as the two sets of negative charge on suramin. If the drug was bound to such enzymes via these hot-spots, the substrate would be sterically hindered from reaching the active site of the enzyme.

A number of antisporozoan drugs block mitochondrial electron transport, most likely at complex III, including **clopidol**, **decoquinate** (c.f. the antibacterial quinolones which are DNA gyrase inhibitors), **parvaquone**, **buparvaquone**, **primaquine** (after metabolism) and **robenidine**. In contrast, the anthelmintic phenol derivatives (**dichlorophen**, **niclosamide**, **nitroxynil**, **oxyclozanide**, **resorantel**) act as uncoupling agents, both of the respiratory chain and also of fumarate reductase.

Many of the anthelmintic drugs which block the functioning of the neuromuscular system appear to work by interfering with that part of the system which uses acetylcholine as the neurotransmitter. **Levamisole**, **morantel** and **pyrantel** almost certainly actually interact with the acetylcholine receptor. The organophosphorous compounds (**bromophos**, **dichlorvos**, **haloxon**, **metrifonate**, **naphthalophos**) are potent inhibitors of acetylcholine esterase, the enzyme which breaks down acetylcholine. **Bephenium**, an analogue of acetylcholine, probably works in a similar fashion. **Piperazine** and **diethylcarbamazine** paralyse nematode muscle, probably by a reversible curare-like effect on neuromuscular junctions. There is little information on the mechanism of action of **oxamniquine**: what there is suggests an effect on the neuromuscular system. **Ivermectin** was thought originally to interfere with the part of the nematode neuromuscular system involving gamma-aminobutyric acid (GABA) as neurotransmitter; in particular it was thought to be a GABA agonist. More recent work, however, in Merck's own laboratories (see Campbell, 1989), suggests that it is not possible to assign a single mechanism of action for this drug in the wide variety of model systems that have been studied. In target organisms, its effect is mediated via a specific high affinity binding site. The physiological response to such binding is an increase in membrane permeability to chloride ions, which is independent of GABA-mediated chloride channels. Gamma-aminobutyric acid-gated chloride ion channels may also

interact with ivermectin, but only at higher drug concentrations. Effects on chloride ion channels are now also thought to underly the anthelmintic effects of praziquantel.

A number of points can be made about these data overall. The majority of the antiprotozoal drugs affect biosynthetic metabolism. Of those that effect energy metabolism, most are still in human use though toxic because of the absence of a satisfactory alternative (e.g. **melarsaprol**, **sodium stibogluconate**, **suramin**), or are veterinary drugs where the degree of selective toxicity required is less than that for human drugs. This suggests that even in protozoa, there are rather more differences in biosynthetic metabolism between them and vertebrates than there are in differences in energy metabolism.

In contrast to antiprotozoal drugs, most of the anthelmintic drugs affect energy metabolism or neuromuscular function. This should not be taken as an indication that helminth metabolism in these areas is distinct from that in vertebrates. It is much more likely to be a reflection of the lack of vulnerability of non-growing mature worms (except for egg production) to inhibitors of biosynthetic processes. The vulnerability of worms to disruption of neuromuscular function in most cases can be related to the fact that they need to maintain their position in the gut (e.g. gastro-intestinal nematodes) or circulatory system (e.g. schistosomes) of the host in order to survive.

9.4.2 Mechanisms of selective toxicity

Investigations with antimicrobial drugs have shown that the establishment of the biochemical mechanism of action of a drug often also provides an explanation as to why only the microbe is killed. This has in many cases also proved to be true for antiparasitic drugs.

The five biochemical mechanisms found to underly the selective action of antimicrobial drugs are: (1) differential uptake or secretion of drug between host cells and microbes; (2) drug activation only in the microbe; (3) the presence of the biochemical target only in the microbe; (4) the biochemical target of the drug differs between host and microbe; and (5) the biochemical target

Table 9.2 Summary of possible mechanisms of selectivity of antiparasite drugs

Differential uptake/ secretion	Drug activation only in parasite	Unique target in parasite	Drug discriminates between target in host and parasite	Pathway blocked more important in parasite than in host
Antiprotozoals				
Amodiaquine	Benznidazole	Dapsone	Amphotericin B	Meglumine antimonate
Chloroquine	Metronidazole	Ethopabate	Amprolium	Melarsaprol
Diminazine	Nifurtimox	Sulphadiazine	Buparvaquone	Sodium stibogluconate
aceturate	Satranidazole	Sulphadoxine	Clopidol	
Halofantrine	Tinidazole	Sulphamethoxazole	Decoquinate	
Homidium bromide		Sulphaquinazoline	Eflornithine	
Isometamidium		Suramin	Parvaquone	
Mefloquine			Primaquine	
Pentamidine			Proguanil	
Quinapyramine			Pyrimethamine	
Quinine			Robenidine	
Tetracycline			Trimethoprim	
Anthelmintics				
			Albendazole	
			Bephenium	
			Bromophos	
			Dichlorvos	
			Diethylcarbamazine	
			Fenbendazole	
			Haloxon	
			Levamisole	
			Mebendazole	
			Metrifonate	
			Morantel	
			Naphthalophos	
			Oxfendazole	
			Piperazine	
			Pyrantel	
			Thiabendazole	
			Triclabendazole	

Unknown (antiprotozoals): diclazuril; halofuginone; lasalocid; monensin; salinomycin. Unknown (anthelmintics): bithinol sulphoxide; bunamidine; closantel; diamphenethide; dichlorophen; niclofolan; niclosamide; nitroxynil; oxyclozanide; resorantel; ivermectin; oxamniquine; nitrosconate; pyraziquantel; rafoxanide.

in the microbe is more critical to its viability and growth than it is to the host cell. Examples of all five mechanisms of selectivity can be found amongst antiparasitic drugs. The distribution of the mechanisms of selective toxicity of currently used key drugs between the five groups is given in Table 9.2.

Several antiprotozoal drugs are concentrated one hundred-fold or more in the parasite compared with the drug concentration in the plasma. Included here are the antimalarial 4-aminoquino-lines, quinolinemethanols and phenanthreneme-thanols (**amodiaquine, chloroquine, halofantrine, mefloquine, quinine**) and **tetracycline**, the anti-protozoal diamidines (**diminazine aceturate, pentamidine**) and the antitrypanosomals, **homidium bromide, isometamidium** and **quinapyramine**. In all cases, selectivity appears to be the result of the absence of such concentrative mechanisms in mammalian cells (see Chapter 6, Section 6.7.3).

Differential drug uptake is also behind the selectivity of antiprotozoal nitro-compounds (e.g.

benznidazole, metronidazole, nifurtimox, satranidazole, tinidazole). It occurs as a consequence of particular factors that activation continually tending to reduce intracellular drug concentrations below plasma levels. The enzyme system involved in **metronidazole** activation has been well-studied in *Trichomonas* and is a pyruvate–ferridoxin oxidoreductase, which has no homologue in mammalian cells (see Chapter 5, Section 5.3).

Sporozoan protozoa generally appear to biosynthesize tetrahydrofolate *de novo* from GTP, *p*-aminobenzoate and glutamate, whereas mammalian cells salvage dietary folates. Dihydropteroate synthase, the target of the sulphonamides (e.g. **sulphadiazine, sulphadoxine, sulphamethoxazole, sulphaquinazoline**), sulphones (e.g. **dapsone**) and *p*-aminobenzoic acids (e.g. **ethopabate**), is thus absent from mammalian cells, providing an explanation for the selectivity of these groups of drugs. Similarly, alpha-glycerophosphate oxidase, most likely the primary target of the antitrypanosomal drug, **suramin**, has no homologue in mammalian cells: NADH is oxidized via the respiratory chain or in the lactate dehydrogenase reaction. The absence of a haem polymerase from mammalian cells probably adds to the selectivity of chloroquine, which results primarily from selective concentration into infected erythrocytes (see above).

One of the classical examples of drugs discriminating between isofunctional targets in hosts and pathogens is provided by the 2,4-diaminopyrimidines (e.g. **pyrimethamine, trimethoprim**) and the **proguanil** metabolite (cycloguanil), and their target, the enzyme dihydrofolate reductase. The latter is involved in both tetrahydrofolate synthesis *de novo* and also in folate salvage, so only this type of mechanism would explain their selectivity (c.f. the sulphonamides, above). There are many other examples of similar selectivities amongst the antiparasitic drugs: (1) **amprolium** appears to discriminate between protozoan and chicken thiamine transport; (2) **amphotericin B** between leishmanial (ergosterol) and mammalian (cholesterol) membrane sterol; (3) **clopidol, decoquinate, parvaquone, buparvaquone, primaquine** metabolite and **robenidine** between complex III of sporozoan and vertebrate electron transport

systems; (4) the benzimidazole carbamates (**albendazole, fenbendazole, mebendazole, oxfendazole, triclabendazole**) between helminth and mammalian tubulin; and (5) the organophosphorous anthelmintics (**bromophos, dichlorovos, haloxon, metrifonate, naphthalophos**), **bephenium, piperazine** and **diethylcarbamazine, morantel** and **pyrantel** and **levamisole** between helminth and mammalian neurotransmission involving acetylcholine. The selectivity of **eflornithine** is more subtle: unlike its trypanosome homologue, the mammalian enzyme turns over rapidly, thus circumventing the blockade.

Glycolytic rates in African trypanosomes are high by comparison with mammalian cells since there is no metabolism beyond pyruvate. This difference is believed to explain, at least in part, the selectivity of the antitrypanosomal arsenicals (e.g. **melarsaprol**). However, lack of metabolism of pyruvate is not apparently a feature of *Leishmania* energy metabolism, and yet the antimonials (**meglumine antimonate, sodium stibogluconate**) are more selective drugs.

Three overall points can be made about these data concerning the mechanisms of selective toxicity of antiparasitic drugs. First, the mechanisms of the antiprotozoal drugs spread evenly across all five categories (Table 9.2). In contrast, those of the anthelmintic drugs, in so far as they are known, are restricted to the target discrimination category. Secondly, few known targets exploited to date have no homologue in mammalian cells. There are only four entries on the list so far, dihydropteroate synthase, alpha-glycerophosphate oxidase, haem polymerase and possibly trypanothione. Thirdly, information in this area for antiparasitic drugs generally, and especially for anthelmintic drugs, is sadly lacking.

9.4.3 Mechanisms of resistance to drugs

Acquired drug resistance has long been a problem in chicken coccidiosis and has tended to appear in the past in the trypanosomiases, whenever chemotherapy has been used exstensively to control them. It is now a major factor in the control of malaria. So far, however, it has proved to be far less of a problem in helminth chemotherapy. Two factors are probably responsible for this dif-

ference: (1) only in very few areas have anthelmintics been used as intensively as for example
in coccidiosis and malaria; and (2) the slow rate of
replication of helminths as compared to protozoa
gives far less opportunity, over a given period of
time, for the generation, selection and therefore
spread into the population of genes giving a drug-
resistance phenotype. Netherless, where intensive use of anthelmintics has occurred, such as
with the benzimidazole carbamates (e.g. **oxfendazole**) for the control of gastro-intestinal nematode infestations in sheep and cattle, drug
resistance has become a problem. This probably
occurred, however, through the selective transmission of pre-existing, drug-resistant worms,
rather then the acquisition of drug-resistance
genes *de novo*.

Studies, particularly with antibacterial drugs,
have led to the recognition of five basic biochemical mechanisms by which microorganisms can
become resistant to drugs: (1) metabolism of the
drug to an inactive form; (2) change in permeability so the drug is no longer taken up or is rapidly
pumped out of the cell; (3) development of an
alternative metabolic pathway so the metabolic
lesion is bypassed; (4) alteration in biochemical
target so the drug no longer binds to it as well;
and (5) elevation in the amount of target enzyme
so high levels of inhibition no longer have metabolic consequences. The basic genetic mechanism behind these developments mainly involves
mutation, often point mutations. However gene
amplification and, in bacteria, the acquisition of
novel drug-resistance genes as a result of the
transfer of genetic factors between individual
cells, populations and strains, also play key roles.

Little experimental work has been done on the
biochemical mechanisms of resistance of antiparasitic drugs or indeed of the genetics behind such
resistance. What is known is summarized below
and discussed in the context of molecular biology
in Chapter 6, Section 6.7.3.

Antitrypanosomal resistance to **pentamidine**
appears to involve alteration in the drug uptake
system. This normally raises the internal drug
concentration many times over that of the external (plasma) concentration. Drug-resistant strains
of trypanosome have a transport system with a
lower maximum velocity transport system (V_{max})

or a higher transport or Michaelis constant (K_m)
or a combination of the two (See Chapter 7,
Section 7.8.3 for more information on this subject).

Resistance to the antimalarial, **chloroquine**,
was originally described as being due to loss of a
high affinity chloroquine-binding site, leading to
a failure to concentrate drug from the plasma. It
appeared in laboratory models of malaria to be
a stable genetic factor, which was inherited in
a simple Mendelian fashion, underwent genetic
recombination with other markers and probably
arose by mutation and selection in the presence
of drug. More recent work, focusing on multi-
drug resistant (mdr) strains of the human malaria
parasite, *Plasmodium falciparum*, suggested that
in some of them, resistance is due to the presence
and/or amplification of one of two genes, pfmdr1
on chromosome 5 or pfmdr2 on chromosome 14,
which are homologues of the mdr gene found in
many drug-resistant tumours. In drug-sensitive
tumours, the protein expressed by this gene is
thought to be involved in pumping natural products out of the cell; in mdr strains, in which the
gene is mutated and/or amplified, it pumps out
drugs, preventing their accumulation to cytocidal
concentrations. It seemed likely that similar mechanisms operated in malaria parasites resistant/
multi-drug resistant to the 4-aminoquinoline,
quinolinemethanol and phenanthrenemethanol
antimalarial drugs (**chloroquine**, **quinine**, **amodiaquine**, **halofantrine**, **mefloquine**). Certainly, experimental drugs (e.g. verapamil), that inhibit the
functioning of this pump, reversed such resistance. However, it subsequently emerged that
not all chloroquine-resistant strains contain
either pfmdr1 or pfmdr2 and where one or other
was present, it was not necessarily amplified. All
chloroquine-resistant parasites, however, contain
a gene on chromosome 7 which is now thought to
play the major role, as yet undefined, in the
resistance mechanism. Most likely, this normally encodes a protein involved in chloroquine
concentration. The mechanism by which drugs
such as verapamil reverse resistance is not explained by this current hypothesis. It may simply
be that they have some antimalarial activity
in their own right which is potentiated by
chloroquine.

Laboratory-induced, **pyrimethamine**-resistant

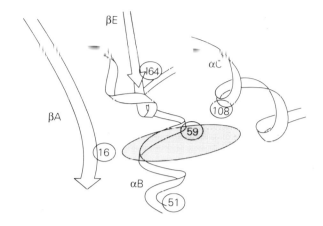

Pyrimethamine resistance*	Resistance to both	Proguanil resistance†
Ser 108 → Asn	Ser 108 → Asn	Ser 108 → Thr
Cys 59 → Arg	Isoleu 164 → Leu	Ala 16 → Val
Asn 51 → Isoleu	Cys 59 → Arg	

* Ser 108 → Asn only gives low level resistance.
† As its metabolite, cycloguanil.

Fig. 9.2 Diagram showing the positions of the point mutations in *Plasmodium falciparum* dihydrofolate reductase that border the active site of the enzyme (shaded portion). Individual point mutations are located on the C α-helix, the B α-helix and the E β-strand of the enzyme. (After Peterson, Milhous & Wellems, 1990, *Proceedings of the National Academy of Sciences of the U.S.A.*, **87**, 3018–3022.)

strains of malaria contain either amplified amounts of dihydrofolate reductase and/or, more commonly, an altered dihydrofolate reductase. In the latter case, the K_m tends to increase, along with the inhibition constant (K_i). In *P. falciparum*, resistance to **pyrimethamine** and **proguanil** (as its active metabolite, cycloguanil) in field isolates arises through point mutations at a limited number of positions along the dihydrofolate reductase/thymidylate synthase gene, which occurs on chromosome 4. Examples are in Fig. 9.2.

Resistance to the anthelmintic benzimidazole carbamates (e.g. **oxfendazole**) and **levamisole** have been linked respectively with a modified tubulin and an altered acetylcholine receptor.

9.5 DISCOVERY AND DEVELOPMENT OF NEW DRUGS

The overall objective of the drug-discovery process is the identification of a novel chemical entity with the potential to produce in an infected patient or animal a therapeutically useful response with an acceptable level of adverse reactions. Since drugs already exist for most parasitic diseases, such an entity is unlikely to be developed unless it has one or more of the following advantages over the existing therapies: better efficacy, lower toxicity, activity by a more preferred route, activity against drug-resistant strains and shorter withdrawal period (drugs for farm animals only).

The drug-discovery process is basically the same in all the anti-infective therapeutic areas, including the antiparasitic area. It begins with an idea and this idea may be of a chemical or

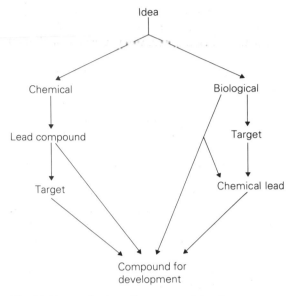

Fig. 9.3 Routes for drug discovery and development.

biological nature (Fig. 9.3). Chemical ideas centre around possible lead compounds and can be processed to molecules for development with or without knowledge of the cellular target through which they exert their effects. The discovery of most of the antiparasitic drugs in use today can be traced back to this approach.

In contrast, biological ideas tend to focus on a cellular target for which a chemical lead is later identified. However, it is possible, e.g. when the biological approach involves screening of natural products and especially of herbal medicines, that a chemical lead or even a compound ready for development will be identified directly without knowledge of the target site. Such approaches have led to the identification of a number of the antiparasitic drugs (e.g. the antinematode compound **ivermectin**), and without doubt they will be successful again. They are, however, no longer representative of how most of the pharmaceutical industry seeks to discover novel chemical entities. Therefore this chapter will concentrate on the biological targeting approach. There are two types of molecule which can be useful here: (1) medicinal chemicals; and (2) therapeutic proteins, especially monoclonal antibodies. The latter are still highly experimen-

tal and so this section will deal only with medicinal chemicals.

The drug-targeting approach to antiparasitic chemotherapy can be divided up into a number of phases. This section will deal first with targets and their validation, second consider the identification of a chemical lead and then will go on to summarize the overall development process. Finally, some of the novel antiparasitic drugs currently in the development pipeline will be discussed. Some of the more molecular aspects are also considered in Chapter 6, Section 6.7.

9.5.1 Targets and their validation

Most people tend to think about drug targets in terms of the enzymes (e.g. dihydrofolate reductase) involved in cellular metabolism. Other proteins in the cell can, however, serve the same purpose, including transporters and pumps, e.g. the receptors involved in neurotransmission in helminths and structural proteins such as tubulin. Other macromolecules can also serve this purpose, including DNA and cholesterol.

The identification of such targets arises either from comparative studies of the molecular biology, biochemistry and physiology of the target parasite and host or from investigation of the mechanisms of action of drugs already known to be active. It is not clear which is the best approach. The problem with the first is that it is not known that one is on to a potential winner until the study is complete. The problem with the second is that the target has already been exploited and might therefore already be associated with problems of drug resistance. For either approach, some clues as to the areas of metabolism to be investigated can be gleaned from consideration of the targets of antimicrobial drugs currently in use, especially those of antiparasitic drugs (see Section 9.4.1).

A critical part of characterizing a potential target or validating a mechanism of action is to establish a target-site assay which requires readily-obtainable amounts of parasite material, is quick and easy to operate and has a high throughput. Subsequently it can then be used to test for drug activity at the target site and can thus play a key role in the identification of a chemical lead.

The ultimate objective of the study of the comparative biochemistry or mechanism of action is to demonstrate that the enzyme concerned represents a target for chemotherapeutic attack. This requires two types of evidence: (1) demonstration that binding of ligands to the target leads *in vivo* directly or indirectly to a static or cidal consequence, that is that drug interaction with the target is associated with efficacy; and (2) evidence either that the target has no counterpart in the host or that it differs significantly from that in the host, that is that the target is associated with selectivity and thus a low level of toxicity. Whether it is best for the consequence of interaction to be static or cidal is not clear. The former relies on the immune system to clear the infection and thus leaves the host with some measure of immunity. The latter does not require the host to have a competent immune system.

9.5.2 Identification of a chemical lead

Having identified and validated a target, the critical next step is to match it with a chemical lead. This must have activity at the target site and be amenable to novel chemical manipulation. Ideally, it should also show some selectivity relative to the isofunctional host enzyme and have some activity in *in vitro* and/or *in vivo* whole organism assays.

The problem with identifying a chemical lead is, of course, not in defining the properties sought but in identifying a compound which fulfils such criteria. There are in fact a number of approaches which can be followed in seeking to achieve this objective and to some extent they overlap. They are detailed in the next five paragraphs.

1 Random screening of compounds in target-site assays is the least rational of all the approaches and as such is likely to require many thousands of compounds to be evaluated. Its attraction in the 1990s is that it might reveal a completely unsuspected activity in a particular compound.

2 A somewhat more rational variant of this approach involves the screening of a carefully selected representative set of several hundred compounds, chosen to provide a wide range of chemical types of diverse physico-chemical properties. Most pharmaceutical companies have now identified and maintain stocks of such a set of compounds.

3 A much more rational approach is to survey the scientific literature for described inhibitors from other systems. Since microbial biochemistry generally and bacterial biochemistry in particular has been studied in so much more depth than parasite biochemistry, it is rare that such a search will not be fruitful for the parasite biochemist.

4 A sure-fire rational approach to getting a chemical lead is of course to make analogues of the substrates or products of the enzyme target. Simple analogues are likely to be competitive inhibitors, with all the potential problems of their effects being overcome by a build-up of substrate. However, such potential problems can be overcome by the design of suicide inhibitors based on a knowledge of reaction mechanisms. The biggest problem is in fact to achieve selectivity since, although isofunctional proteins from mammalian and parasites often vary widely in amino acid sequence, very often there is conservation at the active site, just where, by definition, the competitive inhibitor will bind.

5 This leads to the ultimate in rationality, the *de novo* design of a lead on the basis of target structure using computer graphic techniques. The overall approach is summarized in Chapter 6, Fig. 6.8. The critical activity is the determination of the three-dimensional structure of the target enzyme. This usually requires an X ray crystallographic analysis of the protein, since most targets are too big for two-dimensional NMR analysis. There is thus a need for good-quality crystals of the protein. This is one of two Achilles heels of the approach, since it is not easy and, ironically, is tackled essentially empirically. The other Achilles heel is that, although one might end up with an excellent inhibitor at the target site, such design involves little consideration of the other major requirement for a successful chemotherapeutic agent, a pharmacokinetic and metabolic profile which will ensure that some at least of an administered dose of the compound will be in the vicinity of the target site in a chemically intact state for a reasonable period of time. The reason that this is not built in totally from the beginning is that at the moment we know too little about the ground rules that apply to enable us to do so in a rational way. For this reason, it must be stressed that at this time, such *de novo* design activities should be seen as lead generation exercises, though no doubt in time, totally rational design will be possible.

9.5.3 Overall development process

Once a chemical lead is established for the intended target, the next step involves the synthesis of analogues of the lead compound and their testing in the target-site assays of both parasite and host and in *in vitro* and *in vivo* screens. The objective is to maximize activity at the parasite target site, *in vitro* and *in vivo* and to minimize activity for the host, in other words, using today's 'jargon', to establish the structure activity relationships (SAR). There was a time when this used to involve only medicinal chemists and biologists. Now, however, the physical chemists have become increasingly involved, especially in the analysis of the biological results in terms of the physico-chemical properties (e.g. molecular weight, pka, lipophilicity) of the molecules. The whole process has therefore become increasingly quantitative, hence quantitative structure activity relationships (QSAR).

On the basis of the QSAR, it should be possible to draw up a short-list of perhaps six compounds for detailed study. The criteria for selection of such a short-list are likely to include: (1) the degree of novelty; (2) the level of efficacy, especially against drug resistant strains; (3) the degree of selectivity at the target site; and (4) the absence of acute toxicity.

The idea then is to study the compounds on the short-list in depth. Such investigations are likely to include: (1) for human drugs, efficacy in primate models, if available; (2) for veterinary drugs, efficacy studies in the target animal species; (3) detailed analysis of patentability, unless the whole series is novel; (4) analysis by development chemists of potential problems of scale-up; (5) consideration by pharmaceutical development scientists of formulation issues; (6) preliminary study of the absorption, distribution, metabolism and excretion (ADME) in at least one animal species; (7)

analysis of the acute (usually maximum-tolerated dose) and preliminary chronic toxicity of the compounds, generally in mice; (8) analysis of mutagenicity, usually an *in vitro* Ames test; (9) assessment of general pharmacology; and (10) for molecules which have chiral centres or show *cis/trans* isomerization, analysis of the efficacy, toxicity and ADME of each isomer, since nowadays an individual isomer would be developed. At the end of such a period of detailed assessment, one is hopefully in a position to select a compound for development. If not, consideration can be given to investigating, in detail, other compounds in the series not included on the original short-list, or even the synthesis of new members of the series.

The key activities usually associated with the development of a novel chemical entity are: (1) chemical development and large batch synthesis; (2) pharmaceutical development (the production of compound suitably formulated into capsules, tablets, suspensions, injectables, etc.); (3) development of analytical methods to support quality assurance procedures; (4) chronic toxicological studies in two animal species, reproductive toxicology and investigation of carcinogenicity; and (5) human or veterinary clinical trials to assess safety, tolerance, pharmacokinetics and efficacy. The objective of the first three activities is to provide a drug for the subsequent toxicological and clinical studies. For drugs intended for human use, if all goes well, a dossier of the data thus obtained will be assembled into a marketing authorization application (MAA) for submission to the regulatory authorities.

Such drug discovery and development activities are expensive, time-consuming and speculative. The numbers now put on such statements are of the order of 1 in 10 000 compounds synthesized on an empirical basis, 10–12 years and £100 million (including the costs of the failures). Because patent life (where it exists at all) was until recently restricted to 20 years, and because there is a need to recoup the discovery and development costs as well as manufacturing and distribution costs and the costs of previous failures, new drugs are expensive. This situation is creating a major impasse in the human antiparasitic drug therapeutic area. There is a clear need for new drugs but most of the patients that need them are in the Third World and cannot afford to buy them. Therefore, increasingly, the industry is not developing them. Discussions are going on within the industry, between the industry and the charities and between the industry and the international agencies, such as the WHO, to try and resolve this problem. If this is not done soon, there will be a dearth of new antiparasitic drugs introduced for human parasitic diseases in the twenty-first century.

9.5.4 Drugs in the development pipeline

There are two drugs/drug series in clinical trial for human diseases caused by kinetoplastid protozoa, *allopurinol* and azoles such as *ketoconazole*. The first has long been used to treat gout and the second are used as an antifungal drugs. Thus none of them has attracted a full set of development costs.

There are reports in the clinical literature that orally administered *allopurinol* is efficacious in leishmaniasis, especially if used as an adjunct to other antileishmanial agents. No doubt these will be followed up in the next few years. It may also have utility in Chagas disease. Allopurinol works, after conversion to the ribonucleotide equivalent, by blocking nucleic acid synthesis. Selectivity is a consequence of the fact that mammalian enzymes cannot carry out this conversion. *Ketoconazole* is in clinical trials for Chagas disease and leishmaniasis. Other, safer follow-up antifungal azoles (e.g. itraconazole, fluconazole) are also being tested. These compounds work by inhibiting the *de novo* synthesis of the parasite specific sterol, ergosterol.

In contrast to these studies, there have been no recent investigations of the efficacy of allopurinol riboside in leishmaniasis and the development of a liposomal formulation of sodium stibogluconate was terminated because of problems of toxicity. No decision appears to have been made yet about the development of the 8-aminoquinoline, *WR 6026*, which has good activity in experimental leishmaniasis. No animal trypanocide nor any drugs active against 'anaerobic' protozoa are known to be in development.

However, a number of compounds with activity against sporozoan protozoa are in the develop-

ment pipeline or are about to enter it. Included here are: (1) *azithromycin*, a presumed (by analogy with related antibacterial drugs) protein synthesis inhibitor, which has promising experimental activity *in vivo* against toxoplasmosis; (2) cotrimoxazole for the suppressive prophylaxis of toxoplasmosis in AIDS patients; (3) *566C80* (generic name, atovaquone) a protozoan-specific inhibitor of mitochondrial electron transport at complex III, which is in clinical trials for malaria and toxoplasmosis and now marketed for PCP; (4) *artemisinin*, a herbal medicine of unknown mechanism of action, long-used in China for malaria, and semi-synthetic analogues (e.g. arteether, artemether, artesunate) now under development in the western world, which look promising, especially for cases of cerebral malaria; (5) totally synthetic artemisinin analogues, designed using molecular graphic techniques (e.g. the bicyclic peroxides and the 1,2,4-trioxanes) for potential use in malaria and PCP; (6) *WR 238,605*, an 8-aminoquinoline with sufficient experimental activity for it to be considered as a possible replacement for primaquine in the prevention of malaria relapses; and (7) mdr transporter inhibitors (e.g. verapamil and desipramine) which it is hoped would reverse chloroquine resistance, though those tried so far are not efficacious at non-toxic doses.

No new series are known to be in development for parasitic diseases caused by digeneans or cestodes. A number of series related to Merck's **ivermectin** are, however, at various stages of development in various pharmaccutical houses for the various gastro-intestinal nematode infestations of farm animals. Whether such compounds will be able to carve a niche for themselves in a market-place dominated by ivermectin remains to be seen. In the meantime, **ivermectin** is now in clinical trials for human lymphatic filariasis and yielding promising results. *CGP 6140*, of unknown mode of action, is under clinical investigation as a macrofilariacide for onchocerciasis.

APPENDIX

The chemical structures of antiparasite drugs are shown on pages 236–241. The names of key currently used drugs are in **bold type**; those in development are in *italics*.

Albendazole

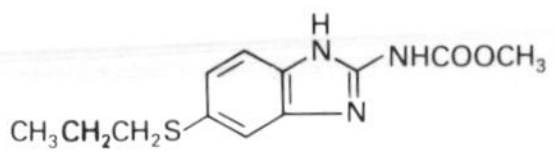

Allopurinol

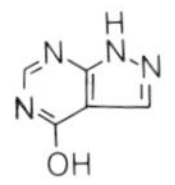

Amodiaquine

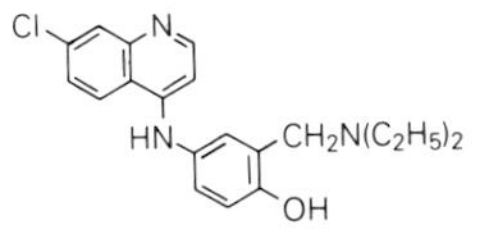

Amphotericin B

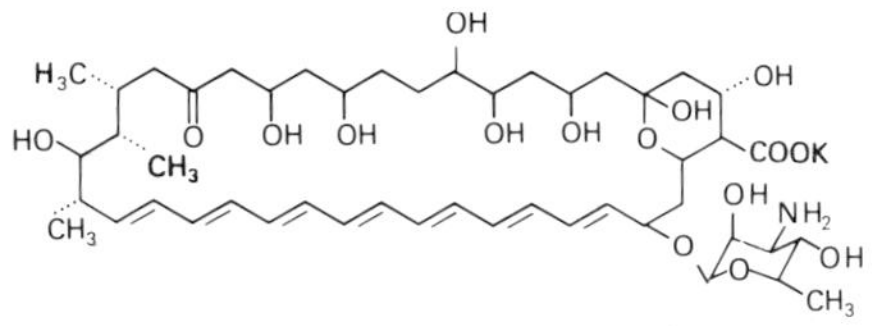

Amprolium

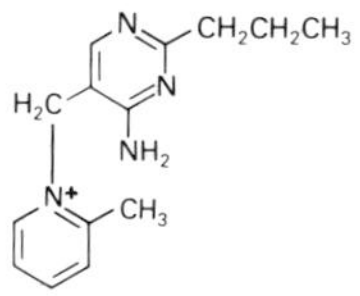

Artemisinin

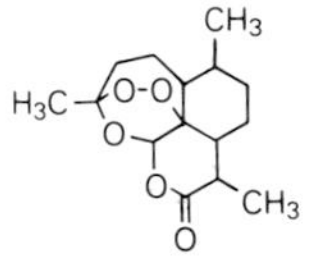

Azithromycin

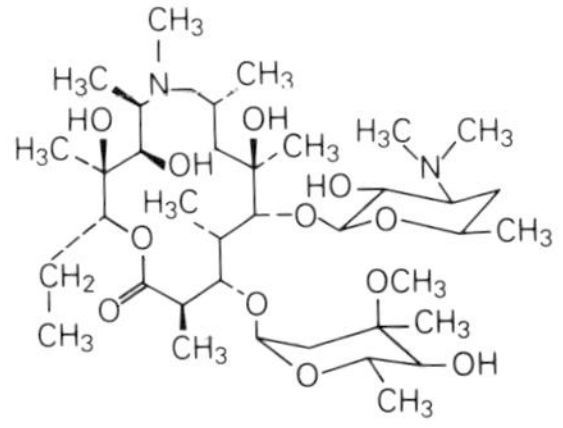

Benznidazole

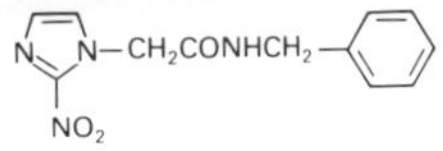

Bephenium

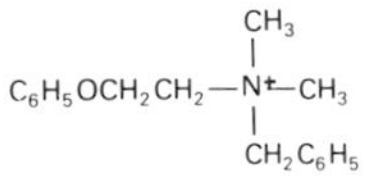

Bithinol

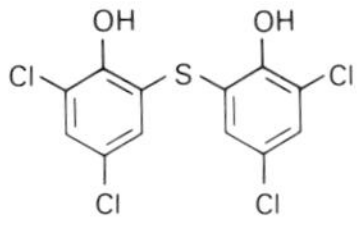

Bromophos

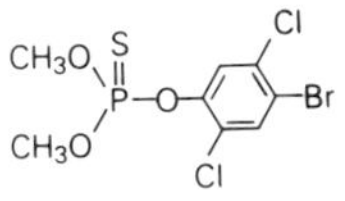

Bunamidine

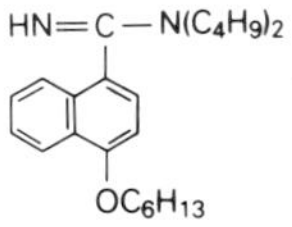

Buparvaquone

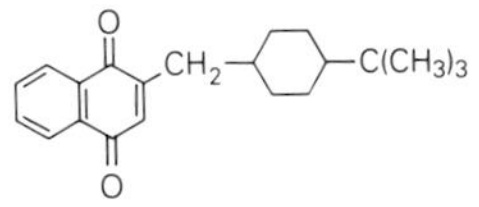

Chloroquine

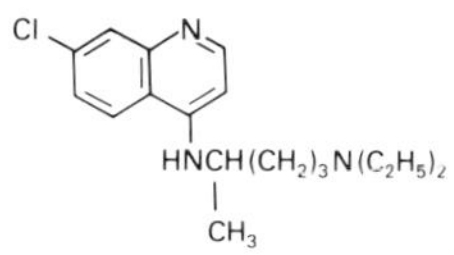

Clopidol

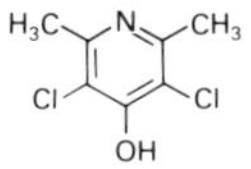

Closantel

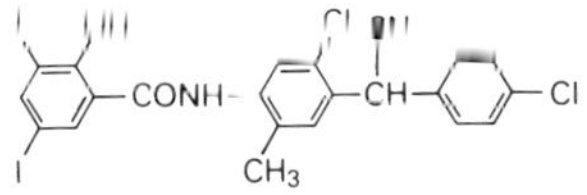

CONH–

CH₃

Cl

Dapsone

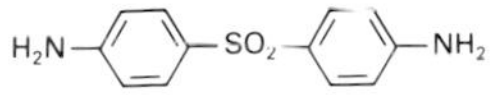

H_2N—〈 〉—SO_2—〈 〉—NH_2

Decoquinate

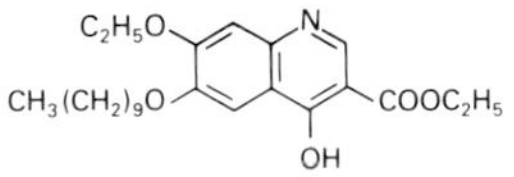

C_2H_5O

$CH_3(CH_2)_9O$

N

$COOC_2H_5$

OH

Diamphenethide

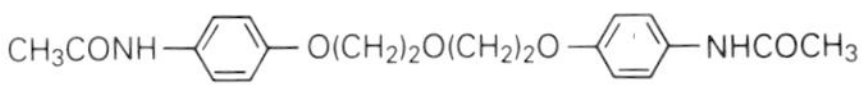

CH_3CONH—〈 〉—$O(CH_2)_2O(CH_2)_2O$—〈 〉—$NHCOCH_3$

Dichlorophen

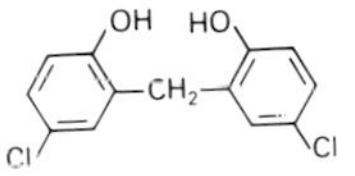

OH HO

CH_2

Cl Cl

Dichlorvos

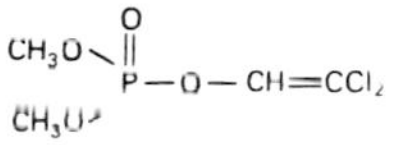

CH_3O

CH_3O

O
||
P—O—CH=CCl₂

Diethylcarbamazine

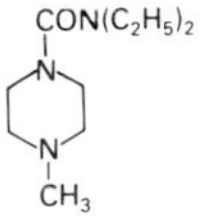

$CON(C_2H_5)_2$

N

N

CH_3

Diminazine aceturate

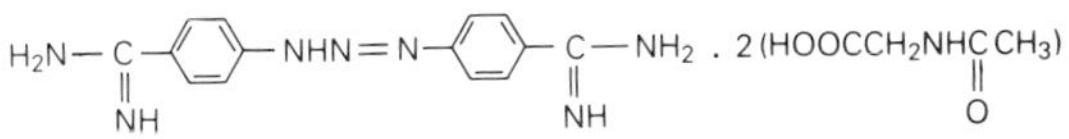

H_2N—C—〈 〉—NHHN=N—〈 〉—C—NH_2 . 2 (HOOCCH₂NHC CH₃)

NH   NH   O

Eflornithine

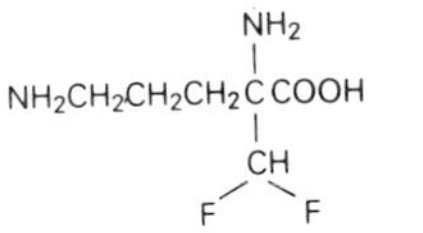

NH_2

$NH_2CH_2CH_2CH_2C$—COOH

CH

F   F

Ethopabate

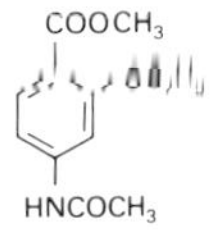

$COOCH_3$

$HNCOCH_3$

Fenbendazole

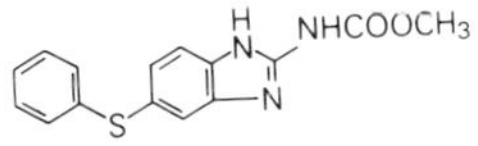

H
N

$NHCOOCH_3$

S

N

Halofantrine

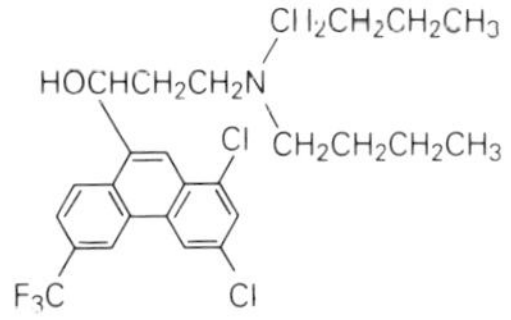

$CH_2CH_2CH_2CH_3$

$HOCHCH_2CH_2N$

$CH_2CH_2CH_2CH_3$

Cl

F_3C

Cl

Halofuginone

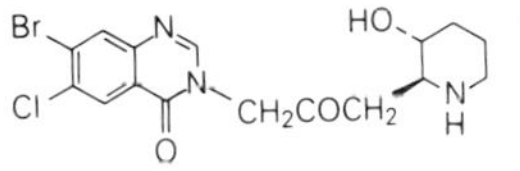

Br

Cl

N

N

O

HO

CH_2COCH_2

N
H

Haloxon

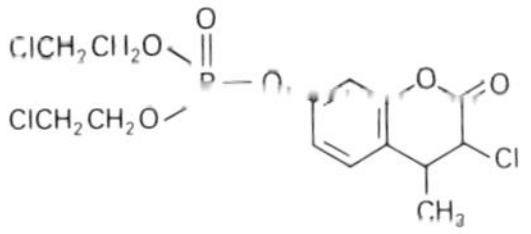

$ClCH_2CH_2O$

$ClCH_2CH_2O$

O
||
P—O

O   O

Cl

CH_3

Homidium bromide

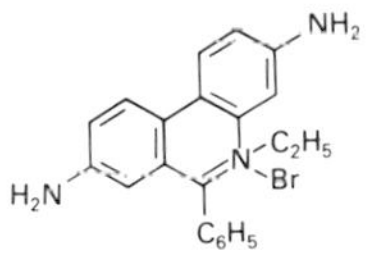

NH_2

H_2N

N

C_2H_5

Br

C_6H_5

Isometamidium chloride

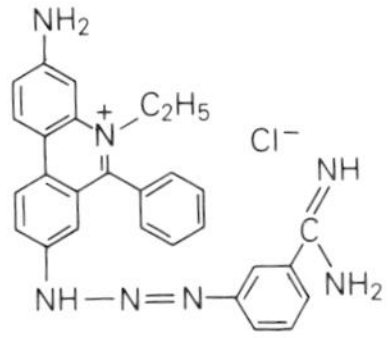

NH_2

+
N

C_2H_5

Cl⁻

NH—N=N

NH

NH_2

Ivermectin

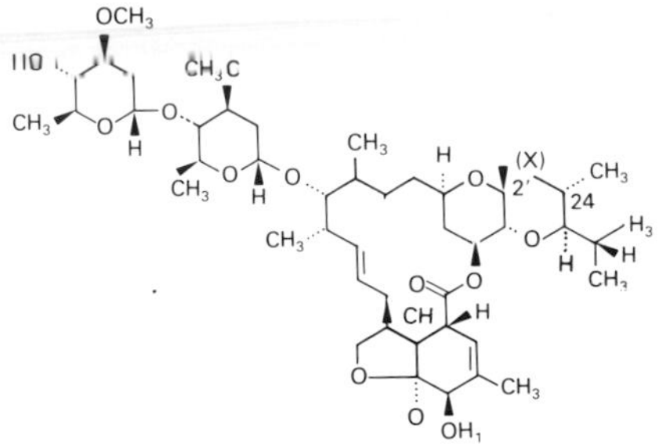

Ketoconazole

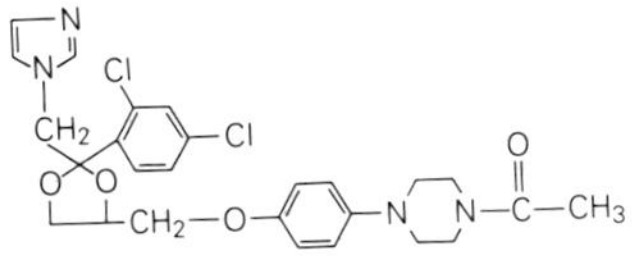

Lasalocid

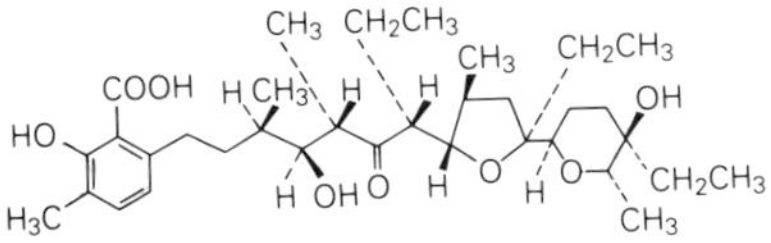

Levamisole

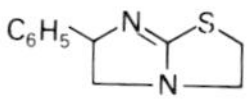

Mebendazole

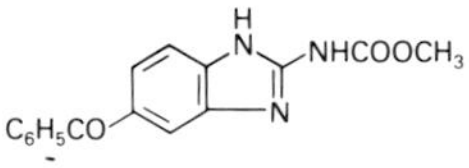

Mefloquine

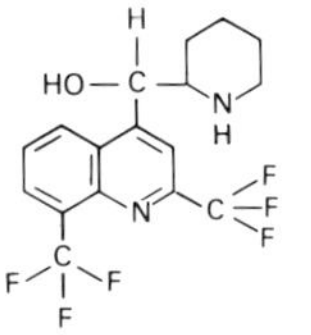

Meglumine antimonate

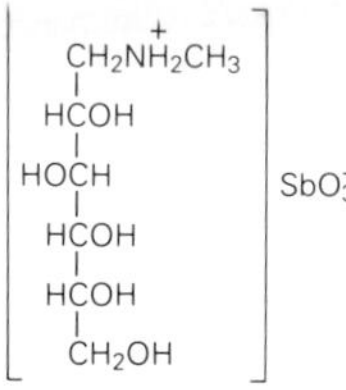

Melarsaprol

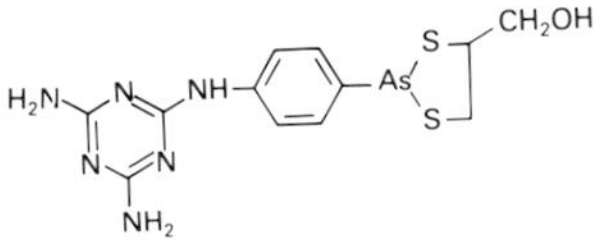

Metrifonate

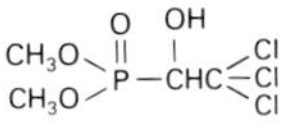

Metronidazole

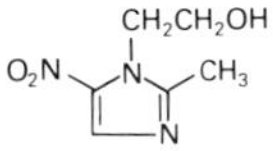

Monensin

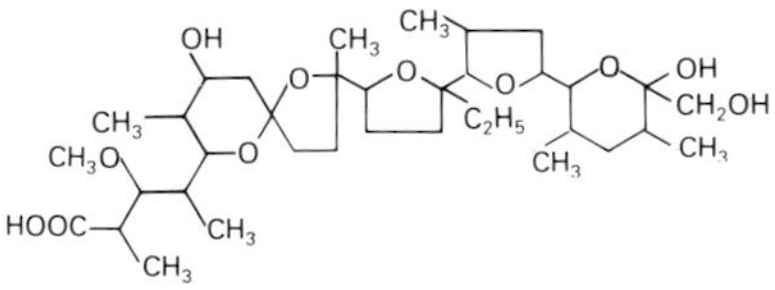

Morantel

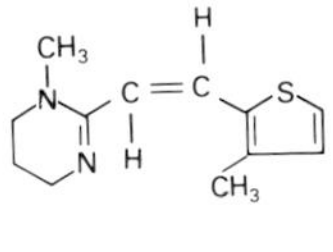

Naphthalophos

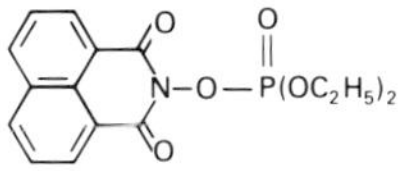

Niclofolan

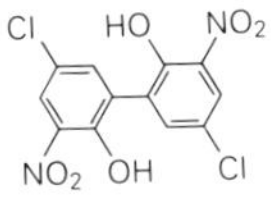

Niclosamide

Nifurtimox

Nitrosconate

Nitroxynil

Oxamniquine

Oxfendazole

Oxyclozanide

Parvaquone

Pentamidine

Piperazine

Praziquantel

Primaquine

Proguanil

Pyrantel

Pyrimethamine

Quinapyramine

Quinine

Rafoxanide

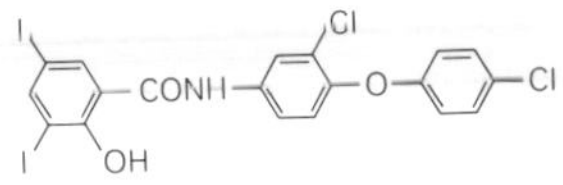

Sulphamethoxazole

Resorantel

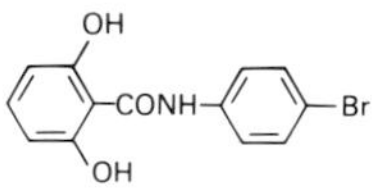

Sulphaquinoxaline

Robenidine

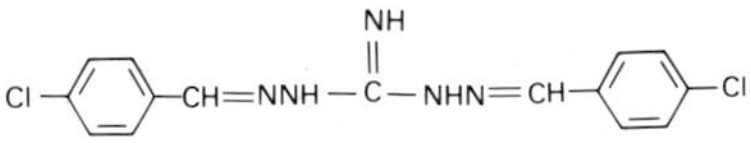

Suramin

Salinomycin

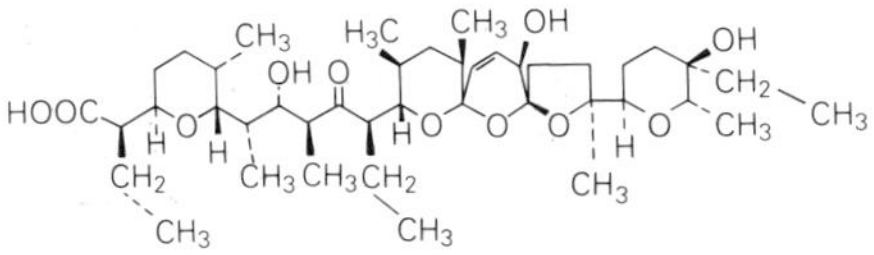

Tetracycline

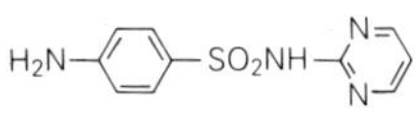

Satranidazole

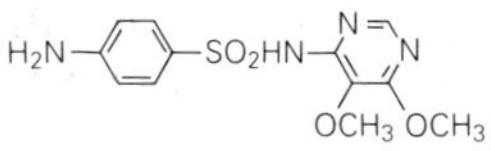

Thiabendazole

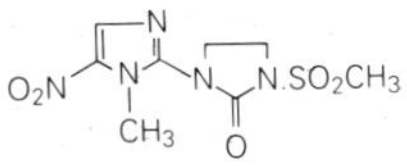

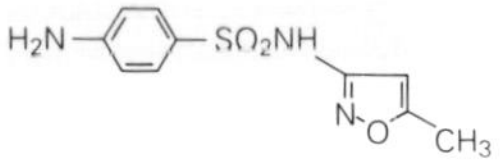

Sodium stibogluconate

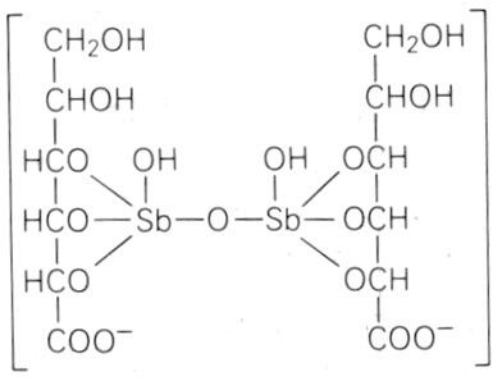

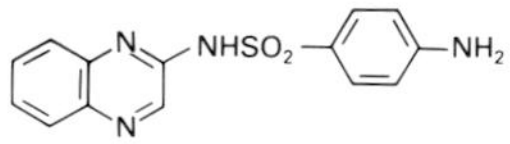

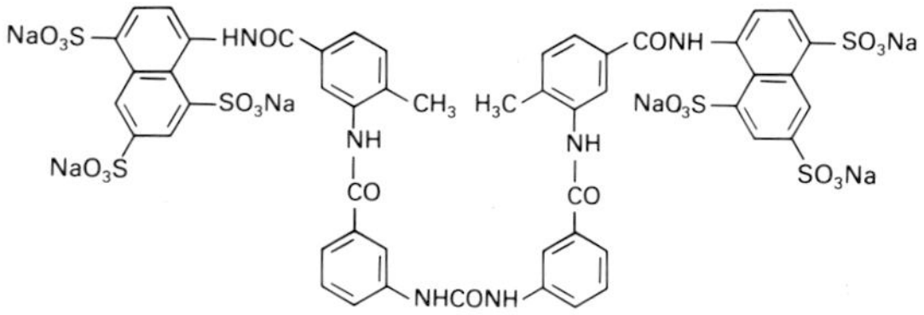

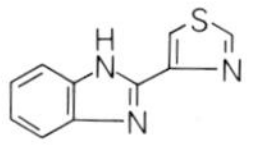

Tinidazole

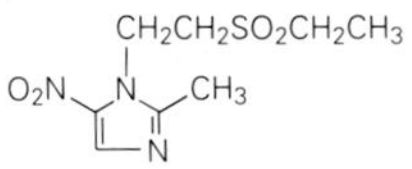

Sulphadiazine

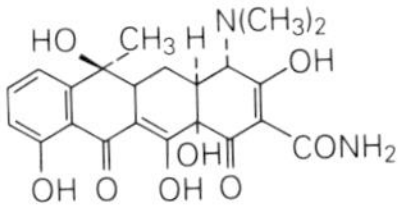

Triclabendazole

Sulphadoxine

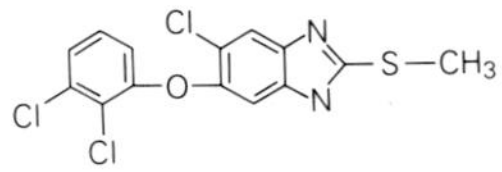

Trimethoprim

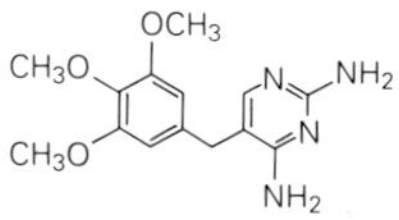

BW 566C80

WR 6026

CGP 6140

O_2N—⬡—NH—⬡—NHCSN⬠N—CH_3

WR 238,605

REFERENCES AND FURTHER READING

Alving, C.R. (1986) Liposomes as drug carriers in leishmaniasis and malaria. *Parasitology Today*, **2**, 101–107.

Ash, C.P.J. (1989) Onchocerciasis and the Mectizan donation programme. *Parasitology Today*, **5**, 63–64.

Ash, C.P.J. (1990) Benzimidazole anthelmintics. *Parasitology Today*, **6**, 105–136.

Barrett, J. (1981) *Biochemistry of Parasitic Helminths*. London, Macmillan.

Campbell, W.C. (1989) *Ivermectin and Abamectin*. New York, Springer-Verlag.

Cowman, A.F. (1991) The P-glycoprotein homologues of *Plasmodium falciparum*: are they involved in chloroquine resistance? *Parasitology Today*, **7**, 70–76.

El-On, J., Jacobs, G.P. & Weinrauch, L. (1988) Topical chemotherapy of cutaneous leishmaniasis. *Parasitology Today*, **4**, 76–81.

Fairlamb, A.H., Carter, N.S., Cunningham, M. & Smith, K. (1992) Characterisation of melarsen-resistant *Trypanosoma brucei brucei* with respect to cross-resistance to other drugs and trypanothione metabolism. *Molecular and Biochemical Parasitology*, **53**, 213–222.

Franklin, T.J. & Snow, G.A. (1989) *Biochemistry of Antimicrobial Action*, 4th edn. London, Chapman & Hall.

Gale, E.F., Cundliffe, E., Reynolds, P.E., Richmond, M.H. & Waring, M. (1981) *The Molecular Basis of Antibiotic Action*, 2nd edn. London, John Wiley.

Ginsberg, H. (1990) Antimalarial drugs: is the lysosomotropic hypothesis still valid? *Parasitology Today*, **6**, 334–337.

Greenwood, D. & O'Grady, F. (1985) *The Scientific Basis of Antimicrobial Chemotherapy*. Cambridge, Cambridge University Press.

Gutteridge, W.E. (1987) Available trypanocidal drugs. *La Medicina Tropicale*, **3**, 118–121.

Gutteridge, W.E. (1989) Parasite vaccines versus antiparasite drugs: rivals or running mates? *Parasitology*, **98** (supplement), S87–S97.

Gutteridge, W.E. (1989) Current trends in parasite chemotherapy. *Verhandlungen der Deutschen Zoologischen Gesellschaft*, **82**, 111–119.

Gutteridge, W.E. & Coombs, G.H. (1977) *Biochemistry of Parasitic Protozoa*. London, Macmillan.

Hart, D., Langridge, A., Barlow, D. & Sutton, B. (1989) Antiparasite drug design. *Parasitology Today*, **5**, 117–120.

Hughes, W.T. (1987) Treatment and prophylaxis for *Pneumocystis carinii* pneumonia. *Parasitology Today*, **3**, 332–335.

Hyde, J.E. (1990) *Molecular Parasitology*. Milton Keynes, Open University Press.

James, D.M. & Gilles, H.M. (1985) *Human Antiparasitic Drugs. Pharmacology and Usage*. Chichester, John Wiley.

Krogstad, D.J., Schlesinger, P.H. & Gluzman, I.Y. (1992) The specificity of chloroquine. *Parasitology Today*, **8**, 183–184.

Long, P.L. & Jeffers, T.K. (1986) Control of chicken coccidiosis. *Parasitology Today*, **2**, 236–240.

Meshnick, S.R. (1990) Chloroquine as intercalator: a hypothesis revived. *Parasitology Today*, **6**, 77–79.

Peterson, D.S., Milhous, W.K. & Wellems, T.E. (1990) Molecular basis of differential resistance to cycloguanil and pyrimethamine in *Plasmodium falciparum* malaria. *Proceedings of the National Academy of Sciences of the U.S.A.*, **87**, 3018–3022.

Slater, A.F.G. (1992) Malaria pigment. *Experimental Parasitology*, **74**, 362–365.

Stryer, L. (1988) *Biochemistry*. New York, W.H. Freeman.

Vande Waa, E.A. (1991) Chemotherapy of filariasis. *Parasitology Today*, **7**, 194–199.

Warren, K.S. (1990) An integrated system for the control of the major human helminth parasites. *Acta Leidensia*, **59**, 433–442.

Wellems, T.E. (1991) Molecular genetics of drug resistance in *Plasmodium falciparum* malaria. *Parasitology Today*, **7**, 110–112.

Wellems, T.E. (1992) How chloroquine works. *Nature*, **355**, 108–109.

Wernsdorfer, W.H. (1991) The development and spread of drug-resistant malaria. *Parasitology Today*, **7**, 297–303.

White, N.J. (1988) The treatment of falciparum malaria. *Parasitology Today*, **4**, 10–14.

Windholz, M. (ed.) (1989) *The Merck Index. An Encyclopedia of Chemicals and Drugs*, 11th edn. Rahway, Merck and Company Inc.

World Health Organization (1985) *The Control of Schistosomiasis*. WHO Technical Report Series No. 728. Geneva, WHO.

World Health Organization (1986) *Epidemiology and Control of African Trypanosomiasis*. WHO Technical Report Series No. 739. Geneva, WHO.

World Health Organization (1987) *Prevention and Control of Intestinal Parasitic Infections*. WHO Technical Report Series No. 749. Geneva, WHO.

World Health Organization (1987) *WHO Expert Committee on Onchocerciasis*. WHO Technical Report Series No. 752. Geneva, WHO.

World Health Organization (1990) *Practical Chemotherapy of Malaria*. WHO Technical Report Series No. 805. Geneva, WHO.

World Health Organization (1990) *WHO Model Prescribing Information. Drugs used in Parasitic Diseases*. Geneva, WHO.

Chapter 10 / Control

D. H. MOLYNEUX

10.1 INTRODUCTION

The success of control measures against any disease is dependent on a knowledge of its aetiology (causative agent) and natural history (epidemiology), including its mode of transmission, e.g. whether it is vector-borne, water-borne, or transmitted by orofaecal, aerosol or venereal methods.

Whatever the method of control employed it is necessary to relate this to the development of an overall strategy for that particular disease based on the best available biological information, allied to practical considerations such as methods and logistics of delivery, relative cost-effectiveness and unit costs, ability to integrate control into existing programmes, potential ecological damage and acceptability to the target community. Before the implementation of a control programme, however, the disease must be identified by the local Ministry of Health as one which merits a priority rating, and funds allocated or obtained. In addition, the potential of ongoing research to produce alternative methods in an acceptable period of time must be assessed.

10.2 COMPONENTS OF CONTROL

The components of control activities fall into various well-defined categories.
1 Disease surveillance and diagnosis; definition of the problem; reporting.
2 Definition of strategy by selection of methodology.
3 Planning and resourcing.
4 Implementation of method, e.g.
  (a) chemotherapy and chemoprophylaxis – drugs
  (b) immunoprophylaxis – vaccines
  (c) vector or reservoir control
  (d) environmental management
  (e) health education
5 Evaluation.

The cost-effective integration of a combination of the above methods (see Table 10.1) fulfils the principles of an integrated control strategy. Consideration as to how such a programme could be integrated into existing activities must also be given.

10.2.1 Integrated control of parasitic infections

Parasitic infections are frequently associated with particular remedies and methods of control but it is appropriate to view control of an agent or a vector, or the relief of the symptomatology of a disease caused by that agent, as part of an integrated control strategy. This is particularly important if that strategy is aimed at alleviating disease on a community basis, at whatever level community is defined – local, regional or national, rather than at the level of the individual. Integrated control strategies are based on coordinated long-term planning and combine knowledge derived from diverse scientific disciplines (e.g. see Table 10.1).

Integrated control strategies are dependent on planning and execution of a high order and can be quantified in terms of social and economic benefits to the human or livestock population. However, although the definition of the problem is a prerequisite to disease control in the community, it is initially dependent on accurate diagnoses. Diagnosis of parasitic infections is sometimes difficult and whilst parasitic infections cannot be diagnosed on clinical grounds alone, there is currently, however, a drive towards establishing the value of self-diagnosis or community diagnosis. The difficulties of provision of appropriate diagnostics at the peripheral levels of the health service remains a serious impediment,

Table 10.1 Components of parasitic diseases control – considerations for developing a strategy

---

COMPONENTS OF CONTROL PROGRAMMES
1 Diagnosis – recognition and definition of problem
2 Field studies – determine extent, severity, prevalence/incidence
3 Evaluate socio-economic consequences of disease control (health econmics)
4 Selection of strategy
5 Search for funding:
  National
  Multilateral/international agency
  Bilateral
6 Plan of operations
7 Continuous evaluation of results of programme
8 Development of programme based on results of control

CONTROL ACTIVITIES
1 Diagnosis methods and approaches
  1.1 Parasitological
  1.2 Serological
  1.3 Clinical
      1.3.1 Community based
      1.3.2 Field teams
      1.3.3 Passive v. active surveillance
      1.3.4 Hospital based
      1.3.5 Link surveillance to other programmes, e.g. leprosy
            chemotherapy, immunization programmes

2 Immunotherapy
  2.1 Cold chain
  2.2 Integration with national EPI (expanded programme on immunization)
  2.3 Logistics
  2.4 Epidemiological evaluation

3 Chemotherapy
  3.1 Selection of drug on the basis of cost, toxicity (side-effects), method of
      delivery
  3.2 Dosages
  3.3 Monitor development of resistance

4 Vector control
  4.1 Ecology of vector
  4.2 Environmental management
      4.2.1 Habitat modification
      4.2.2 Housing, engineering
  4.3 Insecticides
      4.3.1 Method and choice
      4.3.2 Community participation
      4.3.3 Environment
      4.3.4 Cost
      4.3.5 Resistance
      4.3.6 Genetic and biological control

Table 10.1 *Continued*

5 Reservoir control
   5.1 Dogs – hydatid
   5.2 Rodents – *Leishmania*

6 Health education
   6.1 Methodology relevant to social structure and education level of community
   6.2 Health structures – committee; attitude of population of such committees; involvement of schools
   6.3 Education via communication via TV/video/posters/radio/schools

not only to treatment but to disease monitoring and epidemiology. The correct diagnosis is essential for any individual requiring treatment which involves a level of risk. Diagnosis in individuals enables a problem to be defined on a community basis. Serological tests can be used for screening communities to detect suspects before parasitological examination determines whether antibody levels are indicative of active disease or simply previous exposure to antigen. Serological results must be treated with caution and interpretations should be based on the known and well-tested levels of sensitivity and specificity of the particular reagents. Many serological tests, particularly those developed commercially, do however show high correlation between disease and antibody levels and commercially available diagnostic kits also include control positive and negative sera. Methods of antigen and DNA detection for diagnosis are already becoming available and as these are more appropriate and specific than tests based on antibody detection, it is anticipated that a range of such tests will become widely available. Some diseases may require highly sophisticated instruments to confirm diagnosis of a condition, e.g. cysticercosis, hepatic amoebiasis and hydatid disease, where the use of computer assisted tomography can confirm clinical presentations. Such procedures can, however, never be envisaged as routine but can be of value in clinical research, and benefit individuals although they are of limited relevance to disease control in the community.

10.2.2 Objectives of control

Alleviation of discomfort or pain in an individual patient, or larger scale intervention which is properly planned and managed, will contribute to reducing transmission amongst the population. However, such extensive control measures usually have more than a humanitarian objective and are often directed to socio-economic benefit and community well-being. It is only recently that the social and economic value of large-scale measures have been quantified. The Onchocerciasis Control Programme (OCP) in West Africa had a development objective for the Volta Basin where, through river blindness, abandonment and failure to resettle the more fertile valleys was inhibiting agricultural development. Large-scale projects to control malaria, filariasis and schistosomiasis are now recognized as producing both humanitarian and socio-economic benefits to communities through increased agricultural productivity and crop diversification. This is manifested by a move away from purely staple-crop production into a cash-crop agriculture with resultant benefits not only to local communities but also to the balance of payments to endemic countries. Control of animal disease has benefits in terms of reduced mortality as well as increased productivity such as the control of coccidiosis in chickens by coccidiostats in feed and the use of acaricides for tick control which reduces the burden of tick-borne disease such as theileriosis, East Coast Fever, babesiosis and anaplasmosis.

The need for rapid intervention on humanitarian grounds alone is clear if epidemics arise, e.g.: during a sleeping sickness outbreak in Uganda during the 1980s when several thousands of people per year were infected with a virulent *Trypanosoma brucei rhodesiense* infection; epidemics of malaria, such as in Madagascar or Sri Lanka, threaten communities which have lost the immunity which had previously provided a level of protection from epidemics; and migrant populations who are immunologically naive move into areas of agricultural development where new infections become established, e.g. malaria and schistosomiasis associated with dams, irrigation and rice production.

10.2.3 Control versus eradication

It is essential to avoid confusion in the use of these terms. Control of any disease by the reduction of parasite load or vector population, thus reducing transmission, implies recognition that there is likely to be a continuing problem and hence commitment. The objective is to reduce the disease to a level at which it is acceptable to the community or the stockholder by reducing morbidity to levels commensurate with socio-economic development or livestock productivity. Future control activities should aim to sustain the reduced incidence (rate of acquisition in unit time) of the disease but at a cost that is economically justified. Eradication, which is rarely an objective, implies that the disease is to be eliminated and will not recur. The best recent example of eradication is that of the viral disease smallpox. Eradication of some other diseases or vectors in limited geographical areas has also been achieved, e.g. yellow fever in Central and South America, *Anopheles gambiae* from Brazil, *Simulium naevei* in Kenya, animal trypanosomiasis from north-east Nigeria and malaria in Italy. Eradication, therefore, often refers to the total removal of a vector.

10.2.4 Diagnosis

The identification of the cause of disease is fundamental to the solution of the problem presented by the individual, the community or the herd. In larger scale mass-control programmes, or when prophylactic measures are employed, diagnosis of the disease in the individual is of little importance as it is the accepted objective that a strategy of mass treatment, or large-scale vector control, would be to reduce overall the prevalence and hence mortality/morbidity of the disease. Diagnosis is particularly important when dealing with: (1) organisms which are difficult to detect parasitologically; (2) where treatment is risky due to the toxicity of a drug and frequency of side-effects; and (3) when initial problems require to be defined.

10.2.5 The role of primary health care in parasite disease control

The goal of *Health for all by the year 2000* was a declaration by the Alma-Ata UNICEF/WHO Conference of 1978. The objective was to raise the level of health in the broadest sense by a political commitment to the poorest and most deprived. It was recognized that the resources that were inadequate were trained manpower, facilities, equipment and finance. These limited resources have to be used appropriately to give the most relevant and cost-effective intervention to reduce the burden of morbidity and mortality and must be targeted, in particular, on those diseases which are most important in each community. Parasite disease control must be integrated, therefore, with primary health care if that is technically possible. The health-care system must be aware of, capable of responding to and controlling, parasitic infections; preferably integrating control into the health-care delivery system and providing education concerning the tools to control important parasitic diseases in the most feasible and cost-effective way.

An example of the appropriate approaches to disease control is provided in relation to different levels of authority and organization in sleeping sickness (Table 10.2). In this example the role of the different levels of the health-care system are identified. The integrated approach to and organization of vector control of filariasis in India within a primary health-care system, and the relationship between control activities and vector research, are illustrated in Figs 10.1 and 10.2.

Table 10.2 Role of different organizational levels of health services in sleeping sickness control. (After World Health Organization, 1986.)

Level	Activities
Country (ministry level)	Situation analysis
	Programme designs
	National coordination
	Allocation of personnel
	Purchase of equipment and supplies
	Training
	Distribution of equipment and supplies (reagents, drugs, trapping devices, and/or insecticides)
	Distribution of technical documents
Regional (intermediate level)	Active case detection (mobile teams)
	Confirmation of diagnosis
	Treatment
	Participation in follow-up
	Technical supervision of vector control
	Data collection and analysis
	Reporting to ministry
	Distribution of diagnostic reagents and material for vector control
	Vehicle maintenance
District (dispensary health centre)	Passive detection
	Support to mobile teams
Village (primary health care worker)	Identification of suspects
	Active surveillance of population at risk
	Follow-up after treatment
	Coordination of vector control activities undertaken by the village community

10.2.6 Geographical, economic and political variations in control objectives

Considerable variation in the epidemiology of particular diseases occurs from one geographical area to another. In addition, there is also variation in the economic, political and social conditions which prevail within a particular country and region. The WHO has recognized these differences in the case of malaria and recommends that depending on the local circumstances objectives will vary. These so-called tactical variants have graded objectives:

1 the reduction and prevention of malaria associated mortality;

2 reduction and prevention of malaria associated mortality and morbidity, particularly the latter in high-risk groups;

3 as in (2) with reduction of human reservoirs;

4 national control with view to long-term eradication (see Table 10.3).

Similar approaches have been developed for the control of human sleeping sickness, again dependent on government commitment, available human and financial resources, and the existence of an acceptable infrastructure on which to build via appropriate in-service training courses. The ideal situation is rarely achieved in terms of coverage of the population at risk by effective active surveillance for *T. b. gambiense* on a regular basis, hence compromise fall-back strategies are adopted of necessity. However, in accepting this

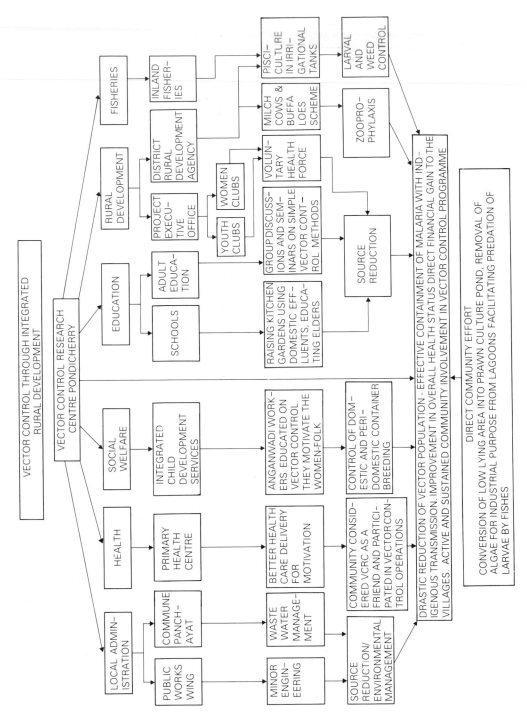

Fig. 10.1 Vector control through integrated rural development. (After Rajagopalan *et al.*, 1987.)

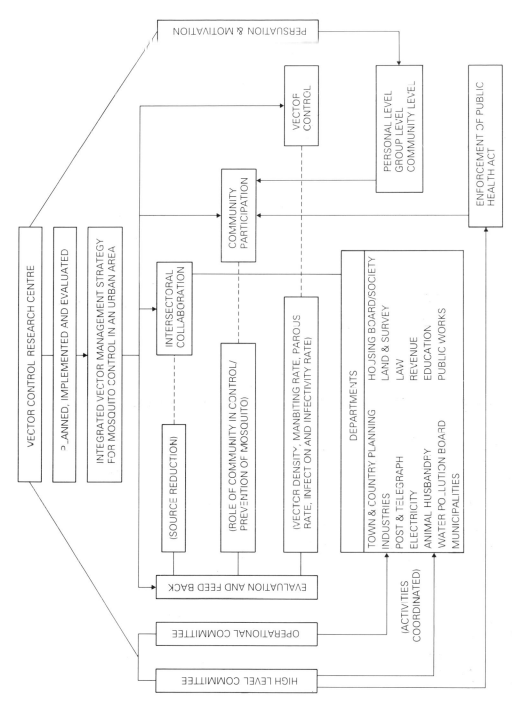

Fig. 10.2 Filariasis integrated vector management strategy. (After Rajagopalan *et al.*, 1987.)

Table 10.3 Tactical variants in malaria control strategies

	Characteristics	Activities

Variant 1
Sub-Saharan Africa,
Central America,
South East Asia
	High prevalence
	Severe clinical disease
	Limited financial resources
	No experience of control
		Develop expertise through training
		Standardize drug therapy and transfer to
		  rural periphery
		Health education – particularly early
		  recognition of symptoms

Variant 2
Richer communities of
area of Variant 1 South
East Asia
	More available resources than Variant 1	Focus drug distribution on high risk
		  groups – prophylaxis for pregnant women
		  and those breast feeding – as above

Variant 3
Indian subcontinent,
Middle East
	Long-term national committment
	Malaria programme with trained
	  workers exists
	Health infrastructure available to plan,
	  implement and evaluate
		All available resources to be employed;
		  including vector control, in depth
		  epidemiological and entomological
		  studies, distribution of resistance to drugs
		  and insecticides. Information flow.
		  Surveillance system for case detection –
		  Health education stresses personal
		  protection measures such as house
		  screening and bed nets. Control may
		  involve community. Training expanded
		  and improved

Variant 4
Middle East, North
Africa
	Communities with considerable
	  experience in control
	Firm government support
	Ample human and financial resources
	Well-informed population
	Strong base of community participation
		As above, but if eradication is to be achieved,
		  complete coverage of all areas necessary.
		  Perfection of surveillance and drug
		  distribution systems so all cases detected
		  and cured

position, government authorities should be aware of the potential risks of compromise and the likely cost if epidemics occur.

10.3 METHODS OF CONTROL

The control of any parasitic disease is a multifaceted operation that must take into account a large number of variables and the need to achieve particular ends, e.g. the control of malaria for a traveller to a malarious area is relatively simple compared with the eradication of the same disease from an endemic area. In this chapter, control at the level of the community has been emphasized but this should not detract from the need of the individual. Many major control schemes are directed at the vectors and these will be considered in Section 10.4. Other methods of control include vaccination and chemotherapy which involve the individual whilst health education and the control of reservoir hosts which, like vector control, is most effective if directed and organized at a higher organizational level with a minimum of individual involvement. In this chapter, vector control has been emphasized but other facets must be mentioned.

10.3.1 Vaccination and chemotherapy

Vaccination

Global smallpox eradication and yellow fever eradication in Central America are examples of the effective use of vaccines. In the developed world, the widespread use of vaccines against childhood diseases has had a dramatic effect on child mortality and morbidity from diphtheria, measles, poliomyelitis and whooping cough. Despite this, lack of acceptance of some of these vaccines (particularly that for whooping cough) has led to the persistence of the disease whilst measles vaccine, which could play a major role in preventing child mortality in the tropics, is not as widely distributed as it should be due to logistic problems. Attempts to develop vaccines for the control of parasitic disease have thus occupied scientists for several decades. The diversity and complexity of protozoa and helminth parasites have, however, presented a multiplicity of problems which have prevented the production of viable and acceptable products with a few notable exceptions. This topic is discussed in more detail in Chapter 8 (Section 8.9) and from a mathematical viewpoint in Chapter 4 (Section 4.6.2).

Chemotherapy

Prophylactic chemotherapy, taking a drug in order to prevent infection, is a well-tried but expensive method of control for the individual. The best known example is malaria for which chloroquine, pyrimethamine or proguanil have protected millions of travellers and visitors to malarious areas. Even with increasing drug resistance, this is still the recommended method. Mass prophylaxis has been practised for the control of malaria in the Gambia and elsewhere and curative mass treatment has been employed for the control of schistosomiasis, to prevent reinfection of snails, whilst large-scale use of mectizan (ivermectin) is now an integral part of OCP where, as a microfilaricide, it is given annually to reduce skin disease and reduce the rate at which blindness develops. In general, mass chemotherapy is expensive and can lead to the development of drug resistance so is not widely used except in epidemic conditions.

For the infected individual, curative chemotherapy is always the action of choice and this contributes to community protection by removing one source of infection. Chemotherapy is discussed in more detail in Chapter 9 and the mathematical aspects of chemotherapy as applied to helminth infections are discussed in Chapter 4 (Section 4.6.4).

10.3.2 Health education

Dissemination of information to populations plays an increasingly important role in disease control. Displays of posters, the use of radio, television and, more recently, video recordings, enable populations to be informed about control activities or bring communities together so that particular remedies, techniques or programmes can be explained and introduced. In many countries, the health infrastructure can be used to disseminate verbally information at a local level or, alternatively, by existing political or local government structures. Poster displays should be targeted in local languages, emphasizing the importance of observing symptoms, or refraining from particular practices, e.g.: not urinating or defaecating into water supplies in the case of schistosomiasis control; correct methods of faeces disposal for geohelminth control; the removal of sheep offal to reduce hydatid transmission; the identification of particular symptoms in sleeping sickness; the need to maintain attendance for drug delivery; and refraining from interfering with tsetse traps.

10.3.3 Control of reservoir hosts and their infections

In parasitic diseases that have proven wild or domestic animal reservoir hosts, elimination of the infection within the host or destruction of that host may have a dramatic effect on the transmission of the infection to humans.

Extensive rodent control in the former USSR has been implemented to reduce cutaneous leishmaniasis due to *Leishmania major*; the reservoir, the great gerbil *Rhombomys opimus*, being controlled by poison (zinc phosphide) or by ploughing up or flooding of gerbil colonies.

Treatment of dogs with the cesticide Prazi-quantel has been used in the control of hydatid disease (*Echinococcus granulosus*) and in some areas feral dog control has also been implemented. Dog control has been undertaken for control of visceral leishmaniasis in China and it is widely practised when rabies is a problem.

Elimination of game animals was practised in Zimbabwe and Uganda as a method of controlling tsetse and hence animal trypanosomiasis; here, however, the principle was to eliminate the preferred hosts of savanna *Glossina* species and hence the fly itself rather than reducing the parasite reservoir directly.

10.4 VECTOR CONTROL

Since the recognition that insects transmit infectious agents and the elucidation of the life cycles of parasites in vectors, the vectors (insects, arachnids and snails) have been targets through which disease control can be achieved. Initial attempts at vector control depended on environmental management to reduce vector populations before insecticides became available and application techniques were developed. Considerable success was achieved by draining swamps to remove larval breeding sites of *Anopheles* mosquitoes (Pontine Marshes in Italy) or by the use of oil to prevent larval mosquito respiration; *Glossina* (tsetse flies) were controlled by the selective destruction of savanna and riverine forest habitats together with control by the destruction of game. The latter method allowed the extensive development of the livestock industry in Zimbabwe. Clearance of riverine vegetation removed the habitat of *G. palpalis*, contributing to the control of sleeping sickness in West Africa. Trapping of *Glossina* for control, which has become an important part of present day control strategy, was introduced by the Portuguese in the early part of this century when plantation workers wore sticky materials on their backs to trap *G. palpalis* attacking them. As early as 1921 trap development in Southern Africa also demonstrated that large numbers of tsetse flies could be attracted to traps.

10.4.1 Insecticides

The advent of insecticides in the 1940s, with the widespread use of dichlorodiphenyltrichloroethane (DDT), resulted in less emphasis on environmental and biological methods of control and the reliance, for a period of two decades, on insecticides. It was only after the publication of Rachel Carson's *Silent Spring* in 1962, describing the proven deleterious environmental effects of the persistent insecticides and the development of insecticide resistance, that a more integrated and rational approach to vector control resulted.

The use of DDT, which was the basis of the WHO Malaria Eradication Campaign of the 1950s, achieved eradication in several subtropical regions and controlled malaria transmission on much of the Indian subcontinent for several years. Gradually, however, DDT resistance developed and alternative insecticides were required; organophosphates, carbamates, and more recently synthetic pyrethroids have been introduced as the spectrum of resistance has widened. DDT was used in other vector control campaigns. The blackflies, *Simulium naevei* and *S. damnosum*, were eradicated from the river systems in Kenya and Uganda, and tsetse (*Glossina*) eradicated from parts of Southern and Eastern Africa and parts of Nigeria.

In recent years the development of the insecticide *Bacillus thuringiensis israeliensis* Serotype H14 (BTI), a toxin derived from the spores of the bacterium *Bacillus thuringiensis*, has been extensively used in blackfly and mosquito control. BTI acts specifically on larvae of blackflies and mosquitoes by a cytolytic effect on larval midgut cells. The advantage of such a product is its specificity; it can, however, only be used as a larvicide but is ideal for blackfly control when, as has occurred in the Onchocerciasis Control Programme, resistance to the organophosphate, Temephos, has occurred. However, it is necessary to develop appropriate formulations of BTI to match operational needs for river-flow rates, adsorption onto particular materials and carry-distance of insecticides in large rivers.

10.4.2 Vector ecology in relation to control

Widespread application of persistent insecticides was initially undertaken with little regard for environmental consequences. However, it was acknowledged at an early stage that increased knowledge of the ecology of vectors was required to assist in developing more cost-effective methods of application, e.g. information on resting sites of *Glossina* in relation to insecticide application by ground-spraying teams, the behaviour of mosquitoes in relation to biting habits and resting habits (indoor or outdoors) and the frequency of feeding on humans (degree of anthropophagy) compared with animal hosts (zoophagy) which led to the development of the techniques of blood-meal analysis to determine the host on which insects feed. The need to measure the degree of resistance to a particular insecticide led to the development of simple test kits for detecting resistance. The location of larval breeding sites, the degree of dispersal of a vector population and the basic ecology of the populations (mean longevity, reproductive rate, parity, biting cycles) led to techniques of age-grading and more refined and efficient trapping and collecting methods. Such studies have led to the discovery of sibling species or species complexes of major vectors, such as the *Anopheles gambiae* and the *S. damnosum* complexes, groups that are morphologically identical, or nearly so, and can only be reliably distinguished by examination of polytene chromosomes or by biochemical methods such as isoenzymes, DNA probes or cuticular hydrocarbon analysis. Members of such complexes have different behaviour patterns such as the degree of anthropophagy (human-biting), resting sites (indoor/outdoors), capacity to migrate (as in *S. damnosum* complex) and susceptibility to parasites, all bearing on their importance as vectors.

10.4.3 Insecticide application methods

Various methods of delivery of insecticides have been developed to maximize the cost-efficiency of insecticide application. The methods of application employed vary and depend on whether the insecticide is to be applied as a residual (persistent) application (effective over a period of weeks or months) or as a non-residual application (effective over a short time-scale). Non-residual insecticides kill target insects which are currently exposed, e.g. adult mosquitoes or *Glossina*, but do not remain effective sufficiently long to kill the proportion of the population which is immature (e.g. larvae or pupae). Insecticides used as residual applications are applied to the inside walls of homes for control of mosquitoes, sandflies or triatomine bugs or to vegetation (twigs, tree trunks) for tsetse control. Such applications are usually undertaken by knapsack spraying techniques. Vehicle-mounted sprayers have been used for residual tsetse control, as have helicopters which can deposit residual insecticides onto tsetse resting sites. Residual insecticide spraying depends on the application of high doses of insecticide as large droplets. Non-residual spraying, however, is dependent on space spraying with smaller droplets ($<50\,\mu m$ diameter) at reduced dosages; the insecticides are not targeted on any particular surface but are present as aerosols for sufficiently long to kill those insects on the wing. Such ultra low volume (ULV) applications can be made by 'fogging' machines which can be hand-held, mounted on vehicles or deposited from aircraft. It is important that, particularly in the case of aerial application, meteorological conditions are optimal to ensure that the insecticide reaches the target populations. Non-residual applications of aerosols of insecticide have been used sequentially to control savanna *G. morsitans* and fogging is an essential tool to control urban mosquito populations such as *Aedes aegypti*, particularly to stop transmission in epidemics of dengue and yellow fever, or for *Culex* and *Anopheles* control.

Insecticides can be used in other ways to reduce disease transmission by control of vectors. Over the past decade the synthetic pyrethroids have been incorporated into tsetse traps (deltamethrin) with attractant odours (octenol, acetone, urinary phenols) to control *Glossina*, into mosquito nets (permethrin) to reduce malaria transmission, into canisters for control of triatomine bugs and as pour-ons for tick control.

Table 10.4 Insecticides use in mosquito control

	Residual dose (g per m²)	Residual effect (months)
ADULTICIDES		
Organochlorines		
DDT	1–2	>6
HCH (and BHC, lindane)	0.2–0.5	3–6
Organophosphates		
Malathion	1–2	2–3
Fenitrothion	1–2	>3
Pirimiphos methyl	1–2	>3
Carbamates		
Bendiocarb	0.4	2–3
Propoxur	1–2	2–3
Pyrethroids		
Permethrin	0.5	2–3
	Dosage (ppm)	Duration of effect (months)
LARVICIDES		
Organophosphates		
Temephos* (abate)	1	2–4
Chloropyrifos (dursban)	1–10	2–10
Fenthion	1–10	2–10

* Temephos is also used in *Simulium* larvae control and control of *Cyclops*, for control of *Dracunculus*, the guinea worm, and can be drunk safely.

10.4.4 Mosquito control

Conventional insecticides such as organochlorines, organophosphates, carbamate and synthetic pyrethroids have been the insecticides of choice for control of mosquitoes by indoor house spraying for *Anopheles*, fogging for control of *Aedes* or the larviciding. Insecticides used are listed in Table 10.4. However, many alternative approaches to mosquito control have been tested, based on an improved knowledge of the ecology and physiology. This is due to the development of insecticide resistance and heightened environmental concern, as well as the need to develop integrated control strategies compatible with cost-effectiveness, foreign exchange consideration and sensitive environmental management.

The list below details the range of approaches available in mosquito control:

1 Environmental management
  (a) swamp drainage and land levelling;
  (b) removal of human-made breeding site (containers, tyres, tins);
  (c) water management and engineering
   • variation in water levels and changes in water flow rates
   • control of aquatic vegetation
   • specific constructions to reduce area of shallow water around shoreline – impoundment.

2 Larviciding by surface treatment
  (a) oil – kerosene, used engine oil;

(b) special larvicidal oils and aliphatic amines;

(c) Paris Green (copper acetoarsenite);

(d) expanded polystyrene beads in *Culex* breeding sites, synthetic monolayers and foams.

3 Bacterial toxins

(a) BTI. Endotoxin crystal derived from spores. Genetically engineered toxins now formulated;

(b) *Bacillus sphaericus*.

4 Physiological methods of control

(a) inhibitors of chitin synthesis – insect growth regulators (IGR) – diflubenzuron, dimilin;

(b) juvenile hormone mimics – altosid, methoprene.

5 Biological control through predators and parasites

(a) predators

• predators of larvae. Mosquito fish, *Gambusia* and *Lebistes*

• predatory larval mosquitoes, *Toxorhynchites*; *Dugesia* (a planarian), *Hydra*, *Notonecta*;

(b) parasites

• fungi – *Coelomomyces*, *Lagenidium*

• protozoa – microsporidians, *Nosema*

• nematoda – mermithids.

6 Genetic control

(a) sterile male release;

(b) introduction of genes rendering mosquitoes insusceptible to infection.

7 Netting and traps

(a) window screens;

(b) covering of containers where larvae develop, to prevent oviposition, and vents from sanitation system;

(c) bed nets – including permethrin impregnated nets;

(d) light traps or exit traps for latrines.

8 Repellents. Specific chemical applied to skin (diethyltoluamide) or impregnation of clothing; mosquito coils, evaporating healing plates; churai (bark) burning.

9 Zooprophylaxis. Location of alternative hosts (cattle) to divert mosquitoes from man.

Insecticides used as both adulticides and larvicides have been the mainstay of control since the mid-1940s. Table 10.4 lists the insecticides which have been used, their chemical class, duration of effect and dosage rate. The strategy for control of *Anopheles* mosquitoes is based on residual house spraying due to their tendency to feed on humans (anthropophily) and to enter homes to rest (endophily). However, the development of resistance to DDT and subsequently to other classes of insecticide and heightened environmental concern led to a more measured approach to the problem which aims for control, including the reduced use of organochlorine compounds and greater use of biodegradable chemicals as well as other appropriate measures locally (mosquito fish, bed nets, larviciding through BTI).

10.4.5 Tsetse control

Prior to the advent of insecticides, control of *G. palpalis* was based on the destruction of linear riverine vegetation. For *G. morsitans*, extensive clearing of savanna, the elimination of preferred food hosts, by shooting game animals or by the creation of barrier zones and fences to keep game and cattle in separate areas, were employed. Tsetse 'pickets' were placed at each junction where a road crossed a fence to prevent flies entering such cattle zones on vehicles; this method was ineffective and costly in terms of labour and fence maintenance.

Persistent DDT applications were made against tsetse flies on a large scale and there were successful programmes in Nigeria, Zimbabwe, Uganda and Kenya. They were based on ground spraying of insecticide to preferred resting sites. Spraying activities against tsetse flies are carried out during the dry season and are dependent on precise planning in order to increase the efficiency of the campaign – disposition of supplies of food, fuel and insecticide; adequate camping and communications; precise mapping – all are prerequisites for success.

Aerial spraying against *G. morsitans* is undertaken by fixed-wing aircraft and involves sequential application of insecticides, e.g. endosulfan and synthetic pyrethroids such as deltamethrin, as a non-residual spray (ULV) spraying to kill those tsetse living as adults. The sequential spraying technique is dependent on an intimate knowledge of tsetse biology, as a further application must be applied before females, that emerge after an application, have had time to larviposit. The last application must be made after those puparia, deposited as larvae immediately before the first spray, have all emerged, as the maximum duration of pupal life can be up to 60 days, and the minimum larval incubation period around 12 days, hence 5–6 sequential applications are usually applied.

Sequential application has been used in Southern and Central Africa (Botswana, Zimbabwe, Zambia) for savanna habitats and night flying has been introduced for flat terrain with the development of sophisticated navigation aids. Aerial spraying against *Glossina* must be undertaken when appropriate weather conditions occur as it is necessary to have a temperature inversion to permit the droplets to reach the ground. These conditions occur from around 17.00 to 09.00 hours; after approximately 09.00 hours ground temperatures are higher than the ambient air temperature and convection currents will ensure no insecticide reaches the ground resting sites. Aerial application requires considerable organization, entomologists to monitor reproductive cycles and age of female *Glossina*, aerial spray contractors, and experts in meteorology and physico-chemical assessment of droplet behaviour.

Residual application of insecticides by helicopter have also been undertaken for tsetse control in West Africa. Insecticide is deposited from low altitude, a few metres above riverine vegetation, to resting sites. Helicopter operations, like fixed spraying, can be undertaken only in inversion conditions. As it is unsafe to fly helicopters at night, spraying is limited to the early morning and just before dark. Residual applications (using larger droplet size) are applied at approximately the same insecticide dose per hectare as for ground spraying. Applications must be undertaken as soon as possible after the end of the rainy season in order to provide the maximum duration of insecticide effect before the next rains, when the effectiveness of the insecticide will be rapidly reduced. Helicopter applications are more expensive than those from fixed-wing aircraft.

Application of deltamethrin as a pour-on product to cattle has also been shown to be successful for tsetse control. Formulations are applied to the back of cattle, spread over the hair of the cow and remain effective for several months, killing *Glossina* as they alight to feed.

Most recent advances in tsetse control have been in the development of trapping devices. Parallel studies in West Africa, on traps based on the biconical trap, and odour-baited traps in Zimbabwe have resulted in widespread use of traps, targets and screens for control. Riverine *Glossina* control is based on deltamethrin-impregnated blue biconical traps, or modifications such as monoconical or pyramidal versions. *Glossina morsitans* group control is based on deltamethrin impregnation of targets baited with acetone, octenol and either cattle urine or synthetic phenols which have been identified as attractants in bovine urine. In some situations no insecticide is necessary as non-return devices are constructed which allow flies to die in such traps. The advantage of traps is that in many parts of Africa local communities have become actively involved in the manufacture, deployment, maintenance and servicing of these traps. It appears that they are the cheapest form of tsetse control.

Sterile male-release programmes have also been evaluated for *Glossina* control. Mass rearing of *Glossina* is undertaken with the objective of continuously releasing sterile males that are competitive with wild males. As female tsetse only mate once, if that mating is sterile no offspring will be produced. The wild population, however, must be initially reduced by approximately 90%, using non-residual insecticide or traps, to permit released sterile males to swamp the remaining wild population by a ratio of 9:1.

No effective methods of biological control of *Glossina* have been applied although predators of both pupae and adult stages do occur.

10.4.6 Sandfly control

Sandfly control (of *Phlebotomus argentipes*) has been achieved in India by the application of persistent insecticides (DDT) during indoor house spraying against *Anopheles*. House and animal-shelter spraying has also been undertaken in parts of South America specifically for the control of sandflies; clearing vegetation around settlements in South and Central America, accompanied by insecticide treatment of parts of the vegetation, e.g. trunks or larger trees, may have a temporary effect on transmission of *Leishmania guyanensis* by *Lutzomyia umbratilis*. Habitat destruction of reservoir host colonies (*Rhombomys opimus*) by ploughing or flooding has been undertaken in the former USSR with the accompanying destruction of sandfly habitats. No biological control measures are available for sandflies. Control of sandflies is difficult when they are not associated with human habitations. Normally, transmission of leishmaniasis is associated with sylvatic or desert habitats where vector control is neither easily undertaken nor likely to be effective. The limited knowledge of preimaginal stages precludes any attack on larvae. Any sandfly control campaign should be evaluated not only by the estimation of sandfly abundance but by the impact of changes in incidence of the disease.

10.4.7 Triatomine bug control

Chagas disease control is dependent on elimination of triatomine species which have invaded houses where transmission of *Trypanosoma cruzi* takes place. The most important vectors, *Triatoma infestans*, *Rhodnius prolixus* and *Panstrongylus megistus* readily adapt to living in close proximity to humans, although in Brazil, another 30 species of bugs have been found to infest the home.

Control of Chagas disease vectors is currently undertaken in only a few countries where the disease is endemic. In Brazil, where an extensive national programme has been developed, the insecticide used in house spraying was benzene hexachloride (BHC) applied as a residual insecticide. In recent years BHC has been replaced by the synthetic pyrethroids, deltamethrin and cypermethrin, as they are more cost-effective, more acceptable to home owners, of lower mammalian toxicity and of greater persistence, through greater stability, on wall surfaces. Pyrethroids also allow savings in costs of application compared with BHC as deltamethrin requires only one application and has a longer persistence, hence, allowing a wider area to be treated in a given time. Insecticide application has met with considerable success but such measures need to be maintained in view of the capacity of the vectors to invade homes.

A more permanent solution would clearly be the improvement of rural housing to reduce the availability of bug habitats in walls and roofs within traditionally built dwellings. Replacement of palm thatch with metal roofing allied to insecticide spraying has a significant impact on infestation with bug elimination recorded in some areas of Venezuela. Introduction of new types of building materials and more extensive plastering of existing dwellings, to reduce the numbers of cracks, will reduce the habitats available for bug colonization. Current strategy involves improving existing dwellings rather than building new ones, but if the house structure is too poor to improve an alternative strategy is to build a new house using compressed building blocks designed not to develop cracks.

Paints containing slow release formulations of insecticides provide alternative methods for controlling domiciliary vectors. Initial costs are high because of the quantity of insecticide incorporated to permit release over long periods but in the long-term they become economically feasible and reduce the numbers of visits required to treated houses.

Recently fumigant cans for triatomine control have been developed as a result of the demonstration of synergistic effects of combinations of insecticides on triatomes. Fumigant cans containing dichlorvos, lindane and deltamethrin have been proved to have good residual effect and a high mortality in exposed bugs as well as excellent acceptance by the community with no toxicological effects.

Although several candidate biological control organisms and compounds have been tested, e.g. Mermithid nematodes, the flagellates *Blastocrithidia* or *Trypanosoma rangeli*, and *Metarhizium* fungi, or juvenile hormone mimics, control of triatomines will depend on insecticides and housing improvement programmes for the foreseeable future.

10.4.8 Blackfly control

Simulium are controlled in the larval and pupal stages rather than as adults, as the preimaginal stages are localized in well-defined breeding sites, whereas adults are widely distributed. The necessity to apply insecticides in aquatic habitats poses environmental problems which require monitoring.

Early blackfly control campaigns used DDT applied to breeding sites. These campaigns resulted in the eradication of *S. damnosum* from the Nile in Uganda and *S. naevei* in western Kenya.

The environmental hazards of organochlorine compounds led to the use of organophosphates as the initial insecticides of choice for the OCP in West Africa. This programme, presently the biggest vector control programme in operation, seeks to alleviate blindness caused by *Onchocerca volvulus*, and to provide the opportunity for development of the fertile valleys of West Africa where settlements have been inhibited by high densities of *S. damnosum* and fear of blindness.

The OCP vector-control activities have been based on weekly applications of the larvicide (temephos) to blackfly breeding sites in up to 18 000 km of river. Applications are made by helicopter or fixed-wing aircraft by a rapid release of insecticide upstream of the breeding site. The amount of insecticide released is calculated to take into account the amount of river discharge over the breeding site. During the rainy season larger numbers of breeding sites require to be treated in view of increased flow, and consequently greater quantities of insecticides are used.

Development of insecticide resistance to temephos has required the rotational use of alternative insecticides. At present five insecticides are employed to combat development and spread of resistance. These are temephos, phoxim pyraclofos (organophosphates), BTI, permethrin (synthetic pyrethroid) and carbosulfan.

The strategy for rotational use of these compounds is complex and decisions are made based on environmental data, cost of insecticide, formulation in relation to amount of river discharge, possibility of development of cross-resistance amongst the organophosphates and resistance in blackflies. This strategy has so far prevented the spread of organophosphate resistance.

Blackfly control has successfully prevented *O. volvulus* infection in children born in the OCP area since 1974/75. This has been accompanied by a reduction in microfilarial load in those that were already infected and reduced the risk of blindness. The programme is now extending westwards to treat blackfly breeding sites from where invasion takes place into the original programme area. The strategy of the programme is not to eradicate blackflies but to control them to reduce transmission levels and hence reduce blindness. Ivermectin, a microfilaricide, has been introduced into the programme; this drug, which is given annually, rapidly reduces the microfilarial load and will, in parallel with blackfly control, reduce transmission and blindness rates.

Blackfly control using insecticides has also been undertaken in Central America (Guatemala) where onchocerciasis is transmitted by *S. ochraceum*. The breeding sites are in small mountain streams with vegetation cover where control is by slow-release formulation of temephos in briquettes.

Environmental management has contributed to blackfly control in temperate regions where blackflies are a nuisance problem, breeding in dam spillways or other outfalls from artificial water bodies. Control can be achieved by raising, lowering or ceasing water flow, killing preimaginal stages through desiccation. Regular application of this strategy has reduced blackflies as a serious nuisance problem and as vectors of *Leucocytozoon*, a protozoan parasite of turkeys in North America.

10.4.9 Midge control

Midges (Ceratopogonidae) are serious nuisances as well as being vectors, and control is undertaken to reduce their irritating bites. This is achieved by environmental management – draining swamps and marshy areas. Application of insecticides to breeding sites can be successful, the most effective being dieldrin which, however, has a potential for side-effects on non-target organisms. Rain is required to wash insecticides through larval breeding sites. Adult control is of little long-term use. The use of repellents provides some protection.

10.4.10 Control of ticks

Tick control using dip tanks containing acaricides has been widely used in South Africa, with considerable success, for the control of theileriasis (East Coast Fever) Cattle should be dipped at least once a week and this strategy is allied to fencing to prevent contact with wild game reservoirs. Currently, an integrated approach to tick control is being adopted allowing cattle to be kept in areas where tick-borne diseases are present. It is important to maintain the resistance which cattle display to ticks whilst preventing epidemics by movement of non-immune animals into areas where they may be more susceptible and to prevent the introduction of new infection through movement of livestock into areas hitherto free.

Acaricides used are organochlorines (toxaphene, lindane), organophosphates (dioxathrion, oxinothiophos, chlorofenvinphos), carbamates (carbaryl) and pyrethroids. They can be applied in dip tanks, spray races or by hand spraying. Pour-on formulations of deltamethrin are being utilized as well as ear tags impregnated with permethrin. Ivermectin by inoculation is also effective against ticks.

Ticks develop resistance to acaricides and control is less effective when cattle graze adjacent to wildlife which maintain the tick species that infest cattle.

A recent approach to control of ticks has been the development of tick-resistant cattle following the observation of acquired immunity to tick infestation, hence reducing challenge from disease. Immunization procedures, using tick antigens to prevent or reduce tick attachment, feeding and egg production are being developed in Australia. Genetically engineered *Boophilus* midgut antigens produce antibodies in cattle which, when *Boophilus* ticks feed, kill the ticks by interfering with the normal digestive process.

10.5 SNAIL CONTROL

Control of snails using chemicals is known as mollusciciding. Early snail control relied on inorganic chemicals such as copper sulphate. Compounds specifically developed for snail control include niclosamide, which is still in use. Disadvantages of molluscicides are that they are adsorbed onto particulate material, inactivated by light and are toxic to non-target invertebrates.

In addition, some snail vectors of digeneans are amphibious and hence a proportion of the population may not be accessible at the time of application. Effective long-term control of snails by molluscicide requires repeated treatment. Prevention of reinvasion of snails over large areas and any large scale programme for schistosomiasis control must incorporate appropriate initial studies of snail ecology and faecal transmission sites. Studies in St Lucia in the Caribbean, Sudan and on the Volta Lake in Ghana suggest that focal mollusciciding accompanied by removal of aquatic weeds may reduce snail numbers and, when accompanied by controlling water supplies and chemotherapy, may reduce schistosomiasis incidence. However, mollusciciding at selected points must be undertaken every 2–4 weeks. In any programme the cost-effectiveness of the activities must be evaluated. Mollusciciding by itself is inadequate for schistosomiasis control but could contribute to successful integrated programmes. The best-known example of an integrated control programme directed against schistosomiasis, caused by *S. japonicum*, was undertaken in China during the cultural revolution. This involved massive drainage schemes, redirecting waterways, filling-in of infested streams, removal of vegetation and hand destruction of snails which entirely eliminated the snail vector from considerable areas. This was accompanied by mass treatment of infected individuals and followed up by encouraging farmers and peasants with financial inducements to look out for snails. The scheme has been a complete success but such an intensive and single-minded campaign is unlikely to be repeated elsewhere. Biological control by introduction of competitive snails *Marisa* and Echinostome trematodes, that compete with human pathogenic trematodes within the snail, are approaches which have been tried with limited success as has the use of plant-derived molluscicides.

10.6 CONTROL OF ECTOPARASITES

Control of lice, fleas and mites that live on the skin of humans or domestic animals is undertaken by topical application of insecticides as

Table 10.5 Methods in ectoparasites control

SCABIES		
Sarcoptes scabei	Family treatment	20–25% benzyl benzoate emulsion. Tetmosol Tesitgal (sulphur compound)
HEAD LICE		
Pediculus humanus capitis	Family treatment	Dust – DDT, malathion. Shampoos
BODY LICE		
P.h. corporis		10% DDT dust or organophosphate dust
CRAB LICE		
Phthirus pubis		As above
FLEAS		
		Insecticidal dusts, applied to body of human or animal. Dust around rodent holes

powders, soaps or shampoos or other chemicals, e.g. benzyl benzoate for the scabies mite *Sarcoptes scabei* (Table 10.5). Hygiene clearly plays an important part in reducing ectoparasite loads and head lice infestations are also more prevalent in children, particularly in institutional accommodation. Control should therefore be directed at groups or families. Body lice infestations are particularly prevalent in populations living at high altitudes who wear several layers of clothing which is infrequently removed.

10.7 SANITATION

The gastro-intestinal helminth parasites of humans (*Ascaris*), hookworms (*Ancylostoma* and *Necator*) and *Trichuris* are, because of the absence of reservoir hosts and insect vectors, amenable to control by imposing sanitation and encouraging hygienic habits via health education. *Schistosoma haematobium* and the prevalence of the gut-dwelling protozoa *Entamoeba histolytica* and *Giardia duodenalis* can also be controlled by these methods.

Eggs and cysts are resistant for many months (*Ascaris* up to 3 years) and faecal material must be stored for long periods or treated chemically before it can be used. The most effective method of control of gut parasites is the provision of simple latrines and the population should be educated not to use human excreta 'night soil' directly as fertilizer, nor should the authorities use excreta removed from latrines for that purpose. Chemicals such as sodium nitrite and calcium superphosphate are ovicidal and, in combination, destroy *Ascaris* and *Trichuris* eggs.

The extent and prevalence of gastro-intestinal infections throughout the world poses a massive problem and one which has not improved over recent decades. Only an integrated approach via mass treatment, safe disposal of waste and provision of latrines will make an impact on this problem. The role of the community and health education and particularly involvement of school-age children as well as the peripheral health sector will be a vital element in any success.

Sterilization of the water supply will also prevent transmission of protozoan cysts of *Entamoeba* and *Giardia* although *Cryptosporidium* is very resistant to normal sterilization procedures. For the control of all water-borne diseases the prevention of contamination of water sources and distribution systems is essential. Precipitation and filtration of urban water will eliminate cysts but few individuals obtain water from such sources in the developing world.

Environmental hygiene by protection of food from cockroaches and flies will prevent the spread of eggs and cysts of parasites and washing and sterilizing vegetables can play a role in control but few deprived communities practice these methods of food preparation. In the case of amoebiasis, a further aspect of control is the detection of cyst carriers among those who handle, distribute or sell food. Street vendors are common foci of gut infections when foods are constantly handled or have been grown where human faeces are used as a fertilizer.

APPENDIX:
INTEGRATED APPROACHES
TO CONTROL OF
SOME PARASITIC DISEASES

This appendix emphasizes the complexity of integrated control and the range of options that are available in defining an appropriate intervention. The list of examples used is selective and the appendix should be read in conjunction with the tables and figures which provide additional information. The development of an integrated control programme will not only involve technical and scientific inputs but at government/regional level will require political, environmental, planning and economic consideration before a programme is financed.

Animal trypanosomiasis (Trypanosoma vivax, T. congolense, T. brucei)
1 Chemotherapy
2 Chemoprophylaxis
3 Vector Control
   3.1 Insecticide application (endosulfan, DDT, synthetic pyrethroids)
      3.1.1 Ground spray
      3.1.2 Aerial application
         3.1.2.1 Fixed wing sequential ultra low volume
         3.1.2.2 Helicopter – residual
   3.2 Deltamethrin impregnated traps and targets incorporating host odours acetone, octenol and urinary phenols (synthetic) or bovine urine itself
   3.3 Sterile male release
   3.4 Deltamethrin pour-on – also used for tick control
4 Development of trypanotolerant cattle
5 Appropriate use of land resource in concert with knowledge of tsetse distribution, trypanosome challenge and need for game for tourist industry
6 Animal trypanosomiasis control linked to livestock development plan including tick-borne disease control

Human trypanosomiasis (T. brucei gambiense and T.b. rhodesiense)
1 Surveillance – serological tests, e.g. CATT (card agglutination test for trypanosomiasis)
2 Detection of parasite for definitive diagnosis
3 Treatment of individual to remove human reservoir
4 Chemoprophylaxis (pentamidine) rarely used and only in T.b. gambiense areas
5 Vector control in epidemics
   5.1 Ground spraying of resting sites: endosulfan, DDT, pyrethroids
   5.2 Deltamethrin impregnated traps and targets

LEISHMANIASES (*LEISHMANIA* SPECIES)

Old World zoonotic cutaneous leishmaniasis (L. major)
1 Reservoir and sandfly control by destruction of burrows, gerbil destruction, ploughing, zinc phosphide
2 Vaccination of exposed populations with avirulent *L. major* vaccine
3 Chemotherapy
4 Health Education

Visceral leishmaniasis (L. donovani, L. infantum)
1 Surveillance and diagnosis to reduce human reservoir (India)
2 Vector control by house spraying (DDT). Impregnated bed nets (permethrin) may be of value
3 Control of dogs or other reservoirs in areas except India (e.g. China, Latin America)

New World cutaneous and mucocutaneous leishmaniasis (L. mexicana, L. guyanensis, L. brasiliensis)
1 No methods are available for control of vectors and reservoirs except short-term reduction of *L. umbratilis* vectors of *L. guyanensis* by insecticides or clearing forest around settlements
2 Passive case detection and treatment

Theileria and Babesia control
1 Tick destruction
   1.1 Dipping or spray races to control ticks by acaricides
   1.2 Pour-on formulations
   1.3 Prospects of antitick vaccines against genetically engineered midgut cell antigen
2 Chemotherapy (see Chapter 9)
3 Chemoprophylactics – long-acting tetracylines
4 Vaccination and treatment (in trials)
   4.1 Vaccine of cryopreserved *Theileria parva* sporozoites or *T. annulata* (passage attenuated schizonts) and long-acting tetracyclines
   4.2 Attenuated *Babesia bovis* blood stage vaccine

5 Maintenance of enzootic stability of immune live-stock populations to ticks to prevent acute disease
6 Prevent contact with game if feasible

South American trypanosomiasis (Trypanosoma cruzi)
1 Surveillance of population
2 Testing of blood in blood banks and treatment of blood (Gentian violet)
3 Treatment of acute cases
4 Vector control
   4.1 Insecticide spraying of houses; BHC, deltamethrin
   4.2 Insecticide cannister paint
   4.3 Improved housing construction; improved bricks and wall plaster, cement or tile floors, tin or tile roofs

Malaria (Plasmodium Species)
1 Reduction of human infection
   1.1 Chemotherapy (see Chapter 9)
   1.2 Chemoprophylaxis (see Chapter 9)
2 Mosquito control (see list 10.4.1)
   2.1
      2.1.1 Adults – house spraying by residual insecticides (see Table 10.4)
      2.1.2 Larvicide – to breeding sites
      2.1.3 Bednets – impregnated with permethrin if possible
   2.2 Environmental and biological control
      2.2.1 Drainage of breeding sites (source reduction)
      2.2.2 Mosquito fish

Coccidia (Eimeria Species)
1 Chemotherapy (see Chapter 9)
2 Attenuated and genetically engineered vaccines under development

HELMINTH DISEASES

Schistosomiasis (Schistosoma haematobium, S. mansoni, S. japonicum)
1 Environmental engineering
   1.1 Construct irrigation canals to prevent vegetation growth
   1.2 Increase flow or intermittent flow in canals
   1.3 Appropriate waste disposal facilities
2 Snail Control
   2.1 Molluscicides – bayluscide; frescon; copper sulphate; plant molluscicide (endod from *Phytolacca*)
   2.2 Removal of vegetation from edges of water contact point
   2.3 Competitor snails *Marisa, Thiara*
3 Chemotherapy (see Chapter 9)
4 Health education

   4.1 Reduce water contact
   4.2 Prevent urination and defaecation around water courses
   4.3 Construction and use of latrines

Onchocerciasis (Onchocerca volvulus)
1 Larviciding
   1.1 Application of selected larvicides (Temephos, BTI, pyraclofos phoxim, permethrin, carbosulfan). Selection dependent on resistance, environmental effects, cost and water flow
   1.2 Temephos slow release briquettes. Small streams in Central America
   1.3 Changing water levels at dam spillway sites
2 Chemotherapy
   2.1 Ivermectin as microfilaricide for improved clinical condition and to reduce transmission
   2.2 No suitable macrofilaricide available

Gastro-intestinal helminths
Hookworm *Ancylostoma, Nectator*
Ascariasis *(Ascaris)*
Trichuriaisis *(Trichuris)*
1 Sanitation and health education
   1.1 Provision of latrines
   1.2 Education in latrine use
   1.3 Avoidance of use of 'night soil' – human faeces for fertilizer without chemical treatment (sodium nitrate and calcium superphosphate)
   1.4 Disinfection of vegetables
2 Chemotherapy
   2.1 Individual
   2.2 Mass chemotherapy

Filariasis
1 Chemotherapy (see Chapter 9)
2 Appropriate vector control techniques (see Figs 10.1 & 10.2)

Hydatid disease (Echinococcus granulosus)
1 Mass treatment of dogs
2 Dog destruction
3 Prevent dog access to sheep offal
4 Sheep quarantine
5 Education programmes

REFERENCES AND FURTHER READING

Briceno-Leon, R. (1987) Rural housing in control of Chagas disease in Venezuela. *Parasitology Today*, **3**, 384–387.

Caincross, S. (1987) Low cost sanitation technology for control of intestinal helminths. *Parasitology Today*, **3**, 94–98.

California Agriculture (1980) *Special Report on Mosquito Research*, Vol. 34, 1–43

Curtis, C.F. (ed.) (1990) *Appropriate Technology in Vector Control*. Boca Raton, CRC Press.

Curtis, C.F. (1991) *Control of Disease Vectors in the Community*. London, Wolfe Publishing.

1988, J.C.P. (1987) Control of Chagas disease in Brazil. *Parasitology Today*, **3**, 336–341.

Dolan, T.T. (1987) Immunisation to control East Coast Fever. *Parasitology Today*, **3**, 4–6.

Gemmell, M.A., Lawson, J.R. & Roberts, M.G. (1987) Towards global control of cystic and alveolar hydatid diseases. *Parasitology Today*, **3**, 144–151.

Greenblatt, C.L. (1985) Vaccination for leishmaniasis. In K.P. Chang & R.S. Bray (eds) *Leishmaniasis*, pp. 163–176. Amsterdam, Elsevier.

Lindsay, J.W. & Gibson, M.E. (1989) Bednets revisited – Old idea, new angle. *Parasitology Today*, **4**, 270–272.

MacLaren, D.J. (ed.) (1989) Vaccines and vaccination strategies. *Parasitology*, **98**, (supplement), S1–S100.

MacLaren, D.J. & Terry, R.J. (1989) Anti-parasitic vaccines. *Transactions of the Royal Society of Tropical Medicine and Hygiene*, **83**, 145–148.

Marinkelle, G.C.J. (1980) The control of leishmania. *Bulletin of the World Health Organization*, **58**, 80–7, 181.

Peters, W. (1992) *A Colour Atlas of Arthropods in Clinical Medicine*. London, Wolfe Publishing.

Rajagopalan, P.K., Panicker, K.N. & Das, P.K. (1987) Control of malaria and filariasis vectors in South India. *Parasitology Today*, **3**, 233–241.

Rozendaal, J.A. (1988) Impregnated mosquito nets and curtains for self protection and vector control.

Tropical Diseases Bulletin, **86**, R1–R41.

Walsh, J.A. & Warren, K.S. (1986) *Strategies for Primary Health Care: Technologies Appropriate for the Control of Disease in the Developing World*. Chicago University of Chicago Press

Weiser, J. (ed.) (1991) *Biological Control of Vectors*. Chichester, Wiley.

World Health Organization (1984) *Report of an Expert Committee on the Leishmaniases*. Technical Report Series No. 701.

World Health Organization (1985) *Ten Years of Onchocerciasis Control*. OCP/GVA/85.1B.

World Health Organization (1986) *Report of an Expert Committee on the Epidemiology and Control of African Trypanosomiasis*. Technical Report Series No. 739. Geneva, WHO.

World Health Organization (1987) *Report on an Expert Committee on Onchocerciasis*. Technical Report Series No. 752. Geneva, WHO.

World Health Organization (1988) *Guidelines for Leishmaniasis Control at Regional and Sub-Regional Levels*. Parasitic Disease Programme. WHO/LEISH/88.25. Geneva, WHO.

Youdeowei, A. & Service, M.W. (eds) (1983) *Pest and Vector Management in the Tropics*. London, Longman.

Young, A.S., Groocock, C.M. & Kariuki, D.P. (1988) Integrated control of ticks and tick-borne diseases of cattle in Africa. *Parasitology*, **96**, 403–432.

Zerba, E. (1988) Insecticidal activity of pyrethroids on insects of medical importance. *Parasitology Today*, **4**, (No. 7 supplement), S3–S8.

Index